S.F. Singer

SINGER

Environmental Impact of Stratospheric Flight

BIOLOGICAL AND CLIMATIC EFFECTS OF AIRCRAFT EMISSIONS IN THE STRATOSPHERE

CLIMATIC IMPACT COMMITTEE
National Research Council
National Academy of Sciences
National Academy of Engineering

Henry G. Booker

NATIONAL ACADEMY OF SCIENCES
WASHINGTON, D.C. 1975

NOTICE: The project which is the subject of this statement was approved by the Governing Board of the National Research Council, acting in behalf of the National Academy of Sciences and National Academy of Engineering. Such approval reflects the Board's judgment that the project is of national importance and appropriate with respect to both the purposes and resources of the National Research Council.

The participants of this study selected to undertake this project and prepare this statement were chosen for recognized scholarly competence and with due consideration for the balance of disciplines appropriate to the project. Responsibility for the detailed aspects of this statement rests with that study committee.

Each report issuing from a study committee of the National Research Council is reviewed by an independent group of qualified individuals according to procedures established and monitored by the Report Review Committee of the National Academy of Sciences and the National Academy of Engineering. Distribution of the report is approved, by the Presidents of the Academies, upon satisfactory completion of the review process.

The activities of the Climatic Impact Committee leading to this report were supported by the Department of Transportation under Contract No. DOT-OS-00035, Task Order No. 21.

ISBN 0-309-02346-7
Library of Congress Catalog Card Number 75-7701

Available from

Printing and Publishing Office, National Academy of Sciences
2101 Constitution Avenue, Washington, D.C. 20418

Printed in the United States of America

Preface

Supersonic aircraft are now being put into service, and an increasing number of subsonic aircraft are flying in the stratosphere. High-flying aircraft had been discharging effluents into the stratosphere for years before much thought was given to the consequences. It was only when serious proposals were made to fly supersonic commercial airliners in the stratosphere that people began to worry. During the main debate in the United States Congress concerning the proposal to invest public funds in the development of an American supersonic transport (SST), concepts of the atmospheric changes that it might cause at cruising altitudes were rudimentary. It was only during the final attempt to revive support for the SST in Congress that a serious concern about a possible reduction of stratospheric ozone came to the fore. This concern is the principal feature of this report. However, the possibility that stratospheric aircraft at cruising altitude might have an influence on climate and agriculture has also been receiving attention recently and is treated in this report as far as present knowledge permits.

Previous reports of the Academies that are relevant are *Summary Statement of the Ad Hoc Panel on* (NO_x) *and the Ozone Layer* (NRC Division of Physical Sciences, 1971) and *Biological Impacts of Increased Intensities of Solar Ultraviolet Radiation* (Environmental Studies Board, 1973).

The wisdom of building supersonic airliners is not discussed in this report from the viewpoint of either economic viability or utilization of world energy resources. Neither does the report address the problems of sonic boom and of low-level atmospheric pollution in the vicinity of airports. The report does not explicitly treat military aircraft, but, based on information received from the U.S. Air Force, the Federal Aviation Administration, and the Civil Aeronautics Board, it is believed that their effects in the stratosphere are not a problem at this time. The report does not discuss in any detail the possibility of radical new fuels for aircraft. The report treats the Space Shuttle in only a preliminary way even though the effects of the Shuttle on the stratosphere may be important. The same is true of nonvehicular effects on the stratosphere, such as those associated with emissions from aerosol cans.

Soon after Congress made its decision not to invest public money at that time in the development of a U.S. SST, it authorized the allocation of funds for research into the scientific facts concerning possible modification of the stratosphere caused by high-flying aircraft. As a result, the Climatic Impact Assessment Program (CIAP) of the Department of Transportation (DOT) was instituted in 1971 and charged with reporting to Congress by the end of 1974 on the possible physical, biological, social, and economic effects that might result from future aircraft operations in the stratosphere. To advise the DOT on CIAP, the National Academies of Sciences and of Engineering established the Climatic Impact Committee in 1972. From the beginning of CIAP, it was the desire of the DOT that a Committee of the Academies would issue an independent report on the subject by the end of 1974. With this in view, the Governing Board of the Academies' National Research Council agreed to sponsor a summer study in 1974. What follows is the report of the Academies' Climatic Impact Committee, based on the summer study and backed by Committee hearings extending over a two-year period.

After an initial survey chapter entitled "Issues and Findings," the report divides into two parts. The first part concerns the physical and biological problem of increased engine emissions in the stratosphere. This includes the reduction of stratospheric ozone and the possible consequent biological effects at ground level and the effects that engine emissions could have on climate and agriculture. The second part addresses the source of these emissions and considers how oxides of nitrogen can be reduced by suitable engine redesign and how sulfur dioxide can be reduced by fuel treatment. This part also discusses the difficulties of assessing the economic, social, and political implications of stratospheric modification that would be caused by aircraft operations and suggests one rationale for arriving at national and international policy.

For those who desire an abbreviated treatment, the crucial arguments are summarized in the Issues and Findings, the prime conclusions being in Section VII. For those who wish to delve into the science, appendixes are provided that summarize the basis of the report and provide a guide to key papers in the scientific literature. The Committee's major recommendations appear at the front of the report.

Because this report ranges over such diverse fields it is not feasible for every member of the Climatic Impact Committee to accept responsibility for every aspect of the study. Different authorship has therefore been attributed to the part dealing with physics and chemistry, the part dealing with engineering, the part dealing with biology and medicine, and the part dealing with economics. The statements for which the Climatic Impact Committee takes collective responsibility appear in the Issues and Findings chapter and in the major recommendations stated at the front of the report.

In preparing this report, the Climatic Impact Committee and its Panels have had the benefit of frequent and prolonged discussions with DOT personnel and contractors associated with CIAP and access to preliminary draft copies of the CIAP Monographs and, very recently, a copy of the Executive Summary of the CIAP Report of Findings. The Climatic Impact Committee wishes to acknowledge the great help that it has received from CIAP over more than a two-year period. Nevertheless, it should be understood that the Climatic Impact Committee has not received complete and final versions of most of the CIAP documents and is not formally reviewing them at this time.

HENRY G. BOOKER, *Chairman*
Climatic Impact Committee

Climatic Impact Committee

HENRY G. BOOKER, *Chairman;* University of California, San Diego

KENNETH J. ARROW, Harvard University

RODERICK K. CLAYTON, Cornell University

ANTONIO FERRI, New York University

THOMAS B. FITZPATRICK, Harvard Medical School

HERBERT FRIEDMAN, Space Science Division, U.S. Naval Research Laboratory

JAMES P. FRIEND, Drexel University

DONALD M. HUNTEN, Kitt Peak National Observatory

HENRY D. JACOBY, Massachusetts Institute of Technology

FRANCIS S. JOHNSON, University of Texas at Dallas

FREDERICK KAUFMAN, University of Pittsburgh

HELMUT E. LANDSBERG, University of Maryland

RENE H. MILLER, Massachusetts Institute of Technology

HANS A. PANOFSKY, Pennsylvania State University

FREDERICK D. ROSSINI, Rice University

RICHARD B. SETLOW, Brookhaven National Laboratory

WILLIAM C. BARTLEY, *Executive Secretary*

Participants

CLIMATIC IMPACT COMMITTEE
SUMMER STUDY
WOODS HOLE, MASSACHUSETTS, JULY 1974

*HENRY G. BOOKER, *Study Chairman;* University of California, San Diego

*WILLIAM C. BARTLEY, *Study Director;* National Research Council

*KENNETH J. ARROW, *Chairman, Panel on Decision Analysis—Monitoring and Surveillance;* Harvard University

WOLFGANG BAIER, Canadian Department of Agriculture

LEO BREWER, University of California, Berkeley

WARREN L. BUTLER, University of California, San Diego

JOSEPH W. CHAMBERLAIN, Rice University

RODERICK K. CLAYTON, Cornell University

EDWARD S. DEEVEY, JR., Florida State Museum

ANTONIO FERRI, New York University

THOMAS B. FITZPATRICK, Harvard Medical School

HERBERT FRIEDMAN, Space Science Division, U.S. Naval Research Laboratory

JAMES P. FRIEND, *Vice Chairman, Panel on Atmospheric Physics and Chemistry;* Drexel University

ARTHUR C. GIESE, Stanford University

DOUGLAS GORDON, University of Queensland, Australia

* Member, Study Steering Committee

FRANCIS T. HAXO, Scripps Institution of Oceanography, University of California

DONALD M. HUNTEN, Kitt Peak National Observatory

HENRY D. JACOBY, Massachusetts Institute of Technology

*FRANCIS S. JOHNSON, *Chairman, Panel on Atmospheric Physics and Chemistry;* University of Texas at Dallas

HAROLD S. JOHNSTON, University of California, Berkeley

FREDERICK KAUFMAN, University of Pittsburgh

HELMUT E. LANDSBERG, University of Maryland

JERRY MAHLMAN, Princeton GFDL/NOAA

RENE H. MILLER, Massachusetts Institute of Technology

*HUGH ODISHAW, University of Arizona

HANS A. PANOFSKY, Pennsylvania State University

*FREDERICK D. ROSSINI, *Chairman, Panel on Engines and Fuels;* Rice University

EDWARD C. RUNGE, University of Missouri

ELIZABETH L. SCOTT, University of California, Berkeley

*RICHARD B. SETLOW, *Chairman, Panel on Biological and Medical Effects;* Biology Department, Brookhaven National Laboratory

ROBERT M. SOLOW, Massachusetts Institute of Technology

A. MICHAEL SPENCE, Stanford University

JOHN W. TUKEY, Bell Laboratories and Princeton University

FREDERICK URBACH, Skin & Cancer Hospital, Philadelphia

RICHARD J. ZECKHAUSER, Harvard University

Consultants to the Study

MARTYN M. CALDWELL, Utah State University

RALPH D'ARGE, University of California, Riverside

J. MORLEY ENGLISH, University of California, Los Angeles

IRVIN GLASSMAN, Princeton University

ROBERT HUDSON, Johnson Spacecraft Center, NASA

CECIL E. LEITH, National Center for Atmospheric Research

* Member, Study Steering Committee

MARTIN MENTER, Haffer & Meiser, Washington, D.C.

S. S. PENNER, University of California, San Diego

ELMAR R. REITER, Colorado State University

GEORGE D. ROBINSON, Center for Environment & Man, Inc.

RICHARD A. RUDEY, Lewis Research Center, NASA

STEVEN SHAVELL, Massachusetts Institute of Technology

MILTON C. WEINSTEIN, Harvard University

Department of Transportation Representatives

ROBERT ATHERTON, Systems Control, Inc.

RICHARD BRENNAN, Systems Control, Inc.

ANTHONY BRODERICK, Transportation Systems Center

SAMUEL CORONITI

KENNETH FORNEY

REYNOLD GREENSTONE

ALAN GROBECKER

JOSEPH GWIAZDOWSKI

THOMAS M. HARD, Transportation Systems Center

RONALD LAU, Systems Control, Inc.

FREDERICK MARMO

JAMES PATMORE, Systems Control, Inc.

ROBERT SCHAINKER, Systems Control, Inc.

JOSEPH E. WIRSCHING, Systems Control, Inc.

Institute for Defense Analyses Representatives

ERNEST BAUER

PYTHAGORAS CUTCHIS

ALEXANDER FLAX

ROBERT OLIVER

Canadian Government Representative

B. BOVILLE, Atmospheric Environmental Service

Science and Technology Policy Office Representative

RUSSELL C. DREW

Contents

Major Recommendations of the Climatic Impact Committee

We RECOMMEND that national and international regulatory authorities be alerted to the existence of potentially serious problems arising from growth of future fleets of stratospheric airliners, both subsonic and supersonic. The most clearly established problem is a potential reduction of ozone in the stratosphere, leading to an increase in biologically harmful ultraviolet light at ground level. (p. 85)

We RECOMMEND that research and development on combustors and other engine components be increased immediately with a view to reducing the effect of engine emissions on stratospheric ozone. We RECOMMEND that the development program be aimed toward a reduction in the emission of oxides of nitrogen by a factor of 10 within the next decade, and that the research program be aimed at substantial further improvement in emissions to be realized in practical engines during the ensuing decade. (p. 72)

We RECOMMEND that nations seek the adoption by international regulation of whatever constraints may be necessary on aircraft emissions and fleet sizes in the stratosphere to prevent undesirable changes in the intensity of ultraviolet light at ground level and undesirable changes of climate. In particular, we RECOMMEND that national and international regulatory authorities adopt standards and policies designed to encourage the entry into stratospheric service of only those new types of aircraft, subsonic or supersonic, that are equipped with engines for which maximum effort has been made to improve emissions—and only if the resulting improvements are judged to be adequate. (p. 85)

We RECOMMEND that biological and medical studies of the effects of changes in ultraviolet radiation upon living organisms be increased. In particular, we RECOMMEND that studies of skin cancer be accelerated to obtain better quantitative understanding of the cause of the disease. (p. 50)

We RECOMMEND that international organizations concerned with atmospheric science, biological science, medical science, and aerospace science devote increased attention to studies concerned with the stratosphere and with possible inadvertent modification thereof. We further RECOMMEND that, to prepare for formal international negotiations looking toward regulation of stratospheric flight, international organizations concerned with the environment and with aerospace flight form international commissions of government-appointed scientists and engineers to review and agree on the state of knowledge concerning the effect of engine emissions on the stratosphere and the means available for modifying them. These scientific understandings should include the effect of stratospheric modification on the intensity of ultraviolet light at the ground, on climate, and on agriculture. They should be updated at frequent regular intervals. (p. 90)

We RECOMMEND that international organizations concerned with the environment and with aerospace flight organize and coordinate the monitoring of any changes in the stratosphere that may be caused by high-altitude flight and that monitoring begin promptly. (p. 34)

We RECOMMEND that studies of physics, chemistry, and transport phenomena in the atmosphere be continued. We endorse the concept that the United States National Aeronautics and Space Administration undertake research on stratospheric phenomena, including inadvertent modification. (p. 34)

We RECOMMEND that it be recognized that control of the sulfur content of fuel for stratospheric airliners will be required in the future. The degree of sulfur content should become a matter of international regulation in the light of further research on the effect of stratospheric flight on climate. (p. 75)

We RECOMMEND that the study initiated by the United States National Aeronautics and Space Administration concerning the effect on the stratosphere of the Space Shuttle be continued on high priority. (p. 34)

We RECOMMEND that an immediate quantitative assessment be made of the possibly large stratospheric effects of stable compounds, such as the chlorofluoromethanes involved in aerosol spray cans and refrigerants. This assessment should include the stratospheric effect of carbon tetrachloride, methyl chloride, and analogous bromine compounds. (p. 34)

1

Issues and Findings

I. The Problem

As one ascends in the atmosphere from ground level, the temperature normally decreases. In latitudes near the poles this decrease may continue to a height of about 8 kilometers (26,000 feet). Near the equator, the decrease of temperature continues further to a height of about 16 km (52,000 ft). The region of decreasing temperature is normally well mixed and is known as the troposphere. Above the troposphere, the temperature at first becomes roughly independent of height and then increases with height up to a level of about 50 km (roughly 160,000 ft). This is the region that is known as the stratosphere (see Figure 1). Subsonic airliners have been flying in the stratosphere up to 12 km (about 40,000 ft) for some time, and plans exist to increase this height to nearly 14 km (45,000 ft). The Anglo-French Concorde and Russian TU-144 supersonic airliners are expected to cruise at between 16 and 17 km (about 54,000 ft), while a next generation of supersonic transports might fly at about 20 km (roughly 66,000 ft). There could ultimately be a hypersonic transport that would fly above 30 km (roughly 100,000 ft).

A region of the atmosphere in which the temperature fails to decrease with height is known as a region of temperature inversion. The effect of a temperature inversion in inhibiting vertical mixing is well known to the inhabitants of the City of Los Angeles, where a temperature inversion at a low level in the troposphere frequently traps the products

This chapter is a report of the Climatic Impact Committee.

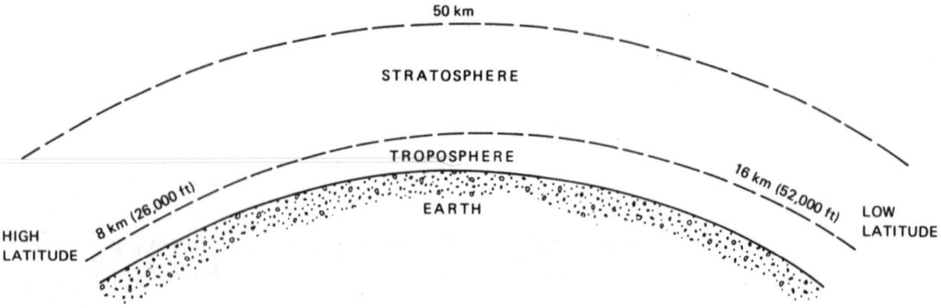

FIGURE 1 Graphic illustration of the variation of the stratosphere from high to low latitude. Not drawn to scale.

of combustion close to ground level. In the stratosphere there exists a permanent inversion of temperature involving a rise of roughly 65°C (about 120°F) over a height interval of roughly 35 km. This temperature inversion permits products of combustion produced in the stratosphere to escape only slowly. They may remain in the stratosphere for years until they are ultimately transported downward into the troposphere and removed by various processes, including rain. The stratosphere can be likened to a city whose garbage is collected every few years instead of daily. Since man does not control the cleaning operation in the stratosphere, he has no alternative but to live with several years' accumulation.

This raises the question of the rate at which man can afford to introduce effluents into the stratosphere. Subsonic airliners frequently fly in the stratosphere at present, and the next generation of subsonic airliners will probably do so more frequently. The Concorde and TU-144 will fly there, and a next generation of supersonic airliners would almost certainly fly higher in the stratosphere, where the cleaning operation is slower. A fleet of aircraft flying in the stratosphere in the future would leave behind there several years' accumulation of the effluents that they produce.

While vertical spreading of the products of combustion in the stratosphere is slow, horizontal mixing is faster and notably extensive. Spreading in the north–south direction as far as the pole and the equator takes months rather than years, while spreading in the east–west direction to all longitudes takes only about a week. The horizontal spreading does not respect national boundaries. It does respect the equator to a considerable extent, so that most stratospheric effluents produced in the northern hemisphere will remain there. Stratospheric modification will be a little greater near the latitude of a heavily traveled route (such as the North Atlantic route), but apart from this, all countries in the northern hemisphere will be affected more or less equally by what each country in the northern hemisphere does with regard to stratospheric flight. There is little that a nation can do individually to protect the stratosphere above it—cooperation with other nations is essential.

In the stratosphere, molecular oxygen absorbs ultraviolet solar radiation at wavelengths below about 240 nm (1 nm $= 10^{-9}$ m), and in the process, ozone is formed. Although ozone is only a minor constituent of the stratosphere, it is important to life as it has evolved so far on earth. Ozone absorbs ultraviolet radiation from the sun at wavelengths below about 320 nm. This absorption shields the earth from the full effect of this radiation and virtually cuts it off at wavelengths below about 295 nm. The amount of ozone in the stratosphere varies with time of day, season of the year, and latitude, both regularly and irregularly; there is more ozone at high than at low latitudes. Scientists have been interested in stratospheric ozone for half a century and, as a result, there is now a good deal of information upon which to draw.

Changes in the average amount of ozone in the stratosphere would not only change one of the conditions that has almost certainly influenced the evolution of life on earth so far, but it could also have an influence on climate. The temperature inversion in the stratosphere is created by the absorption of solar radiation by the ozone, and a small change in the amount of ozone would create a small change in the temperature of the stratosphere. By a complicated process that is inadequately understood, this could lead to a small change in temperature at the earth's surface and to a small change in rainfall. Even small global changes in temperature and rainfall can have effects on marginal agriculture that are not necessarily small.

II. The Engine Effluents

For jet engines using hydrocarbon fuels, the major exhaust products that are discharged into the stratosphere are water and carbon dioxide, which are not likely to have important effects on man or the environment. Water is present in the natural stratosphere in amounts of a few parts per million. Although large fleets of SST's could bring about a small increase in the water-vapor content, this would only have a small influence on the radiation balance, because the radiative effect of water is small compared with that of carbon dioxide, whose concentration is a hundred times larger. Moreover, the amount of carbon dioxide added to the stratosphere by even a large stratospheric fleet of airliners would be a negligible fraction of the carbon dioxide already present there. Any effect on climate through the absorption and re-emission of infrared radiation would be unimportant compared with the effects of the large amounts of carbon dioxide produced at the earth's surface by nature and by man.

The minor products of combustion include nitric oxide (NO) and nitrogen dioxide (NO_2), sulfur dioxide (SO_2) and sulfur trioxide (SO_3), carbon monoxide (CO), and soot. The products that give cause for concern are NO_x (that is, NO and NO_2) and SO_x (that is, SO_2 and SO_3). The rates of emission of these products can be modified by engine design and fuel treatment.

NO_x is of concern because it can lead to a reduction in the amount of ozone in the stratosphere and because it absorbs visible sunlight. Reduction of ozone permits an extension to shorter wavelengths of the ultraviolet spectrum of solar radiation reaching the ground, and this has biological consequences. Slight changes in temperature could also ensue, leading to an influence on climate.

Nitric oxide is formed in the combustion process at high flame temperatures in a jet engine. With the materials presently used for turbine blades it is necessary to reduce the temperature of the gas before it passes through the turbine. By redesigning the engines to avoid unnecessarily high temperatures during combustion, the amount of nitric oxide created can, in principle, be reduced substantially. However, if cooling arangements are improved or if new materials become available for turbine blades, higher engine efficiency would be possible by using higher gas temperatures. The higher the gas temperature, the greater the amount of nitric oxide. If an effort were made to increase efficiency in the subsonic fleet by designing engines that operate at higher temperatures, these new engines could increase the production of nitric oxide unless designed to avoid this.

Production of nitric oxide in jet engines leads to discharge of NO_x into the atmosphere. Here NO_x acts as a catalyst in chemical reactions that result in a reduction of the amount of stratospheric ozone. It is only the slow downward transport caused by atmospheric mixing that tends to rid the stratosphere of NO_x.

The other minor product of combustion that is of concern is SO_x, which leads to the production of sulfate aerosols. The aerosols reduce slightly the solar radiation reaching the earth's surface and may have an effect on climate. SO_x arises from sulfur compounds that are present as impurities in the fuel now used. Refining processes are available to remove these impurities.

III. The Consequences

Effects on surface temperature and on rainfall arise from NO_x and SO_x in the engine effluents. We do not at present know whether there would be a net increase in temperature or a net decrease, although the latter is more likely. A net global change of more than a few tenths of a degree seems unlikely for even a large fleet of stratospheric airliners. The change in rainfall is difficult to assess and would vary from place to place. Some parts of the world would benefit somewhat, others would suffer somewhat. It is impossible to assess at this time whether the world as a whole would be better or worse. Marginal farming in some areas of the world might become more marginal or might disappear, while in others it might improve somewhat. Changes would be larger in subpolar regions than in temperate latitudes; they would probably be quite small in tropical regions. In the northern hemisphere much land is settled and cultivated in high latitudes; the climate here might become a little more variable and in some areas the frost-free season might become a little shorter. However, there seems to be little prospect of anything that could be called an ice age.

While the climatic effects of the effluents from aircraft flying in the stratosphere are probably small and difficult to assess, those effects directly related to human health may be estimated with increasing accuracy. Reasonable understanding now seems to exist of the effect of aircraft effluents on the amount of ozone in the stratosphere. It is true that the methods used for calculating mixing in the atmosphere are imperfect. It is also true that past experience in upper atmospheric research has shown that a chemical reaction once thought to be unimportant may later acquire an unforeseen significance. Continued research is the only way to deal with such possibilities. Nevertheless, the Panel on Atmospheric Physics and Chemistry of this Committee has concluded (see Chapter 2, Section III) that stratospheric injection of NO_x will cause a decrease in the ozone content of the stratosphere, and that, for a fleet of stratospheric airliners of known size, engine emission, traffic routes, frequency, and altitude, the magnitude of the decrease can now be reasonably well predicted. From this, the concomitant enhancement in the ultraviolet radiation received at the ground can be predicted (see Table 3 below). What effect would this have on biological systems, and in particular on man?

Biologists divide the ultraviolet spectrum of sunlight at the surface of the earth into three portions: uv-A (wavelengths greater than 320 nm), uv-B (290 to 320 nm), and uv-C (below 290 nm). Since the range of wavelengths particularly affected by changes in stratospheric ozone is approximately 290–320 nm, this study is principally concerned with the uv-B portion of the spectrum. Although radiation from the sun makes life possible on earth, nearly all known effects of uv-B are deleterious.

Life exists in a delicate balance between various harmful influences on the one hand and defensive or restorative measures utilized by living organisms on the other. Biologists believe that the existing balance between damage and protection has been achieved over a long period of time; readjustment to a sudden increase of uv-B may involve natural selection of the better protective devices in a population of organisms and would require many generations of the organism to accomplish.

The duration of the Department of Transportation's Climatic Impact Assessment Program has been far too short to assess the full range of damaging effects of uv-B on life and the long-term responses to such damage. It may be recalled that the principal deleterious effects of smog on vegetation were recognized only after years of exposure; had one attempted to predict the consequences of smog through a research program conducted over a two- or three-year period, it is doubtful whether the seriousness of the problem would have been realized.

It has been inferred from laboratory experiments that an increase of uv-B corresponding to a 50 percent reduction of ozone may be damaging to a number of higher plants. The complex ecosystems of microscopic plants and animals that constitute the ultimate supply of food in the oceans might also be affected.

In human beings, the effect of increased exposure to uv radiation is best known in the forms of skin tanning and sunburn (erythema). Experiments using uv light sources have shown that erythemal response varies with wavelength. The results of these experiments are summarized in what is known as an "erythemal action spectrum." The skin's peak sensitivity occurs at a wavelength of about 297 nm, which is only just within the sunlight spectrum reaching the earth's surface. At wavelengths longer than about 305 nm the skin's sensitivity is appreciably reduced, and in the visible part of the solar spectrum it is negligible. By combining the erythemal action spectrum with the spectrum of the sun's ultraviolet radiation reaching the earth, it is possible to describe the overall erythemal effectiveness of the various wavelengths in the solar spectrum after it is filtered by the atmosphere. The wavelength most responsible for erythemal effects when the sun is at an elevation about 60° usually lies between 300 and 305 nm. This peak effectiveness would be shifted downward in wavelength by about 5 nm and the total impact more than doubled if the amount of ozone in the stratosphere were reduced to one half of its normal value. For smaller changes, the amount of erythemally effective radiation reaching ground level increases by about 2 percent for each 1 percent decrease in stratospheric ozone.

There is considerable natural variability in the ozone content of the stratosphere. In consequence, there is a natural variability in the intensity of solar radiation reaching the ground in the uv-B band. This variability is greater than the present effect of stratospheric aircraft. Nevertheless, the effect of the airliners is systematic and worldwide. For a fleet of stratospheric aircraft of known size, engine emission, traffic routes, and altitude, the percentage increase in the erythemally effective exposure can be calculated to a useful degree of accuracy (see Chapter 2, Table 4). In theory, people could offset this exposure by protective action, such as wearing increased clothing, applying suntan lotion, or spending less time outdoors.

What about biological effects more serious for man than sunburn?

IV. Skin Cancer

Medical and biological scientists are of the opinion that essentially the same spectrum of radiation responsible for sunburn is also responsible for skin cancer.

Two forms of skin cancer are recognized. One variety (nonmelanoma) is found principally in older people, probably indicating the existence of a cumulative effect over many years. A few hundred new cases are reported annually per 100,000 population in the United States, but many cases are probably not reported. If medical help is sought, this variety of skin cancer is easily diagnosed and can be successfully treated with surgery or x rays in virtually all cases. About one death per 100,000 population is reported annually in the United States, but this may be due to errors of classification on death certificates. The disease usually starts in sun-exposed areas of the body and infrequently spreads to other areas. It is a recurring illness that often involves disfigurement.

The more dangerous variety of skin cancer (melanoma) affects a younger segment of the population (30 to 50 years of age) and, when encountered, is a medical problem about as serious as breast cancer in women. Its incidence is small compared with that of the first variety (see Table 1), but it is the most serious disease associated with the skin. It is responsible for a few deaths annually per 100,000 population in the United States. Although fatal forms of skin cancer are not con-

TABLE 1 Incidence and Mortality due to Skin Cancer (per year per 100,000 Population at Risk in the United States)[a]

Type of Skin Cancer	New Cases	Deaths
Nonmelanoma	100–300	About 1
Melanoma	3–9	1–3

[a] See Chapter 3, Figure 3.

fined to those areas of the body that are habitually exposed to sunlight, the disease usually appears initially on parts of the body that are uncovered or that are only lightly covered. It is more common in outdoor than in indoor workers. It is less common for heavily pigmented populations than for light-skinned Caucasians. It is less common in high than in low latitudes. Statistics show that the incidence of this type of skin cancer has increased in the United States in the last decade, possibly because of better reporting but possibly also because of increased leisure time and outdoor recreation.

The above statements suggest a high likelihood that incidence of skin cancer is associated with exposure to solar radiation, but they do not prove that the disease is due to exposure to the uv-B part of the solar spectrum, and it is only this part that could be significantly affected by stratospheric modification induced by high-flying aircraft. The reasons for believing that it is the uv-B portion of the solar spectrum that is largely responsible for skin cancer can be summarized as follows.

First, statistical analysis of epidemiological data shows that the logarithm of the incidence of skin cancer, and the logarithm of the mortality due to skin cancer, are each linearly dependent on the intensity of uv-B radiation. The intensity of uv-B radiation is better correlated with the incidence of skin cancer and with the mortality due to skin cancer than is any other environmental variable.

Second, sunburn is certainly due primarily to the uv-B portion of the solar spectrum, and skin cancers develop in the same tissue as that damaged by sunburn. People who contract skin cancer are often found to be those who sunburn easily.

Third, ultraviolet radiation is known to damage DNA—the genetic material of living cells—and to cause mutations in animal cells. It is a reasonable hypothesis that skin cancer involves the absorption of radiation by DNA. The absorption spectrum of DNA is very similar to the erythemal action spectrum.

Fourth, skin cancer is induced in mice by uv-B radiation and hardly at all by longer wavelength radiation; however, presence of the longer wavelength radiation may accentuate the effect of the uv-B radiation by a mechanism not currently understood.

Although the incidence of skin cancer is influenced by various factors, the evidence is good that the prime cause of the disease is uv-B radiation, and that the biological action spectrum for skin cancer is roughly the same as the erythemal action spectrum. The Panel on Biological and Medical Effects of this Committee has concluded (see Chapter 3, Section II.B) that a reasonable estimate of the percentage increase in skin cancer arising from a fleet of stratospheric aircraft of known size, engine emission, traffic routes, frequency, and altitude may be made by assuming that a 1 percent decrease of stratospheric ozone will cause roughly a 2 percent increase in skin cancer. It should be noted, however, that, while a 10 percent decrease in stratospheric ozone gives about a 20 percent increase in melanoma mortality, a 10 percent decrease in stratospheric ozone appears to give more than a 20 percent increase in the incidence of skin cancer—possibly a 30 percent increase.

V. The Options

To control biological and climatic effects caused by flight operations in the stratosphere, we have available the options of modifying engines and fuel, limiting stratospheric flight, or both. However, the situation is complicated by the necessity for international agreement.

It is unlikely that NO_x emissions can be completely eliminated except through a radical change of propulsion. In the absence of such a development, growth of flight in the stratosphere would have to be severely restricted if one is to avoid completely the risk of any increase in skin cancer or of any perturbation in climate. This would apply not merely to supersonic airliners but also to any increase in the use of subsonic airliners in the stratosphere. The base of the stratosphere varies somewhat from day to day and from place to place. Roughly speaking, it would be necessary to ban flight above about 8 km (26,000 ft) in high latitudes, 12 km (about 40,000 ft) in middle latitudes, and 16 km (52,000 ft) in low latitudes. For a given volume of air traffic, this would tend to crowd the air lanes at the lower altitudes in areas of the world where traffic is heavy. Possible deaths from any resulting accidents should be borne in mind when considering the increased risk of skin cancer that would be avoided.

The engines of present wide-body jet airliners are more efficient than those of the original 707, but they emit between two and three times as much NO_x for the same amount of fuel consumed. Subsonic airliners are now being designed to fly at altitudes up to nearly 14 km (45,000 ft), that is, well into the stratosphere at middle and high latitudes. The engines of these aircraft may emit NO_x at the rate of the present wide-body jet airliners if no preventive action is taken. Moreover, if future engines are designed to operate at higher turbine inlet temperatures, increased engine efficiency will become possible, but the NO_x effluent may increase.

One possibility would be control of the emissions and the number of subsonic airliners permitted to fly at heights above 8 km (26,000 ft) and especially above about 12 km (40,000 ft). Another possible action would be to limit the permissible total annual NO_x emission from the entire fleet of stratospheric airliners. If feasible, this would provide an incentive for designing engines with low NO_x emission. Actions of this sort would have major effects on the operation of airliners and on the design of future aircraft.

The stratosphere is the only feasible place where supersonic airliners could cruise efficiently. The alternatives are to limit the emissions and the number of such airliners flying or to limit the total annual NO_x emission of the fleet.

Theoretical possibilities exist for reducing the NO_x emission of jet engines, and laboratory experiments have been performed confirming the theory. The principle is to prevent the gases in the engine from exceeding a temperature critical to the formation of nitric oxide. At present, this would be unlikely to have a significant effect on engine efficiency because the gas entering the turbine is already limited to temperatures well below critical levels by the materials of which the blades are constructed. This may not continue to be true in the future, however, if cooling of the blades is improved or new materials become available.

If reductions in NO_x emissions as large as those already obtained in laboratory experiments were to be achievable in practical engines, consideration would have to be given to the extent to which reduction of NO_x is appropriate. Because of the effects of other emissions, a situation could be reached for which the overall emissions of the engines would have no net influence on atmospheric ozone. However, we are not within sight of this interesting possibility at this time, and the immediate technical challenge is to reduce the NO_x emissions of practical engines.

How can effective action be taken to reduce the NO_x emission from jet engines? What needs to be done can probably be accomplished in a decade by a technological society dedicated to this objective. Whether it will be done depends largely on the extent to which the public is prepared to press for it and to pay for it.

To minimize possible disturbance to climate by flight in the stratosphere, one needs not only to reduce the NO_x emission but also to control the sulfur content of the fuel. The technology for sulfur reduction is known, but implementation will increase fuel costs.

At this point, however, we have to face the fact that no country can act individually to provide comprehensive protection of the stratosphere over its own territory. It is likely that the spreading of stratospheric perturbations between the northern and southern hemispheres will be inhibited by the hemispherical nature of atmospheric circulation. But, in each hemisphere separately, spreading will be nearly complete. Concerted action by all countries in the northern hemisphere will be required to protect the stratosphere over all countries in the northern hemisphere.

The necessary international organizations already exist. Most nations belong to the International Civil Aviation Organization (ICAO). Each member nation is obligated, under the governing convention on international civil aviation (the so-called Chicago Convention), to conform to standards adopted by ICAO unless, within a stated period, it files a deviation. Nations generally have adopted the ICAO standard as a minimum; when a deviation has been filed it generally has been to advise the Council of the adoption domestically of a more stringent standard than that of ICAO. The ICAO Council has approved the report of its Air Navigation Commission recommending action to establish engine emission standards for stratospheric aircraft, but it concluded, in regard to establishing a permissible level of these emissions, that prime responsibility rests with the World Meteorological Organization (WMO). It would also be the responsibility of the WMO to establish arrangements for monitoring any changes that may take place in the stratosphere.

Different countries will have different concerns about the establishment and implementation of controls on stratospheric flying. Some countries have a greater interest in manufacturing stratospheric airliners than others do. Countries with a heavily pigmented population and countries in high latitudes may have only limited interest in skin cancer but might be vitally affected by any action tending to affect agriculture even slightly. Representatives of all these interests would have to subscribe to a restrictive ICAO standard if stratospheric modification is to be controlled.

VI. The Predictions

What can be expected to happen environmentally in the stratosphere if flight at great altitudes is allowed to develop with or without changes of engines or fuel, and what are people's reactions to the prospects likely to be?

The present study cannot provide a good numerical prediction of the effect of stratospheric flight on climate and agriculture. Nevertheless, small changes in temperature and rainfall could have substantial repercussions economically (see Chapter 8, Section II). The economic consequences are difficult to assess and are both positive and negative, but it would be misleading simply to combine them and to conclude that the overall effect may be slight. The plight of the farmer in one area whose crops fail will not be offset by the increased prosperity of a farmer in another area, even if the two live in the same country. If they live in different countries, the economic, social, and political complications will certainly be accentuated. The trouble will be compounded if the aircraft thought to be responsible belong to yet a third country. A relatively inexpensive but not necessarily complete solution to these difficulties is to control the sulfur content of the fuel. It is predicted that people who understand these considerations would want to buy this insurance. However, this approach will be complicated by the overall problem of world energy resources and the necessity for international agreement.

For skin cancer, the problem of prediction is different. It is not merely a question of deciding that the number of supersonic transport aircraft should not exceed some particular number, whether this be ten, a hundred, or a thousand. Subsonic aircraft are also involved. Moreover, the present Concorde is quite a different aircraft from the formerly projected U.S. SST. Both aircraft are substantially different from any that a manufacturer might build a decade hence. Calculations are radically affected by whether one assumes that (a) no effort is made to reduce the NO_x emission from the engines, (b) enough effort is made to reduce emissions by a factor of 10, or (c) a larger effort is made to reduce emissions by a factor of 20 or more. The results of calculations are appreciably different depending on whether the aircraft is designed to fly at 15 km, 20 km, 25 km, or 30 km.

Predictions need to be made for aircraft that are actually flying or are capable of flying. If the predictions indicate undesirable consequences, restrictions can be placed on the use of those aircraft, and the undesirable consequences can be taken into account in designing new aircraft.

The aircraft that we have considered are (a) the present wide-body subsonic airliner, with an assumed average operational flying height of 10.5 km (about 35,000 ft); (b) a new generation of wide-body subsonic airliner, with an assumed average operational flying height of 13.5 km (44,000 ft); (c) the present Concorde supersonic airliner, with an assumed average operational flying height of 16.5 km (54,000 ft); and (d) the formerly projected U.S. SST, with an assumed average operational flying height of 19.5 km (64,000 ft). We refer to these four types of vehicle as the present subsonic, the projected subsonic, the present supersonic, and the formerly projected supersonic aircraft. ICAO data for the Russian TU-144 indicate that this airliner can be considered roughly as a Concorde-type aircraft for these calculations.

Estimates of the percentage reduction in stratospheric ozone per hundred aircraft operating in the stratosphere are shown in Table 2 for each of the four types of aircraft. The table refers to the northern hemisphere, where roughly the present route structure is assumed. Certain key numbers enter into these estimates:

1. Quantitative values have to be used describing the degree of mixing in the atmosphere at different heights. It is our imperfect knowledge of this transport mechanism that is mainly responsible for the uncertainty factors quoted.

2. Each subsonic aircraft is assumed to emit as NO_x 1.5 percent by weight of its consumed fuel; each supersonic aircraft 1.8 percent. It is this pair of numbers that, in principle, could be reduced substantially by engine redesign. Reduction of the NO_x emissions by a factor of 10 would reduce the percentage figures in Table 2 by the same factor.

3. Each supersonic airliner is assumed to spend 5 hours per day in the stratosphere. The figure takes into account the fact that a supersonic airliner covers a given distance in the stratosphere quickly but nevertheless requires roughly the same turnaround time (approach time and ground time) out of the stratosphere as does a subsonic airliner.

4. Each subsonic airliner operating in the stratosphere is assumed

TABLE 2 Percentage Reduction in Stratospheric Ozone per Hundred Aircraft Operating in the Stratosphere [a]

Aircraft Type	Percentage per 100 Aircraft	Uncertainty Factor
Present subsonic	0.02	10
Projected subsonic	0.2	10
Present supersonic	0.7	3
Formerly projected supersonic	3	2

[a] For further details see Table 4 in Chapter 2.

to spend 7 hours per day there. Note that Table 2 excludes all flights that do not enter the stratosphere because these flights have no significant influence on ozone.

Each 1 percent reduction in ozone will lead to an increase of about 2 percent in erythemally effective exposure. On this basis, one deduces from Table 2 the increases in erythemally effective exposure that are shown in Table 3. The factors of uncertainty for Table 3 are almost the same as those for Table 2. Subject to the additional uncertainties described in Section IV above, use may be made of the numbers in Table 3 as a reasonable indication of the percentage increases in the incidence of skin cancer likely to occur for each hundred of the specified aircraft flying in the stratosphere.

The following points should be noted in Tables 2 and 3:

(a) The last line indicates that a fleet of 300 or 400 of the formerly projected U.S. SST's would in all probability have caused about a 10 percent decrease in stratospheric ozone, and a 20 percent increase in skin cancer. In the long run, in a midlatitude country with a population like that of the United States, a few additional people would probably have died each year, and a few hundred additional people would probably have contracted skin cancer each year, for each formerly projected U.S. SST put into service.

(b) Comparison of the second and third lines of the tables indicates that a new generation of wide-body subsonic airliners might be as effective in decreasing ozone and increasing the incidence of skin cancer as the Concorde/TU-144 airliners if the fleet size for the former is about four times that for the latter.

(c) Comparison of the first and second lines of the tables indicates that, if wide-body jet airliners with existing emission characteristics were to fly at 13.5 km (44,000 ft), that is, 3 km (10,000 ft) higher than the average assumed for present operations, these aircraft would be

TABLE 3 Percentage Increase in Erythemally Effective Exposure per Hundred Aircraft Operating in the Stratosphere [a]

Aircraft Type	Percentage per 100 Aircraft
Present subsonic	0.04
Projected subsonic	0.4
Present supersonic	1.4
Formerly projected supersonic	6

[a] For further details see Table 4 in Chapter 2.

about ten times more effective in decreasing ozone and increasing the incidence of skin cancer. This shows that the biological effects likely to be produced at ground level by stratospheric aircraft are crucially affected by the height at which they cruise.

(d) Since there are about 400 wide-body subsonic jets already in service, the first line of Table 3 indicates that these aircraft can be expected to cause an increase in the incidence of skin cancer of about a sixth of 1 percent. Even if all jet airliners currently in service are taken into account, the figure is probably well under 1 percent. However, the accuracy of calculations for the cruising altitudes now in use is poor.

If one permits substantial numbers of aircraft to fly in the stratosphere, one must be prepared for increases in the incidence of skin cancer comparable with the increases in erythemally effective exposure and therefore of the order of magnitude shown in Table 3. At present, the annual figures for skin cancer in the United States are reported to be a few deaths and a few hundred new cases, each per 100,000 population. These may be regarded as base data roughly representative of the situation prior to a possible major worldwide development in stratospheric flight.

Anyone who has a view about what percentage increase in skin cancer should be risked and who is willing to make an assumption about the mix that is appropriate in the four listed categories of aircraft, can form a judgment from Table 3 concerning how many airliners of each type should be permitted to fly. Moreover, if the percentages in Table 3 are divided by 10, one can also form a judgment as to whether it is worth, say, a 10 to 15 percent increase in the cost of producing a new generation of engines in order to reduce the NO_x emission by a factor of 10.

In using Table 3 in this way it should be remembered that the numbers of aircraft calculated refer only to aircraft flying in the stratosphere. In addition to these, any number of aircraft can fly below the stratosphere (see Figure 1) without affecting significantly the amount of ozone in the upper atmosphere and the incidence of biological consequences at ground level.

VII. The Conclusions

It is concluded that an increase in the number of stratospheric airliners with present engines will diminish the amount of ozone in the stratosphere and consequently will increase the intensity of ultraviolet light at ground level. The potential harmful biological effects of an increase in the intensity of ultraviolet light at ground level are such that the burden of proof that these consequences can be accepted must lie with those proposing increased use of such aircraft. If additional aircraft are put into service in the stratosphere, one must be prepared for an increase in skin cancer unless appropriate international regulations having the effect of limiting the total emission of NO_x by stratospheric airliners, both subsonic and supersonic, are successfully negotiated and implemented. The seriousness of the potential increase in skin cancer must be judged by the public and their representatives in the light of information such as that summarized in Table 3.

Research is needed to place the cause of skin cancer outside the realm of debate in hearings concerning the regulation of stratospheric flight. Developmental work is required to demonstrate that jet engines actually can be built to function satisfactorily under operational conditions with NO_x emissions well below those of contemporary aircraft engines. Research into the biological effects of a change in stratospheric ozone, into atmospheric circulation and mixing, and into atmospheric chemistry should also be continued. All these actions can be taken unilaterally by individual countries. They are needed as a prerequisite to fruitful international negotiations aimed at controlling stratospheric airliners in such a way as to limit the modification that they produce in the stratosphere.

As far as climatic change and agricultural effects are concerned, no clear-cut statement can be made concerning expected changes in temperature and rainfall. Nevertheless, a global change in surface temperature of a few tenths of a degree and an associated change in rainfall are not ruled out. Local changes may be larger, and the economic, social, and political effects of such changes could be substantial. To insure against this, the action required is not only to reduce the NO_x emissions but also to control the sulfur content of the fuel. This is technically feasible. The percentage increase in the cost of the fuel is small, but the provision of special fuel for stratospheric airliners may have to be part of an overall strategy in the handling of world energy resources.

Our understanding of these problems is still far from satisfactory in spite of the great progress made during the last three years in association with the Department of Transportation's Climatic Impact Assessment Program. Some quantitative statements that we now believe to be true may have to be modified somewhat in the light of further work. Continued scientific research and engineering development will be required, combined with a continuing reassessment of progress.

I

THE PROBLEM

2

Effects of Nitrogen Oxides on Ozone and Ultraviolet Radiation

I. Introduction

Potential environmental problems are posed when large numbers of aircraft operate at higher altitudes than most of today's traffic. Subsonic aircraft have operated in the lower regions of the stratosphere for some time. What effect on the environment may occur when aircraft fly at altitudes of 16–23 km (52,000–75,000 ft) instead of the present 10–12 km (33,000–39,000 ft)?

In this chapter, we discuss the consequences of introducing nitrogen oxides *directly* into the stratosphere. Studies have indicated that the injection of nitrogen oxides directly into the stratosphere in sufficient amounts will, by a chain of events, ultimately produce deleterious biological effects on man. These effects could result because the layer of ozone in the stratosphere serves as a shield against ultraviolet radiation, and nitrogen oxides remove ozone and thereby reduce the shielding.

II. The Natural Stratosphere

The stratosphere is a nearly cloudless and remarkably quiescent region of the earth's atmosphere that is susceptible to contamination. The

---

This chapter presents the findings of the Panel on Atmospheric Physics and Chemistry: F. Johnson, *Chairman;* J. Friend, *Vice Chairman;* J. Chamberlain, D. Hunten, H. Johnston, F. Kaufman, H. Landsberg, and H. Panofsky. Consultants to the Panel: R. Hudson, C. Leith, E. Reiter, and G. Robinson.

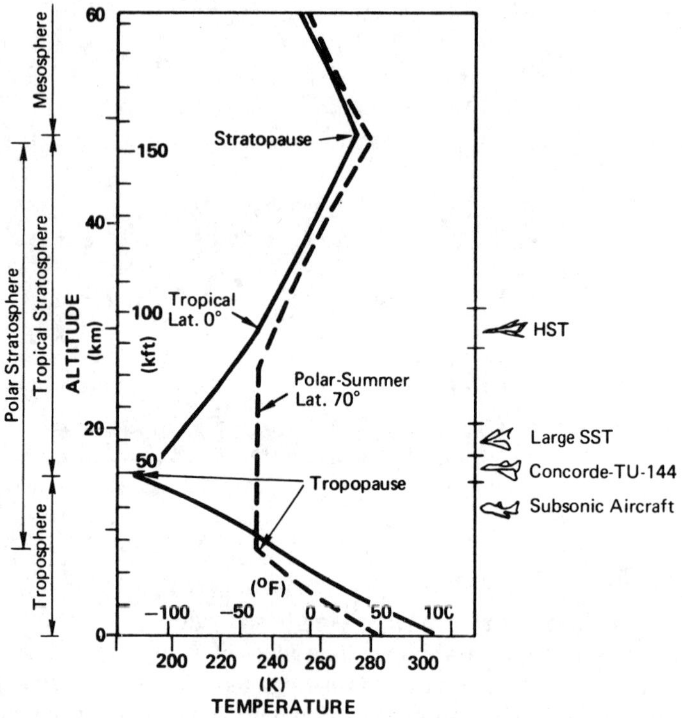

FIGURE 2 The atmosphere's temperature/altitude profile.

stratosphere is characterized by a temperature inversion—a region in which the temperature remains constant or increases with height. As a result, the stratosphere is not stirred by the violent vertical motions characteristic of the troposphere below. The regions called the troposphere (the turbulent region of the atmosphere below the stratosphere), the stratosphere, and the mesosphere (the portion of the atmosphere immediately above the stratosphere) are shown in Figure 2. Also shown in this figure, for comparison, are the cruise altitudes of regular subsonic commercial jet aircraft, the altitudes of the Concorde and Tupolev SST's, the altitudes that were proposed for the 1971 U.S. SST, and projected altitudes for hypersonic transports (HST's). Temperature profiles for the atmosphere vary with latitude so that the bottom of the stratosphere is at a considerably higher altitude at the equator than it is, for example, at latitude 70 degrees; the base of the stratosphere is known as the *tropopause,* and it varies in altitude from about 8 km

(26,000 ft) or less at high latitudes to about 16 km (52,000 ft) over the equator.

The atmospheric pressure decreases exponentially with increasing altitude; it is the weight of the column of air of unit cross-sectional area above any given altitude. Air density is roughly proportional to pressure. At the tropical tropopause it is about 10 percent of that at sea level, and at the stratopause (about 50 km, or 164,000 ft) it is about 0.1 percent.

The special properties of the stratosphere—its temperature inversion and resulting slow vertical mixing—are a consequence in part of the presence of ozone (O_3), which is formed rapidly in the upper stratosphere at altitudes of 30 to 50 km by the photolysis of molecular oxygen (O_2), producing atomic oxygen (O), which in turn recombines with O_2 to form O_3. Below about a 25-km altitude, the ozone concentrations are determined mainly by complicated transport mechanisms that move air toward the poles and downward. These mechanisms are responsible for the latitude distribution of ozone and for the paradox that the total amount of ozone overhead is large at high latitudes where its rate of production is small. The absorption of ultraviolet sunlight (of wavelengths 240 to 320 nm) by ozone shields the earth's surface from harmful ultraviolet radiation and heats the upper stratosphere, thereby producing the temperature inversion. Several natural ozone destruction processes balance its photochemical formation. The most important one, only recently identified, is the catalytic decomposition of ozone by nitric oxide (NO) and nitrogen dioxide (NO_2). These two oxides of nitrogen are referred to collectively as NO_x. The natural concentrations of these oxides of nitrogen are not yet accurately known. They are only present in a few parts per billion, but they are recycled during ozone removal (i.e., they act as a catalyst). The artificial increase of these nitrogen oxide concentrations by effluents from aircraft flying in the stratosphere and the consequent reduction of total stratospheric ozone are the principal subjects of this chapter.

Reactive molecular fragments (radicals), which are formed in the photolysis of water vapor in the stratosphere, mainly at levels above 40 km, also destroy ozone catalytically. Indeed, the initial concern regarding the possibility of ozone reduction by exhaust gases from high-flying aircraft centered around the question of the effects of water vapor. Further study has shown, however, that water vapor tends to reduce the concentration of NO_x by partially converting it to nitric acid. This tends to counteract the catalytic destruction of ozone by radicals derived from water vapor. An increase in stratospheric water could cause either an increase or decrease in ozone, depending on the water distribution. Moreover, the natural concentration of water vapor in the stratosphere (a few parts per million) is so much larger than that of nitrogen oxides that, even though water is a major engine effluent, the fractional change of stratospheric water produced by SST's is much smaller than the fractional change of stratospheric nitrogen oxides.

The stratosphere also contains numerous other minor constituents including carbon dioxide, methane, hydrogen, carbon monoxide, nitrous oxide, and nitric acid. Many of these are important to the chemical processes that occur within the stratosphere. A description of the observed profiles of these various constituents and their significance is presented in Appendix A.

III. Injection of Contaminants

Let us now look a little more carefully at what happens when contaminants are injected into the stratosphere. (A more detailed discussion of stratospheric composition and its chemistry is presented in Appendix A.) This discussion deals mainly with NO_x emitted from engines of aircraft flying in the stratosphere, although this is by no means the only contaminant nor are aircraft the only source. Large volcanic eruptions can inject a considerable volume of gas and debris into the stratospheric region. Also, there were significant man-made injections into the stratosphere from atmospheric nuclear testing in the early 1960's. We believe that contaminants generated at the earth's surface by industry and automobiles have essentially negligible effects on the stratosphere because of the efficiency of rain in removing most contaminants from the troposphere. However, some highly stable gases, such as chlorofluoromethanes (e.g., Freons), may ultimately find their way into the stratosphere.

As aircraft fly through the stratosphere, effluent gases from their engines gradually mix and spread throughout the region. Zonal east–west winds carry contaminants released into the stratosphere around the globe in a period ranging from days to a few weeks. Latitudinal spreading is much slower—requiring the order of a few months to cover a significant portion of a hemisphere—and is brought about by north–south meanderings of the zonal winds. Because of the nature of the general circulation of the stratosphere, there is relatively weak coupling of the motions in the northern hemisphere with those in the southern hemisphere. One manifestation of this weak coupling is slow interhemispheric transport.

Vertical transport in the stratosphere is very slow in comparison with horizontal transport. Yet it is by means of downward transport that matter is lost from the stratosphere. Once in the troposphere, it is subject to various removal processes via precipitation, absorption surfaces, dry decomposition, and chemical transformations at surfaces. These processes, aided by rapid vertical transport in the troposphere, lead to relatively short tropospheric residence times for most trace substances—of the order of about 10 to 20 days. By contrast, the residence times of material injected a few kilometers above the tropopause are of the order of one to a few years.

The distribution of effluent oxides of nitrogen (plus nitric acid) resulting from continual operation of a given fleet of aircraft will ultimately become a quasi-steady pattern. The nature of this pattern is governed by the rate of injection, the dispersion rates within the stratosphere, the transport to the troposphere, and the tropospheric removal rates. The magnitude of the effect of the NO_x on stratospheric ozone depends upon all of these factors and the rates of all the relevant chemical reactions characteristic of the NO_x–HO_x–O_3 system. In order to estimate the effects of the effluent NO_x, recourse must be made to model calculations. Observations of stratospheric radioactivity and experience with numerous models over a 16-year period permit the generalization that the buildup of concentration of the effluents is limited by the vertical transport within the stratosphere and the exchange that takes place between the stratosphere and the troposphere. Also, on the basis of well-established chemical knowledge (see Appendix A), it is known that the effluent oxides of nitrogen are fairly ineffective destroyers of ozone at SST cruise altitudes, where they are recycled slowly, but much more effective as they diffuse upward into the ozone production region. Detailed model calculations are required to provide quantitative estimates of the effects of the NO_x emissions on ozone.

In principle, three-dimensional models of the general circulation of the atmosphere could produce the required estimates. Limited studies of the dispersion of the chemically inert tracer injected to simulate the SST fleet activity have been performed. However, such models representing the stratosphere with suitable computational refinement capable of handling the important chemical reactions have not yet been developed, because they exceed the capacity of available computers.

The most useful models currently available are one dimensional, in which the vertical profile of the eddy-diffusion coefficient is specified. In these models, it is assumed that quasi-horizontal spreading over a hemisphere is rapid compared with vertical spreading of effluents, and the global average vertical transport in the stratosphere is represented by the eddy-diffusion process. These models are capable of including all the known important chemical reactions of the stratospheric NO_x–HO_x–O_3 system.

Calculations from two-dimensional models and from a combination of a two-dimensional and a three-dimensional model have recently become available; these results are in general agreement with those of one-dimensional models. Future improvements in the ability to model the transport of chemically reactive substances would seem to lie in development of detailed multidimensional models, as well as more sophisticated applications of one-dimensional models.

We have considered the effects of several "standard" fleets of aircraft using a one-dimensional model; details of the model calculations and the values of certain model parameters and reaction rate coefficients (which are the best available) are given in Appendix A. We have assumed that, because of the weak interhemispheric coupling, the effects are confined mainly to the hemisphere in which the emissions occur. Some corridor effect—buildup of concentrations near flight routes—is also to be expected, but it is small compared with the interhemisphere difference when the injection is mainly within one hemisphere. Taking the two effects together, we consider the emissions from stratospheric aircraft to be approximated by a uniform distribution over the hemisphere involved. This provides a slight overestimate at low latitudes in the hemisphere concerned, a corresponding underestimate at low latitudes in the opposite hemisphere, and an even smaller underestimate at higher latitudes in the opposite hemisphere. It is a very good approximation at middle and high latitudes in the hemisphere of release.

The results of the model calculations of the ozone reductions for the various fleets of aircraft are shown in Table 4. The fleet emissions should be considered as equivalently defined by the fuel flow and, through an emission index, by the NO_x production. The numbers of aircraft are shown for illustration only. Actual characteristics and usage of aircraft would inevitably differ from values assumed here. For further orientation, the approximate number of wide-bodied aircraft flying in 1974 is shown. Further computations involved in the prediction of ozone reduction and the increase of erythemally effective exposure are described in Appendix A. Reduced emission indices would permit more aircraft to fly in the stratosphere for the same reduction of ozone.

Wide-body subsonic aircraft were assumed to be at cruising altitude 7 hours per day, with a total fuel flow of 10,000 kg/h and an emission index of 15 g/kg. Supersonic aircraft were assigned 5 hours per day, an emission index of 18 g/kg, and fuel flows of 19,000 and 50,000 kg/h for the two sizes. The predicted ozone reductions per hundred aircraft

TABLE 4 Decrease of Stratospheric Ozone and Increase of Erythemally Effective uv Radiation for Illustrative Fleets of Aircraft [a]

Fleet	Height (km)	Fuel Flow (10^9 kg/yr)	NO_x (10^6 kg/yr)	$X(NO_x)$ 10^{-9}	O_3 (% Decrease)	Erythemally Effective uv Radiation (% Increase)
1. 400 wide-body present subsonic	10.5	10.4	156	0.058	0.082	0.16
2. 100 wide-body projected subsonic	13.5	2.6	39	0.113	0.16	0.32
3. 100 present SST	16.5	3.49	62.8	0.52	0.72	1.44
4. 100 large SST	19.5	9.13	164	2.37	3.27	6.5
5. CIAP 1990 upperbound fleet	16.5	1.62	29.2	0.24	11	22
	19.5	30.6	551	7.96		
6. CIAP 2005 upperbound fleet	16.5	8.09	146	1.20	> 30	Large
	19.5	155	2790	40.3		

[a] The first four lines are intended for orientation and are not meant to indicate expected traffic, except for Line 1, which represents long-distance traffic for 1974. Lines 5 and 6 are taken, in simplified form, from the CIAP upperbound fleets. $X(NO_x)$ is the increase in mixing ratio (ratio of NO_x concentration to total atmospheric concentration) in the ozone layer after equilibrium has been reached. The last two columns are estimates of the corresponding percent reduction of ozone and percent increase of erythemally effective uv radiation. Confidence limits are estimated to be a factor of 2 for flight at 19.5 km, a factor of 3 for flight at 16.5 km, and a factor of 10 either way for subsonic aircraft.

are 0.7 percent for present SST's and 3.3 percent for large SST's. The relation between ozone reduction and fleet size is nearly proportional, dropping off somewhat for very large effects. Details are given in Appendix A, Sections I and III. The effect of injection at 19.5 km is almost twice as great as at 16.5 km, reflecting the fact that contaminants have farther to travel downward before being transported out of the stratosphere. Thus the burden corresponding to a constant rate of injection builds up accordingly. Table 4 gives percentage of ozone reduction as a function of the rate of NO_x injection. A factor of 2 greater than the calculated global average was applied to approximate the air-corridor effect and hemispheric confinement mentioned above.

The predictions in Table 4 are based on an eddy-diffusion coefficient profile that was constructed using observed vertical distributions of methane in the stratosphere. This permitted determination of the values of the eddy-diffusion coefficient as a function of height. The ability of this model to describe the average transport properties of the atmosphere has been confirmed by using it to simulate the transport of carbon-14 that was introduced into the stratosphere by nuclear weapons tests prior to 1963 and for which good observational data exist. The overall confidence in predictions of the effects of stratospheric modification has been enhanced by this confirmation and by the good agreement with observation that is obtained in other calculations of total ozone using a similar model (see Appendix A, Section III). Nevertheless, there is a great deal of subjectivity in the estimated uncertainties in the model calculations. The uncertainties are estimated to be a factor of 2 for flights operating at 20 km, a factor of 3 at 17 km, and a somewhat larger factor for those operating at lower altitudes. The uncertainty is larger close to the tropopause because of difficulty in adequately modeling the complex behavior of the atmosphere near the gap in the tropopause at midlatitude (see Appendix A, Figure A.20). This high uncertainty complicates predictions of the effect of the present subsonic traffic that flies in the lower stratosphere. The uncertainties in predictions will be reduced along with the subjectivity of their estimates as our ability to model atmospheric behavior improves.

There are other sources of atmospheric contamination not strictly relevant to high-altitude vehicle effects but of equal concern for understanding the region. The possible effect of chlorofluoromethanes (e.g., Freons) used for many years as refrigerants in air-conditioning systems and now extensively used as propellants in aerosol spray cans has only recently been recognized. Once these materials are released in the troposphere, their only means of removal appears to be transfer to the stratosphere, where they eventually are destroyed by the ultraviolet solar radiation that cannot penetrate to the lower atmosphere. One of the dissociation products, chlorine, is a potential catalyst for stratospheric ozone destruction. The time constant for any such effects in the ozone layer (see chemical reactions described in Appendix A, Section III.D) will be very long—on the order of 100 years for a constant rate of usage—due to the large initial accumulation rate in the troposphere compared with its rate of transfer into the stratosphere. It is therefore virtually impossible to be sure at this time that there is no competitive destruction process to photodissociation in the stratosphere and the subsequent loss of chlorine by downward transport to the troposphere. Preliminary calculations based on a continuation of current release rates suggest an ozone decrease of 1–2 percent by 1975, reaching 14–20 percent sometime in the more distant future. However, a very large uncertainty exists, based on our current level of understanding. It seems clear that a detailed assessment of this problem should be made in view of the possibility of large ultimate effects and the long time that would be required for the atmosphere to cleanse itself of chlorofluoromethanes. Because of the long time-constant due to the tropospheric storage, there is time for interhemispheric transfer to occur, and the problem is worldwide rather than hemispheric in nature.

Effects from the Space Shuttle could also occur. Preliminary estimates indicate that the various chemical effects (principally catalytic ozone reduction by Cl due to HCl in the exhaust gas) will probably be relatively small—an ozone decrease of about 0.4 percent for a traffic rate of 50 flights per year. However, the chemical processes involved are complex, and a more detailed examination of the potential impact of the Shuttle on the stratosphere is needed before a definitive answer can be given. NASA is conducting a study of the matter.

IV. Effects on Ultraviolet Radiation from the Sun

Ozone is a strong absorber of the sun's rays in the ultraviolet portion of the spectrum. Hence, the ozone layer is responsible for the ultraviolet cutoff of the solar spectrum received at the earth's surface. In terms of wavelengths of ultraviolet light received on the earth, the shielding effect of the present ozone layer begins at about 320 nm (nanometers) and becomes increasingly effective at shorter wavelengths—the so-called cutoff being near 295 nm. Variations in the ozone content of the stratosphere shift this cutoff; for example, a reduction in ozone allows more shorter-wavelength ultraviolet radiation to pass through. It is a straightforward matter to calculate the relative change of ultraviolet light at the earth's surface for any given change in ozone. Subtle changes in the amount of ultraviolet light reaching the surface of the earth in this narrow wavelength band at 295–320 nm are not trivial. In particular, it is at these wavelengths that the skin of many people is vulnerable to cumulative damage. It is now well established that this portion of the sunlight causes sunburn—light-induced skin irritation or erythema. Today, many physicians believe that the erythemal action spectrum also applies for skin cancer (i.e., that the solar-induced skin irritation may be a principal stress factor in producing the cancer). Additional evidence supporting this view is provided by the variable incidence of skin cancer with latitude; it varies roughly in proportion to the erythemally effective ultraviolet radiation rather than in accordance with longer wavelength ultraviolet or visible radiation.

Another theory, not incompatible with the above, on the cause of skin cancer is widely held by biologists, namely, that it is related to genetic damage to DNA molecules caused by absorption of ultraviolet light. It happens, moreover, that the wavelengths that produce sunburn are roughly the same ones that cause DNA damage—that is to say, the action spectra for these two hypothetical causes of skin cancer are about the same, which suggests a causal relationship.

It is now believed that the most effective wavelengths in the sun's rays that relate to skin cancer are between 295 and 310 nm; the shorter wavelengths are strongly attenuated by ozone, and the longer wavelengths are intrinsically less effective in producing biological damage. This is based on calculations of the combined effects of the skin's action spectra and the ozone cutoff of the sun's spectrum. Appendix B describes this combined action and presents methods for predicting changes in biological dosage with changes in ozone. Briefly we can say, however, that the percentage increase in biological dosage is about twice the percentage decrease in ozone.

V. Research and Monitoring

In the early 1970's when the possibility of deleterious effects from aircraft flying in the stratosphere was first noticed, little was understood of stratospheric processes. There was less known about this region than about other parts of the atmosphere above and below because it is too high for investigation by many meteorological research aircraft, and remote sensing from satellites had not been adequately developed. As the U.S. Department of Transportation undertook to organize its Climatic Impact Assessment Program (CIAP) and other interested countries took similar steps, it was found that there were few observational data available to check the various theories on the stratosphere that had been developed during the previous decade.

Progress has been made over the last three years in conceiving and developing some of the new instruments needed and deploying them on platforms such as high-altitude aircraft, balloons, and rocket-launched parachutes. Remote sensing from satellites is increasingly a source of data, especially valuable because of its wide geographical coverage. However, the gathering of information on composition and properties has barely begun. It is important to measure not only the stratospheric constituents of direct interest, such as ozone and nitrogen oxides, but also substances that act as tracers and give information about atmospheric transport. Progress in computerized modeling of climate depends crucially on such data for verification and for the generation of new ideas.

Repetitive and intensive sampling and analyses of the various species of interest in the atmosphere can be considered as monitoring. In this sense, two types of monitoring may be distinguished: scientific monitoring and regulatory monitoring. By scientific monitoring is meant obtaining measurements of sufficiently high quality to provide data usable to advance knowledge. From this knowledge will come new ideas and validated models of the complex system of stratospheric chemistry and transport. Regulatory monitoring consists of regular, repetitive measurements of certain quantities for the purposes of ensuring that legally prescribed limits of these quantities are not exceeded. It is not necessary for the regulatory monitoring to produce data of a quality sufficient for scientific use. At the present time there are no statutory or other prescribed limits on stratospheric contaminants. If and when there are, it will be necessary to have a regulatory monitoring program. The question will arise as to what substances should be monitored. The answer to that question will come from the experience that a scientific monitoring program will provide. Appendix A gives a fairly complete list of substances that are thought to be of importance in the stratospheric chemical system. Monitoring should start promptly, before any appreciable man-made modification has occurred.

VI. Conclusions and Recommendations

Based on present empirical data and theoretical models of the stratosphere, we CONCLUDE that one can predict to within a factor of 2 or 3 the extent to which the ozone layer will be reduced for a given injection of NO_x into the stratosphere. With an emission index of about 18 grams of NO_x per kilogram of fuel burned, an average fleet of 100 of the present Concordes and TU-144's, operating at altitude 5 hours per day, can be expected to reduce the ozone layer by about 0.7 percent; the effect of 500 such airplanes would be five times greater.

Because of the natural variability of concentrations of minor constituents of the stratosphere and the problems associated with sampling and analyzing trace constituents, adequate knowledge of the chemical composition of the stratosphere is not easily attainable. The transport properties of the atmosphere are even more difficult to measure on an extensive basis. Yet, we must have this kind of information if we are to reduce the uncertainties in our predictions. Reduction of these uncertainties is important even for assessing the impact of the present subsonic fleets. *We* RECOMMEND *that studies of physics, chemistry, and transport phenomena in the atmosphere be continued. We endorse the concept that the United States National Aeronautics and Space Administration undertake research on stratospheric phenomena, including inadvertent modification.*

Stratospheric modification by SST's is a worldwide problem that calls for close cooperation internationally. We believe that existing international organizations (e.g., the International Council of Scientific Unions, the International Civil Aviation Organization, the World Meteorological Organization) should make a detailed study of the problem. *We* RECOMMEND *that international organizations concerned with the environment and with aerospace flight organize and coordinate the monitoring of any changes in the stratosphere that may be caused by high-altitude flight and that monitoring begin promptly.*

We CONCLUDE that present rates of release of chlorofluoromethanes (e.g., Freons) at the earth's surface may also represent a significant problem in terms of ozone reduction. *We* RECOMMEND *that an immediate quantitative assessment be made of the possibly large stratospheric effects of stable compounds such as the chlorofluoromethanes involved in aerosol spray cans and refrigerants. This assessment should include the stratospheric effect of carbon tetrachloride, methyl chloride, and analogous bromine compounds.*

We also RECOMMEND *that the study initiated by the United States National Aeronautics and Space Administration concerning the effect on the stratosphere of the Space Shuttle be continued on high priority.*

Planning and budgeting by governments concerning research on the stratosphere will determine how rapidly progress will be made. To assist in this area the Panel BELIEVES that a Board or Council should be appointed to review and advise on the activities of various U.S. government agencies engaged in stratospheric research.

3

Biological and Medical Effects of Nitrogen Oxide Emissions

I. Introduction

It is shown in the previous chapter that the addition of NO_x to the stratosphere from aircraft engines or other sources can reduce the total ozone column density with a consequent increase in the intensity of solar ultraviolet (uv) radiation reaching the earth's surface. This chapter addresses the question of whether an increase in ultraviolet flux can pose a problem to the human population, to agriculture, or to any other part of the biosphere.

II. Sun Exposure and Man

Although people have recognized that they should take precautions against painful sunburn, depending on their particular susceptibility, there has been a general belief that sunlight is one of the gifts from nature. The danger of excessive sun exposure is a subject that has only recently received considerable attention. It is now virtually certain that there are a number of negative long-term effects of sun exposure.

There is no immediate warning signal of inadvertent uv exposure

---

This chapter presents the findings of the Panel on Biological and Medical Effects: R. Setlow, *Chairman;* W. Baier, W. Butler, R. Clayton, E. Deevey, T. Fitzpatrick, A. Giese, D. Gordon, F. Haxo, E. Runge, E. Scott, and F. Urbach. Consultant to the Panel: M. Caldwell.

such as occurs with a heat burn. It is surreptitious and invisible, and reaction to it is delayed (even sunburn is a delayed reaction). There is strong evidence that damage is cumulative—the older a person is the more damage he has. Increased exposure to uv is known to increase the rate of skin aging. As one dermatologist put it, "the beaches are filled today with people seeking to improve or maintain those highly coveted *dark golden tans*—unknowingly they are building *elephant skin* for their later years." Only recently have there been public warnings that excessive exposure to sunlight can cause skin cancer.

The major known direct effects of the sun's uv radiation are on the skin and eyes, because uv radiation will not penetrate deeper than the skin itself (see CIAP Monograph 5, Chapter 3.5). At short wavelengths little uv radiation penetrates the ozone layer; at long wavelengths the skin is relatively insensitive. The damage comes from a narrow band of the solar uv spectrum from 295 to 320 nm—a band whose intensity at the earth's surface is most altered by changes in stratospheric ozone concentration. This range nearly coincides with the 280–315-nm band for which the scientific shorthand is *uv-B*.

There are no quantitative data on prematurely aged skin so that it is not possible to quantify its relationship to solar radiation. On the other hand, there are data on skin cancer—the most common of all cancers—that indicate a close association with exposure to sunlight, particularly the uv component of sunlight.

A. EVIDENCE LINKING CANCER WITH EXPOSURE TO THE SUN

Statistics show that skin cancer is more prevalent in geographical locations near the equator than at higher latitudes. The incidence doubles with each 8–11 degree decrease in latitude. This latitude effect is significant because incident uv radiation increases from the poles to the equator. A person living in Texas receives considerably more uv-B than a person living in Minnesota, assuming that both spend the same amount of time in the sun. There are, of course, other factors that must be reckoned with in dealing with these statistics, including local cloudiness, lifestyle and hours of outdoor exposure, hereditary genetic differences, and the amount of natural pigmentation (which serves as a built-in filter). However, the incidence of skin cancer and the mortality due to skin cancer are better correlated with the intensity of uv-B radiation than with any other environmental variable.

Of the two different general types of disease that come under the broad category of skin cancer, malignant melanoma is the most dangerous; the prognosis for malignant melanoma is similar to that for breast and other commonly known cancers. The more common non-melanoma skin cancer, although rarely lethal, is disfiguring, expensive, and ultimately very limiting—if a person develops one skin cancer then he is advised to limit his exposure to sunlight in an attempt to control recurrence. Moreover, the problem of an increase in the incidence of skin cancer will not necessarily be solved by better medical care alone. Prophylaxis is unrealistic—people will simply not protect themselves by staying out of the sun. For example, although Queensland, Australia, has the highest incidence of skin cancer in the world, the population does not protect itself any more than do less susceptible populations.

Because the association of skin cancer with solar uv radiation is strong, it is reasonable and prudent to analyze the epidemiological evidence in terms of incidence (or mortality) as a function of uv flux. An extensive description of skin cancers and the epidemiological data will be found in Appendix C. The following are brief summaries of the biological evidence that leads us to associate solar uv light in the range 295–320 nm with human skin cancer in light-skinned individuals.

Basal- and squamous-cell carcinomas are the most common forms of skin cancer. Their association with exposure to *sunlight* may be summarized as follows:

1. They occur almost exclusively on exposed areas of the body that receive high light intensities: nose, ears, cheeks, back of the neck, etc.
2. There is higher incidence among outdoor workers: farmers, sailors.
3. There is higher incidence among fair-skinned people: Celtic, albino-Indians of Panama.
4. There is higher incidence at low latitudes (see Figure 3).

Their association with the *uv-B component* of sunlight may be summarized as follows:

1. Wavelengths known to be effective in producing erythema (sunburn) are below 320 nm (see Figure 4). Injury of the skin by the sun is most likely the principal stress factor in inducing skin cancer. Sunburn and skin cancer arise in the same tissue, and individuals who sunburn easily have a higher than average probability for developing skin cancer.
2. In numerous experimental studies on mice, wavelengths below 320 nm were effective in inducing skin cancer. Longer uv wavelengths were ineffective, although a recent experiment indicates that they may accentuate the effects of shorter-wavelength radiation.
3. There is good evidence that skin cancer arises from uv-induced

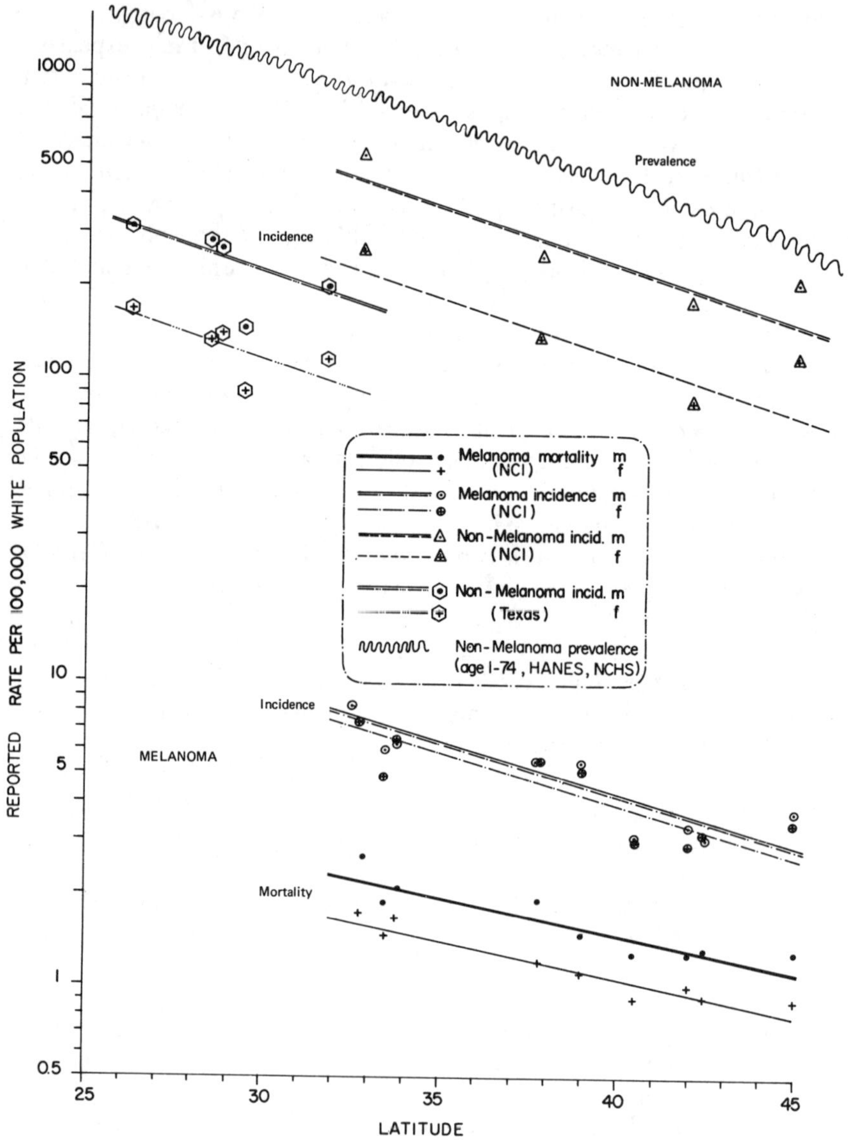

FIGURE 3 Reported skin-cancer rates among whites as a function of latitude. [Sources: melanoma mortality from Mason and McKay (1974), melanoma incidence from National Cancer Institute (1974), nonmelanoma skin-cancer incidence (NCI) from Scotto *et al.* (1974), nonmelanoma skin-cancer incidence (Texas) from Macdonald (1974a), and prevalence of nonmelanoma skin cancer based on preliminary data from the Health and Nutrition Examination Survey of the National Center for Health Statistics (McDowell, 1974).] (The complete references appear in Appendix C.)

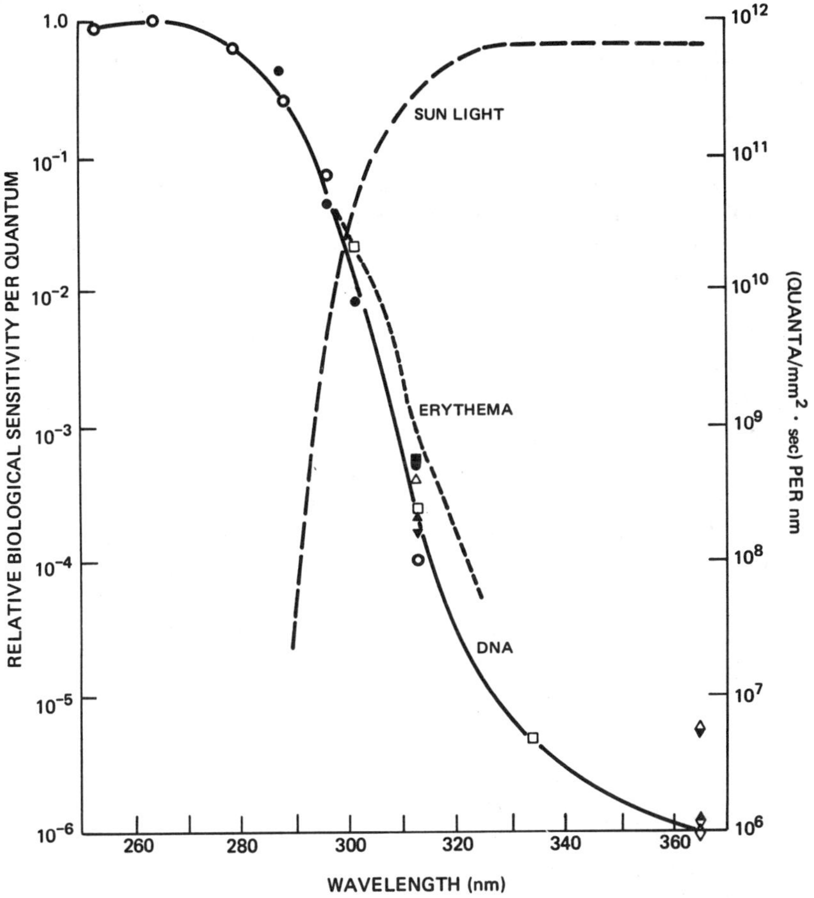

FIGURE 4 The solid curve represents the average action spectrum for affecting DNA. The dotted curve is the long-wavelength portion of a recent erythemal action spectrum normalized to the DNA spectrum at 297 nm. The sun's spectrum at the earth's surface was calculated by Green for Gainesville, Florida, for 2.3 mm O_3 and a zenith angle of 25°. The sun's spectra for different amounts of ozone are shown in Appendix B, Figure B.3, and the range of effective wavelengths (the product of the action spectrum and the sun's spectrum) are detailed in Figure B.4 of that appendix.

changes in DNA—the carrier of genetic information in living cells. Wavelengths below 310 nm are hundreds of times more damaging to DNA than are the wavelengths in excess of 330 nm (see Figure 4). Ultraviolet light is mutagenic, and since many chemical carcinogens are also mutagens (i.e., affect DNA), we expect uv light to be carcinogenic by virtue of its action on DNA. Individuals with the disease xeroderma pigmentosum are extraordinarily susceptible to skin cancer induced by sunlight (in one study a prevalence of greater than 50 percent by age 10). The cells of these individuals are defective in at least one of several molecular mechanisms that repair uv damage to DNA. Hence, there is a strong implication that damage to DNA that is not properly repaired leads to cancer.

The available evidence indicates that the spectral sensitivity for skin cancer is similar to either the action spectrum for erythema production or the spectrum for damaging DNA. The two spectra are similar but not quite identical (see Figure 4).

Malignant melanoma is much rarer than the skin carcinomas mentioned above, but its mortality rate is about one third of the incidence. Its occurrence is also associated with exposure to *sunlight* through the following observations:

1. It occurs on exposed areas of body but *not* exclusively those that receive the highest intensity.
2. There is higher incidence among fair-skinned people.
3. There is higher incidence at low latitudes.

Its association with the *uv-B component* of sunlight can be summarized as follows:

1. Sunburn and melanoma arise in the same tissues, and individuals who sunburn easily have a higher than average probability for developing skin cancer.
2. Although uv irradiation alone does not give rise to melanomas in mice, irradiation of a *chemically induced* benign pigmented lesion does do so.
3. Individuals with xeroderma pigmentosum (see above) have an extraordinarily high prevalence of melanoma compared with the average population. (A recent study observed a prevalence of 7 out of 15 for these individuals compared with about 2 per 10,000 for the average population.)

Although the evidence associating uv-B with malignant melanomas is not so strong as for nonmelanoma, we believe that the only action spectra that we can prudently use for any quantitative estimate of the potential hazard arising from an increase in uv-B are those given in Figure 4.

B. ESTIMATION OF THE EFFECT OF INCREASED uv-B ON THE INCREASE OF SKIN CANCER AMONG LIGHT-SKINNED INDIVIDUALS

There have not been any reports of skin cancer that resulted from therapeutic uv irradiation of skin or from treatment of skin diseases with other agents that also are carcinogenic in mice. However, there have not been any long-range studies, or any comparisons between such treatments and animal experiments, that would permit quantitative estimates of skin-cancer hazards of therapeutic uv irradiation of man.

Even though there is a problem about extrapolating quantitatively from skin cancer in mice to skin cancer in man, we do know that the molecules of mice are similar to those of man and that the photochemical reactions in both mouse and human cells are similar to reactions in solution. The scientific community cannot do laboratory experiments on uv-induced skin cancer in man; and even if it could, it would take too many years for the results to be available. Hence we are forced to rely on the wealth of photobiological data that indicate that uv irradiation is harmful for the cells of all creatures and to use action spectra similar to those in Figure 4, combined with *epidemiological data,* to estimate reasonable values for the increase in skin cancer resulting from an increase in uv-B.

The epidemiological data used by the Panel to Review Statistics on Skin Cancer of the Committee on National Statistics of the National Research Council are summarized in Figure 3. They are described in greater detail in Appendix C. That Appendix also discusses the problems associated with using such data for quantitative prediction. Laboratory and clinical evidence suggests a causal relation between uv-B and skin cancer and therefore suggests an empirical study of the relations between observed incidence and mortality of skin cancer and the intensity of uv-B for a number of locations. Our understanding of the skin-cancer process from an epidemiological point of view is quite limited. We do not know the quantitative answers to many important questions, including which variables in addition to uv flux are important. We have good theories as to the uv wavelengths that are important for skin cancer, but we do not know how the division of total exposure into doses determines the effectiveness of a given total exposure or how the way in which we live determines both total exposure and its division into doses. Without this knowledge, the best approach seems to be to learn what we can from the available data for skin-cancer rates by place, age, and sex.

Among the simple measures of the incidence of skin cancer and of the uv flux, the ones that are mutually related in an almost linear manner are (a) the logarithm of the incidence and (b) the flux, defined as the number of photons per square meter per second per nanometer *averaged over the year* (see Figure 5). Moreover, use of these variables leads to residuals of about the same size for different flux levels. As a consequence, all the analyses reported here are based on linear relations involving these two variables, either alone or in combination with others (see Appendix C).

The simple relation takes the form $\mu=\alpha+\beta F$, where μ is the log incidence (or mortality), α and β are constants to be fitted, and F is a computed uv flux. The fit for 14 combinations of sexes (2) and age-groups (7) is good in the large but not completely satisfactory in detail; individual geographical areas are often consistently above or below the fit for all combinations of age and sex. Clearly there are other variables (perhaps physical or related to lifestyle; see Appendix C) that make definite, but not large, contributions to the observed log incidences. Until these variables are identified—and measured—we have little choice but to use the simple relationship above and let these additional variables contribute to our uncertainties.

Because the other relationships studied are so poorly determined with so few observations, we have chosen to present the results of analyses of log incidence (or mortality) and uv flux alone and to discuss in Appendix C the difficulties that arise when more complicated, but potentially useful, models (i.e., interpolation and extrapolation formulas) are tried out.

Table 5 gives estimates and statistically determined confidence intervals for percentage increase in skin-cancer incidence and mortality for an increase in uv-B associated with a 10 percent ozone reduction. The predictions are for two geographical locations and make use of a wavelength sensitivity intermediate between that for action on DNA and that for sunburn. Further discussion and tables with more detail appear in Appendix C. In view of the strong possibility that other environmental factors influence skin-cancer statistics, the actual uncertainties in the predicted increases in skin cancer are greater than indicated in Table 5.

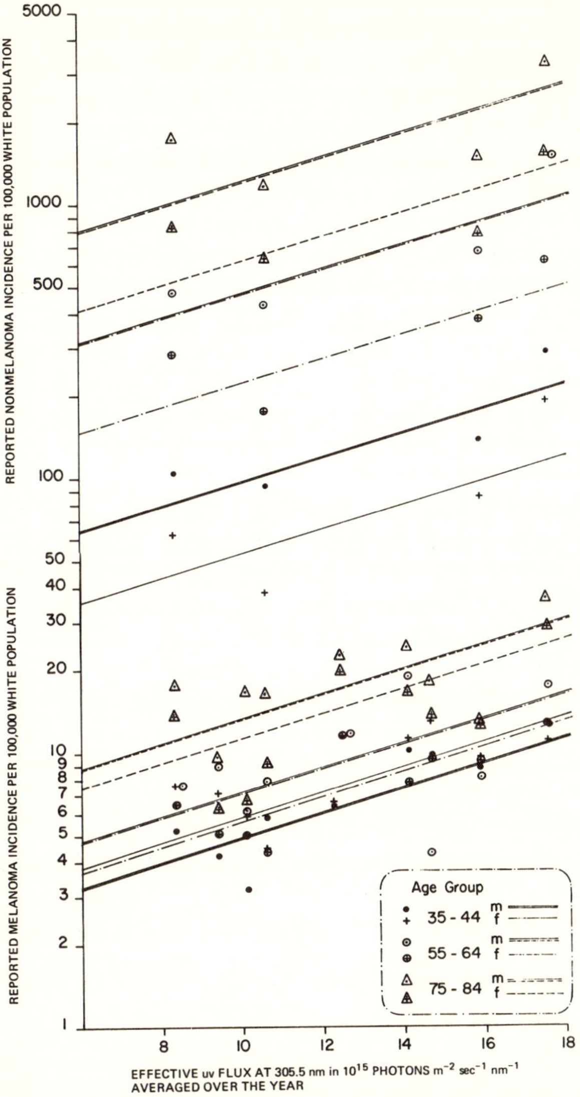

FIGURE 5 Reported skin-cancer incidence rates per 100,000 white population among three age groups as functions of ultraviolet flux. [Sources: melanoma incidence from National Cancer Institute (1974), nonmelanoma skin-cancer incidence from Scotto (1974) and Scotto *et al.* (1974).] (The complete references appear in Appendix C.) The computations of uv flux used to prepare this figure are not adjusted for altitude.

TABLE 5 Estimates of Percentage Increase in Skin Cancer for a 10 Percent Reduction in Ozone [a]

Disease	Rate	Data Source [b]	No. of Areas	Locality	Estimate (%)	90% Confidence Interval
Nonmelanoma (clearly associated with exposure to uv-B)	Incidence	SNCS [d] 1947–48	10	Dallas-Ft. Worth	109	80 to 145
				Minn.-St. Paul	48	36 to 60
	Incidence	TNCS 1971–72	4	Dallas-Ft. Worth	40	9 to 83
				Minn.-St. Paul	19	4 to 36
	Prevalence [c]	HANES [d] 1971–72	35	Dallas-Ft. Worth	73	27 to 143
				Minn.-St. Paul	34	13 to 60
	Mortality			(Data not useful for this purpose)		
Malignant melanoma (less clearly associated with exposure to uv-B)	Incidence	TNCS 1969–71	9	Dallas-Ft. Worth	48	39 to 59
				Minn.-St. Paul	22	18 to 26
	Mortality	M & M 1950–69	9 in TNCS	Dallas-Ft. Worth	30	23 to 38
				Minn.-St. Paul	14	11 to 18
	Mortality	M & M 1950–69	73 in HANES + TNCS	Dallas-Ft. Worth	13	10 to 15
				Minn.-St. Paul	6	5 to 7

[a] Estimates and 90 percent confidence limits of percentage increase in skin cancer among whites in two localities for a 10 percent ozone reduction using the simple model ($\mu = \alpha + \beta F$) and the weighted average effective uv flux corresponding to an action spectrum midway between that for DNA damage and that for erythema. (See Appendix C for a discussion of the epidemiological data and for predictions of percentage increase using other formulas.)

[b] SNCS, Second National Cancer Survey (melanoma and nonmelanoma together); TNCS, Third National Cancer Survey; HANES, Health and Nutrition Examination Survey; M & M, Mason and McKay.

[c] The sum of the prevalences of skin cancer, severe solar keratosis, and moderate solar keratosis used for the estimates.

[d] Slightly different weights of uv flux used.

How much greater is far from clear. Accordingly, the confidence intervals proposed should be interpreted as minimum uncertainties. Nevertheless, although the predictions in Table 5 cover a wide range and are not completely consistent, they provide strong evidence that increases in uv will produce increases in skin cancers. The model predicts that a 10 percent decrease in stratospheric ozone will give rise to about a 20 percent increase in melanoma mortality. However, a 10 percent decrease in stratospheric ozone appears to give more than a 20 percent increase in the incidence of skin cancer—possibly a 30 percent increase.

Several other estimates of the percentage increase in skin cancer that would result from a decrease in ozone are summarized in CIAP Monograph 5, Chapter 3. These estimates employed similar carcinogenic action spectra but different models than reported above. They used many fewer data than used by the Panel on Skin Cancer Statistics. There is good qualitative agreement among the various estimates that a 1 percent decrease in stratospheric ozone would cause a 2 percent increase in skin cancer.

III. Ultraviolet Sun Exposure and Living Organisms Other Than Man

Living organisms withstand the damaging effects of uv-B by four principal means: protective coverings and pigmentation, behavioral adaptation to avoid the sunlight, photoreactivation or photoprotective mechanisms, and dark repair mechanisms. Many forms of life have developed one or more of these protective mechanisms sufficient to tolerate existing levels of uv-B radiation. A significant increase of uv-B can be expected to precipitate a disturbance in the existing balance of life.

A. EXPERIMENTAL LIMITATIONS

It is difficult to determine the effect that increases in uv-B will have on ecological balance. Wide variation in uv levels occurs over the globe naturally; the range of natural variation of uv over latitude alone far exceeds any changes that might be caused by aircraft fleets of moderate size. While species have adapted to uv, they have also adapted to climate and soils, which also vary geographically. One cannot assume that, if uv were increased globally, plants and animals that have learned

how? ?

to survive at high uv levels at low latitudes would then migrate to higher latitudes—to do so would require readaptation in many other ways as well.

Experiments to determine the consequences of reduced or enhanced uv-B in sunlight on living organisms are difficult to perform. Plants and animals that naturally live at high latitudes can be moved to lower latitudes and observed in uv-enhanced sunlight, but the climate must be controlled to simulate the lower temperatures, humidity, and reduced total sun exposure characteristic of higher latitudes. This approach cannot be used with plants and animals that already live in the tropics. Electric lamps can be used to *supplement* the uv of natural sunlight. Unfortunately, the only lamps available today also emit appreciable levels of longer-wavelength uv (e.g., a strong emission line at 365 nm), and there is evidence that presence of other wavelengths may modify the effects of uv-B. Improving lamp technology will require several years. In supplementation experiments attempting to simulate a 50 percent ozone reduction, the adverse effects of increased uv-B tended to be diminished when high intensities of visible light were also present, indicating the significant role of photorepair and photoprotective mechanisms in nature. Similar photoprotection was also observed in insects. Since reduction of ozone changes the intensities of only the shorter wavelengths, it is difficult to use simple supplementation experiments to predict quantitatively the consequences of ozone reduction. The study of the effects of uv-B on aquatic systems has been hampered by the paucity of reliable measurements on the penetration of uv-B into natural waters of the sea.

An alternate approach is to selectively delete uv-B using Mylar plastic filters or take into account the natural filtering effects of seawater for aquatic studies. Such deletion experiments indicate that present ambient levels of uv-B exert an inhibitory effect on photosynthesis in phytoplankton (sea life). Similar experiments on higher plants under field conditions indicate essentially no effect.

B. A FEW CONCLUSIONS ABOUT PLANTS AND uv-B

The conclusions outlined below are derived mainly from studies sponsored by the Department of Transportation. These studies and other relevant data are summarized in CIAP Monograph 5 (especially Chapter 3). Some of the data are given in Appendix D.

Deletion experiments are of questionable validity, because existing protective mechanisms, such as photoreactivation, may counteract the deleterious effects of uv-B. To date, experiments under field conditions have been largely of the deletion type because of the technical difficulties of large-scale supplementary irradiance—the energy requirements alone make supplementation experiments in the field almost prohibitive. However, the Panel believes that deletion experiments on plants, showing little or no detectable effect under field conditions, should not be taken to predict the consequences of *increased* levels of uv-B radiation.

The more useful experiments to date are those that have used supplemental uv-B radiation in growth chambers, in greenhouses, or in the field. In the studies sponsored by the Department of Transportation, supplemental uv-B was designed to simulate the effect of a 50 percent reduction of ozone. This simulation suffered from the fact that light included uv-A (at 365 nm) as well as uv-B. As a control, the light was screened with Mylar allowing only the uv-A to come through. The following results emerged from these studies:

1. Plants exposed to supplemental uv in growth chambers or in greenhouses showed 20 to 50 percent inhibition of growth. The Mylar control (uv-A only) caused less than half this degree of inhibition. Experiments in the field on the effects of supplemental uv on plant growth showed smaller, but statistically significant, decreases for some species.
2. Plants exposed to supplemental uv in growth chambers showed a 10 to 30 percent decline in chlorophyll content, accompanied by a similar decline in their capacity for photosynthesis.
3. Plants exposed to supplemental uv in the field developed degenerative changes in the structure of some of their cells. The frequency of harmful mutations, most easily scored as abnormal or missing stamen hairs, was increased sevenfold by uv-A alone (Mylar control) and twentyfold by the full supplemental uv-A plus uv-B.
4. Seedlings are more sensitive to uv than are mature plants, and single-celled algae are many times more sensitive. A few hours' exposure to natural sunlight is enough to kill many types of single-celled algae, and a 50 percent reduction of ozone is expected to decrease the lethal exposure time by more than a factor of 2.

Although the experiments on plant growth in the field were not completely conclusive, and although it is hazardous to interpolate from the 50 percent reduction level, the research sponsored by the Department of Transportation's CIAP during its short existence has laid valuable groundwork. Much more research is clearly needed before the effects of ozone reduction at levels of 10 percent or less can be predicted with confidence.

C. uv-B AND INSECTS

Some information about the effects of uv-B on insects is available. Mortality of insect larvae in cages is increased by supplemental uv, but in nature these larvae keep themselves sheltered from the sun. The eyes of insects are equipped to detect uv and to see patterns of polarized light that indicate the position of the sun. They are guided in feeding and in pollinating by uv reflectance patterns on flowers. Mate selection, at least in some butterflies, depends on visual cues that include uv reflectance patterns. A single small-scale experiment has been conducted to see what insects might do when the uv illumination on alfalfa plants in the field is increased in the uv-B range corresponding to a 50 percent reduction of ozone; this experiment gave no statistically meaningful results.

D. EFFECTS OF uv-B ON AQUATIC SYSTEMS

Primary production in the ocean depends on the photosynthetic activity of a whole range of species of living organisms. Because of the unique importance of phytoplankton and zooplankton to the ecology of the world's food chain, the sensitivity of these species to enhanced levels of uv-B is of prime concern. Elucidation of this question requires a thorough understanding of how deeply various wavelengths of uv penetrate natural waters. A few preliminary measurements of the penetration of uv-B into natural waters have been made using the Robertson uv sensor, but unfortunately this sensor is incapable of measuring the integral of radiance with a large enough solid angle to take into account diffuse radiation coming from the scattering effect of the atmosphere and of the ocean. Development of new instruments will be required for this purpose. At this time, no precise relationship between turbidity and penetration of uv-B radiation into seawater has been determined. Until this information is available, is related in some way to the vertical distribution of planktonic organisms, and is averaged meaningfully over the major population of plankton in the seas, it will not be possible to predict the effects of ambient or increased uv-B on marine plants and animals.

IV. Conclusions and Recommendations Regarding the Potential Biological Effects of Reducing Stratospheric Ozone

Laboratory experiments have shown that all unshielded cells are highly vulnerable to sunlight and may be killed by relatively short exposure to full sunlight. While such cells and organisms are generally protected to varying degrees in nature so that they experience sublethal doses of radiation, any increase in uv radiation could be considered to increase the pressure against survival. Because of the relationships between species in ecosystems, damage to one species might jeopardize an entire ecosystem. Hence, the potential effects of any elevation of the present uv-B levels of sunlight reaching the earth's surface should be taken most seriously and studied thoroughly.

There are persuasive, almost compelling, biological and medical arguments that uv-B is a cause of skin cancer. Statistical analyses show that uv radiation is associated with skin cancer, but the magnitude of the effect is indicated only within rather wide limits. The only relationship that has been found consistently to be applicable to a number of different data sets from different sites is log incidence (or mortality) linearly dependent on uv irradiance. A 10 percent decrease in stratospheric ozone causes about a 20 percent increase in melanoma mortality. However, while a 1 percent decrease in stratospheric ozone causes roughly a 2 percent increase in the incidence of skin cancer, a 10 percent decrease in stratospheric ozone appears to give more than a 20 percent increase in the incidence of skin cancer—possibly a 30 percent increase. The statistical results and predictions presented in Table 5 are preliminary and may be changed as a result of further analysis. The general pattern of the results is not likely to change, but explicit and useful relationships identifying environmental factors other than uv radiation may be found. Considerably more research is needed to establish the quantitative nature of these relationships.

The effects of uv-B on organisms other than man could also be deleterious, with some organisms being much more sensitive to this radiation than others. The Panel CONCLUDES that considerably more research is needed before the effects of increased levels of uv radiation on living organisms and natural ecosystems can be adequately predicted. Much time will be required not only to determine the direct effects of uv radiation on a particular organism but also to recognize the more subtle and long-term influences on the total ecosystem. The Panel

BELIEVES that any conclusions on the effects of ozone reduction on living organisms must be of limited scope at the present time because there are potentially great deleterious effects as yet unevaluated. The time for biological studies in CIAP has been too short even to design and execute reliable short-term field-scale experiments. It is imperative that follow-on biological studies be undertaken.

We RECOMMEND *that biological and medical studies of the effects of changes in ultraviolet radiation upon living organisms be increased. In particular, we* RECOMMEND *that studies of skin cancer be accelerated to obtain better quantitative understanding of the cause of the disease.* Specific studies needed are (1) the earth-based monitoring program initiated during CIAP should be pursued in conjunction with atmospheric monitoring and should include continuous measurements of uv-B and total sunlight at a number of stations at both high and low latitudes (greater than 55° and less than 10°) for at least five years; (2) a program is needed to gather epidemiologic data on prevalence of non-melanoma skin cancer over a wide span of latitudes and incidence data, including age of onset versus uv-B, sex, occupation, exposure habits, and skin phenotype; (3) a study should be made of the penetration of uv-B into natural waters—uv-B sensors to measure diffuse irradiance under water need to be developed; (4) a research program should be set up to study the effects of uv-B on living organisms from microbial forms to higher plants and animals—a funding commitment should be made over sufficient periods of time (up to five years) for those projects that require a prolonged period of experimentation and data collection.

4

Effects of Exhaust Effluents on Surface Climate

I. Introduction

A second environmental problem identified as a potential consequence of operating large numbers of aircraft in the stratosphere is effects on the earth's climate—temperature, precipitation, or cloudiness. Emissions in the stratosphere of both NO_x and SO_x can affect surface climate. NO_x emissions have been discussed earlier. Sulfur oxides are formed in the combustor from the reaction of sulfur in the fuel with oxygen in the air. Studies indicate that the injection of sulfur oxides in the stratosphere will lead to the formation of *aerosols* (i.e., particles of solid or liquid suspended in the air), which affect the transmission of light to the earth (see Appendix F, Section III).

Climatic effects, such as change in the earth's surface temperature and change in precipitation, could possibly be traced to aerosol formation as well as to changes in NO_x or ozone content, if aerosols occur on a large enough scale.

II. Natural Climatic Variability

Climate involves weather over a long period of time, usually measured in decades or longer; it may focus on a given area of the earth's surface

This chapter presents the findings of the Panel on Atmospheric Physics and Chemistry: F. Johnson, *Chairman;* J. Friend, *Vice Chairman;* J. Chamberlain, D. Hunten, H. Johnston, F. Kaufman, H. Landsberg, and H. Panofsky. Consultants to the Panel: R. Hudson, C. Leith, E. Reiter, and G. Robinson.

or encompass the state of the entire globe. Although climate is often represented by average variables of temperature and precipitation for a given area and period of time, it includes a broad range of conditions and fluctuations from year to year. Occasionally extremes occur—flood-producing rains, droughts, or excessive heat or cold. Climatic indicators generally have a rather wide dispersion; climate fluctuates in irregular rhythms over both short and long intervals for reasons that are poorly understood. These natural fluctuations cover a wide range—since the end of the last glacial period about 11,000 years ago, there have been alternating periods measured in centuries during which the climate was warmer or colder, wetter or dryer than the present. Because of the great variability, transitions from one stage to another are discernable only after considerable time has elapsed following a change in trend.

The enormous masses of the oceans act as a climatic flywheel, smoothing out any impulses that might otherwise lead to a sudden change in climate. There are a multiplicity of such impulses, both terrestrial and extraterrestrial. They are only incompletely understood and only partially accessible to the tools of physical and mathematical simulation.

Present climatic models fall far short of completeness in representing real climatic conditions. They are approaching the stage where average climatic conditions over the globe can be represented numerically with some accuracy; and some of the models seem to simulate dispersion statistics of climate reasonably well, although it remains to be seen if they do so for the right reasons. No model can yet handle in detail the extremes in weather conditions, that is, localized weather as distinct from general synoptic-scale conditions. Present climatic models have very limited capability for establishing the climatic impacts of stratospheric contamination in quantitatively reliable detail, but they do provide a generalized measure, attainable in no other way, of what those impacts might be. Internal meteorological interactions, such as those involving clouds, are not yet realistically incorporated in climatic models; this is an important limitation but not necessarily a crippling one. Present numerical climatic models cannot yet contribute a great deal to the question of whether emissions from high-flying aircraft might influence the climate, but the Panel BELIEVES that it is highly desirable to continue to develop these methods for simulating climate.

Natural variations in climate can conceal for a considerable length of time a small change in temperature—even as large as $\pm 0.5\,^{\circ}\mathrm{C}$ on the global scale. Yet, this is the magnitude of the change that might be the result of man's activities in the stratosphere. It could take many years, if not decades, to discover changes of this level of magnitude, and, in fact, such changes have occurred naturally over the past few centuries.

When these changes do occur, they do not take place uniformly over the globe—changes at low latitudes are small or negligible, while those at high latitudes are quite substantial. At continental locations above 40–50° latitude, a change of 0.5°C on the global scale can seriously affect zones marginal for life processes. Depending on the sign of the change, they cause advances or retreats of forest limits into or from the tundra and have brought about changes in the type of crops that can be grown.

When a climate change occurs, a number of factors are particularly critical to agriculture. For example, changes in the frequency of years with adverse conditions can be much more important than how the average temperature changes. The length of frost-free season is a far more important parameter for crops than average temperature. Thus, no accurate predictions of effects on agriculture can be based on estimates of average temperature alone. Appraisals of the influence on precipitation of subtle changes in the earth's surface temperature (again of the order of a few tenths of a degree) are speculative at the present time. Variability of precipitation and resulting soil moisture are greater than that of temperature, so that trends are still more difficult to discover; in certain marginal agricultural areas rainfall is highly erratic.

The difficulties of discerning in climatic records subtle changes that may be produced by some activity of man must be borne in mind when estimating potential effects of high-altitude aircraft. Before considering such estimates, however, it is useful to make a gross assessment of the effect of other man-made interferences with the atmosphere. Man has interfered with the atmosphere since the Industrial Revolution; substantial contamination has taken place, and some of the end products have made their way into the stratosphere. Yet, these human influences resulting from industry, agriculture, and traffic, even though extending over many decades, have so far been correlated only with changes on a local scale (see Appendix E); on a global scale they are still hidden in the wider swings of natural fluctuations. The earth's climate system has not exhibited any major instabilities, and studies indicate that all man-made effects on local climate to date are reversible in the sense that they would be expected to disappear if the modifying influence were to be removed.

III. Estimates of Effects of Stratospheric Effluents on Climate

Studies have been made of potential effects on climate of several aircraft engine effluents, including CO_2, H_2O, and NO_x, which affect stratospheric ozone or radiative properties, and SO_2, which forms particles that increase the opacity of the atmosphere. Briefly, CO_2 warms the lower layers of air—a process popularly known as the "greenhouse effect." However, CO_2 added to the stratosphere by SST's is likely to remain below 0.1 percent of the amount already there, and hence its effect on the temperature distribution can be neglected; the burning of fossil fuels at the earth's surface contributes more to the increase in stratospheric CO_2 than will high-flying aircraft. Larger injections of water vapor (H_2O) can also produce a greenhouse effect, and recent calculations for currently projected fleets show that global average warming near the ground from H_2O contributed by high-altitude aircraft may be of the order of 0.1 °C—a possibility that should be investigated further.

In the past, it has been speculated that water vapor emitted from high-altitude aircraft might create additional cloud formations. In terms of climate this could be quite significant because clouds radiate different amounts of energy into space as compared with that which is emitted without them, and if they occur over the sunlit hemisphere they also reflect solar radiation. However, stratospheric clouds are rare; only 150 sightings or so have been reported over the past century. The stratosphere is extremely dry, and the air must be cooled to temperatures in the vicinity of −87 °C before clouds can form. Such temperatures rarely occur even in the Arctic winter stratosphere, although they may occur somewhat more frequently in the Antarctic winter stratosphere. It may be that clouds occur more frequently in the stratosphere than the few sightings suggest, because the low temperature required frequently occurs in the polar night, where they cannot easily be seen. In any case, such "nacreous" clouds are generally more or less transparent. Thus, although it has not been proved conclusively, it is believed that stratospheric clouds do not contribute significantly to the earth's heat budget.

In spite of its small concentration, stratospheric ozone plays an important part in producing the earth's climate because it is a major absorber in the ultraviolet and infrared portions of the sun's spectrum and a weak absorber in the visible. As it absorbs energy, ozone also emits infrared radiation, some of which is received at the earth's surface. Although a reduction of ozone would decrease the amount of re-radiation of infrared and have a cooling effect on the earth's surface, reduction of ozone also leads to additional sunlight reaching the surface, which has a warming effect. Since ozone absorbs most of the sun's ultraviolet radiation at wavelengths below 295 nm, an additional effect is the heating of the upper stratosphere; hence, a reduction of the ozone would cool the stratosphere, and many climate models suggest that stratospheric cooling is accompanied by a compensating, but somewhat weaker, tropospheric warming. Finally, NO_2 absorbs visible radiation, thereby heating the stratosphere and cooling the troposphere. Since NO_2 injection decreases the natural ozone, the heating effects of NO_2 tend to compensate the loss of stratospheric heating caused by the ozone decrease. All these various heating and cooling effects are so uncertain that little is known about the net effect. In particular, we cannot assume that it is negligible.

A large fraction of the SO_2 emitted into the stratosphere from aircraft engines is transformed into sulfate particles. Such particles absorb and reflect solar radiation. It is more difficult to evaluate these effects quantitatively than it is to evaluate the effects of absorption by ozone, for reasons described in Appendix F. As with absorption by ozone, absorption of sunlight by particles acts to heat the stratosphere, and this warming can cause changes in the troposphere immediately below. The exact character of such changes is, however, very much in doubt. The eruption of Mount Agung in 1963 seems to have offered a natural experiment. There is some evidence, though controversial, that following this eruption the equatorial stratosphere warmed by some 6°C due to the increased particles in the region; if there was an associated surface cooling, it was too small to recognize among the normal fluctuations. It is generally expected, although without much confidence, that a reduction in surface temperature will result from SO_2 emissions in the stratosphere.

In summary, it is possible that emissions of NO_x, SO_x, and water vapor by high-altitude aircraft flying in the stratosphere may produce long-term climatic changes; such changes may or may not be negligible in terms of effects on life at the earth's surface. Although climatic effects, if they occurred, would be of enormous importance, the link between stratospheric modification and change of climate is very poorly known.

IV. Conclusions Regarding Predictions on Climatic Change

A complete and unambiguous model for studies of climate would be a major scientific breakthrough. Such a model would adequately take into account the circulation of the atmosphere and the oceans and allow one to compute in detail for about a thousand years all important scales of motion and processes such as cloud formation and change in snow and ice cover that influence the radiation balance of the globe. Unfortunately, such a model requires a computing speed and capacity far beyond any available at present or in the foreseeable future. Thus, the practical models today require approximations for the average influence of motions and processes that cannot be treated in detail, and the accuracy of these approximations must be checked against observations of present climate.

This means that they are less valid for *changes* in climate and casts doubt on the predictions of such changes.

The most complete models of the atmospheric part of the atmosphere–ocean system are the three-dimensional atmospheric general circulation models that resolve most of the important horizontal scales of motion and permit simulations of time periods up to about a year with available amounts of computer time. But the vertical-convective transport of heat and moisture that influences cloud formation is still approximated rather crudely. In many models, sea-surface temperatures are prescribed as input rather than calculated; these models are of limited value for studies of climatic change because sea-surface temperature is known to have a regulating effect on climate. There are two-dimensional zonal models that treat the atmosphere and ocean averaged over longitude. This type of model can explicitly treat large vertical circulation cells such as the Hadley cell, which is known to have significant climatic importance in the tropics. However, these models must use approximations for the transport by horizontal eddies, such as the cyclonic disturbances important in the midlatitudes, and these approximations are presently unreliable. There are zonal models that cut down on computation sufficiently to allow simulation of zonal changes during a complete century, including a crude treatment of relatively slowly reacting ocean circulation. But these models sacrifice longitudinal structure and do not allow study of regional climate change. Finally, there are simple energy-balance models in which there is no explicit hydrodynamic transport in the atmosphere or ocean and the latitudinal redistribution of heat is only approximated. Perhaps the largest uncertainty in the predictions made with climatic models comes from the unknown effect of changes in cloud albedo and height, changes that are ignored or estimated unreliably in all such models. A more detailed discussion of climatic models is presented in Appendix E.

The Panel CONCLUDES that specific predictions of climatic change due to the effects of stratospheric contamination are not possible with much confidence today. However, it is the opinion of scientists working in this area that climate changes resulting from aircraft emissions in the stratosphere would not be large. The prevailing opinion seems to be that high-altitude flight operations sufficient to decrease the ozone column by 10 percent or to increase particulates in the stratosphere by 0.1 $\mu g\ m^{-3}$, would produce long-term global surface temperature changes of no more than a few tenths of a degree Celsius, and global rainfall would not be altered by more than perhaps ± 5 percent. The effects of particulates alone would most likely be to decrease global-average temperature by several tenths of a degree Celsius. However, as we have discussed earlier, such global changes will not be uniformly distributed over the earth. Temperature changes in higher latitudes of the continents of the northern hemisphere could reach values several times those estimated for the globe (i.e., perhaps 1 °C). Locally, precipitation could also be considerably more disturbed than the small global changes might indicate, and variations as large as ±10 percent might well occur. Finally, should there be an appreciable cooling in higher latitudes, one must expect greater variability of weather conditions from year to year with all the ramifications that this may have for crops in these latitudes. These effects may be considerably larger than changes in the mean values of temperature and precipitation might suggest.

5

Climate Effects on Agriculture and the Biosphere

I. Introduction

The impact of any climatic changes, be they man-made or natural, must be considered on the basis of *regional* soil–climate resources. Climate and soil provide the natural basis for selection of farming systems, but variation in weather from year to year, and especially the distribution of certain critical meteorological elements during a typical year, determines the long-range success of farming.

It is possible to make a quantitative evaluation of the effects of climate on agriculture for many agricultural systems. In general, there are three aspects of crop production that are affected by changes in weather and climate: (1) length of growing season in relation to the time required to mature the crop and the risk involved that the season will be too short for crop maturation or harvest; (2) level of production for the crop based on available soil and climate resources; and (3) economics of crop production taking into account agricultural technology, production costs, and marketability.

This chapter discusses primarily the potential effects of postulated climatic changes on selected agricultural systems. We will describe the available methodology for such a quantitative evaluation and its limitations.

This chapter presents the findings of the Panel on Biological and Medical Effects: R. Setlow, *Chairman;* W. Baier, W. Butler, R. Clayton, E. Deevey, T. Fitzpatrick, A. Giese, D. Gordon, F. Haxo, E. Runge, E. Scott, and F. Urbach. Consultant to the Panel: M. Caldwell.

There is also the question of the effects of climate changes on ecosystems in general. Here, unfortunately, little definitive can be said. Organisms and ecosystems seem to have what might be called "coping strategies" that respond to the variances of temperature and rain and, within broad limits, may be relatively insensitive to mean values for these parameters. There is evidence that evolutionary "strategies" over geologic time have adapted to changes in mean values for climatic parameters, whereas the response to changes in *variances* has resulted in physiological and biogeographical responses. For example, a small change in the length of frost-free season, if continued for a few years, has the effect of shifting the range of some important organisms or biotic communities by a few degrees of latitude or longitude. In mountainous regions, the response may have involved a change of several hundred meters in altitude. Work is going on under the International Biological Program (IBP) to model large natural ecosystems or *biomes*. The IBP models seem to have inherent limitations, as discussed in Appendix G, but in general, their conclusions seem to be in accord with those of the agricultural models for production of various crops discussed below.

II. Agriculture and Climate Change

In the absence of any firm predictions of surface climate change resulting from high-altitude aircraft (see Chapter 4), potential effects of resulting changes in crop yields are presented here in terms of parametric estimates of the sensitivity of crops to climate changes. The changes considered extend to $\pm 3°C$ and ± 30 percent change in precipitation during the growing season in the particular locality considered; these are not global figures, and the range is greater than is actually likely to occur.

It is not possible to predict variability in rainfall *patterns* during the year nor the resultant change in production. Maximum production of most agricultural crops results from timely operations (e.g., tillage, planting, harvesting) with generally well-distributed rainfall during the growing season. There may be notable exceptions to the latter, however, in marginal farming areas. More variation in climate from year to year or week to week would reduce yield and production of most crops, particularly in marginal agricultural zones. The models discussed below do not include any secondary effects of climatic change such as altered probability of crop disease outbreaks or other pathological considerations. These models cannot take into account the degree to which agricultural technology can accommodate climate change—crop breeding programs have had spectacular success in many situations, particularly in regard to disease resistance and improved yields over the past several decades. The degree to which new varieties of various crop species might be found and selected for yield optimization under new climatic conditions is very difficult to predict, although it is well recognized as a major factor in agriculture.

A simple response to a decrease in temperature is the *movement* of general cropping zones toward the equator. However, this is not always feasible. For example, in North America a southward shift of the corn belt to compensate for decreased temperatures would place the belt on steeper and less fertile soils than those where corn is now grown. This, as well as effects such as agricultural zones crossing coastal and political boundaries, can pose significant problems. Movement of major agricultural zones, particularly during a reasonably short period of a few decades, can also greatly affect total crop production.

III. Modeling Crop Yields as a Function of Climate

Modeling techniques have been applied to agricultural systems based on present crop varieties and agricultural technology to predict changes in yield as a function of perturbations in climate. An example of such modeling techniques is described in Appendix H, and only the results will be presented here.

The primary purpose of most statistical models relating crop or animal production to weather and climate has been to estimate yields from climatological data that reflect the weather patterns over the past years or decades. In these models, a change in the meteorological input does not reflect the overall change in climate that would occur if the temperature would decrease by, say, 1 to 3°C. Although there are no better objective techniques available to evaluate the effects of such assumed climatic changes on agricultural production, it must be realized in the interpretation of the results from such models that a decline in temperature may well cause climatic conditions unlike those that have occurred in the immediate past. Such changes could cause problems in crop and animal production, losses in income for cereal growers, and possibly losses of entire areas for the production of cereals. The possible reduction in quantity and quality of grain on a large scale would aggravate the already critical world food situation.

Results of current model runs indicate that productivity varies strikingly from one crop species to another and to a large extent is highly dependent on geographical location. Productivity changes for shifts in climate have been evaluated for wheat, corn, soybeans, and rice. Planting dates and the maturation dates of most wheat varieties, corn, and sorghum hybrids would be delayed by one to two weeks if mean temperatures were lowered by 1 to 3°C over the Great Plains and agriculturally similar areas. The combined effects of delayed planting and maturing would mean that the effective growing season could be shortened by as much as one month. Therefore, it is likely that it would be necessary to change to shorter season hybrids or varieties or to a shorter season crop. But, the model suggests that the lower number of days of high maximum temperature, coupled with the reduced potential evapotranspiration, would be conducive to greater yields of wheat, corn, and sorghum in most areas. However, this is only true if the growing season is long enough for the crop to continue to grow. It is interesting that the negative effects of decreased precipitation on wheat, corn, and sorghum yields are offset by the beneficial effects of lower mean temperature. This seems to indicate that for nonmarginal areas such as many regions in the United States, these crops would benefit from some cooling trends; the reverse may be true for the growing of crops in marginal areas, for example, in certain parts of Canada and the Soviet Union, where the growing season is barely long enough to accommodate the crop at the present time. The models for all crops studied, except rice and cotton, show that increasing temperature, with precipitation unchanged, decreases yields.

Corn yields are more affected by temperature changes than by rainfall changes. It is estimated that corn yield would increase on the average some 11 percent for each 1°C decrease in local temperature but would only increase some 1.5 percent for each 10 percent increase in rainfall. Changes in yield due to changes in precipitation here assume the same week-to-week distribution of rainfall. Increased variation in rainfall distribution would adversely affect corn yields, and changes in climate would have a greater effect on the corn yield in the western part of the United States than in the eastern corn belt. Other parametric combinations for the U.S. corn belt are given in Table I.1 of Appendix I.

Wheat is basically a cool-season crop, and decreases of temperature generally favor wheat growth, particularly during its midseason development. Three basic production criteria are most likely to be affected by temperature changes: (1) average planting and maturation dates, (2) average crop maturation dates, and (3) grain production. Computations show that local temperatures of 0.5 to 2.0°C lower than normal can result in a 2- to 10-day delay in average planting dates and maturation dates. Finally, grain production is unfavorably affected by high maximum temperatures and favorably affected by low seasonal mean temperature. It is indirectly affected by modification of the soil-water balance through changes in the rate of evapotranspiration according to calculations using the models. This means that decreases in local temperature by 1 to 2°C tend to increase yields by one to two bushels per acre in the major wheat-producing states. If temperatures are increased by the same amount, with precipitation remaining unchanged, wheat yields will decrease by one to three bushels per acre. The influence of rainfall is a bit more complicated to predict because of soil-water storage limits that vary geographically. For instance, a decrease in seasonal precipitation by 10 percent, without a change in temperature, tends to decrease yields by one bushel per acre in the western edge of the midwest of the United States, while tending to favor yields by about the same rate in the eastern edge. Table I.2 of Appendix I gives other parametric combinations for wheat.

Modeling studies have been made for soybeans, rice, and cotton, and these results are summarized in Chapters 4 and 5 of CIAP Monograph 5. No quantitative assessment has been made on the effect of climatic change on the production of beef, swine, and poultry. A study has been made of dairy milk production, and current models suggest that summer milk production should increase for temperature reductions—northern locations benefiting the least and southern locations benefiting the most. However, calculations did not take into account a possible effect of an increase in cold stress that dairy cattle might experience during winter months or any other interactions that could affect animal heat balance. On the other hand, a decrease in precipitation would result in curtailment of range-land productivity and hence of carrying capacity for grazing animals.

Hydrologic implications of climatic changes have also been evaluated for selected locations in North America by means of climatological water budget methods (see, for example, Table I.3 of Appendix I). As might be expected, effects on runoff from agricultural land are largely a function of the slope and nature of the runoff-producing surface and the annual precipitation. In many parts of the United States, drainage below the root zone of agricultural crops is a major source of groundwater recharge. Drainage amounts vary greatly with the rainfall intensity and the soil-water conditions existing at the time of rainfall, plant cover, and temperature. One also cannot ignore the ramifications of climatic change in the irrigated agricultural areas, because changes in yield of water from present watersheds could increase or decrease the area under irrigation and thus have economic consequences.

IV. Conclusions Regarding the Potential Impact of Climatic Changes

The Panel CONCLUDES that the greatest effects from climatic changes will be in marginal areas of crop production, especially if temperature decrease results in the shortening of the growing season for present crops. At the same time, the Panel CONCLUDES that potential climatic changes will probably cause a minimal change in the soil-water balance and drought. But deep percolation and runoff from agricultural lands may be affected in some areas. The ramifications of climatic change in irrigated agricultural areas must also be considered since changes in yield of water from present watersheds could increase or decrease the area under irrigation. The Panel BELIEVES that agricultural technology can accommodate the effects of minor climatic changes to some extent, but any increase in variability of length of growing season and moisture supply during the growing season would reduce crop production, particularly in the marginal agricultural zones, which show unusually large variances in productivity from year to year.

II

APPROACHES TO A SOLUTION

6

Nitrogen Oxide Emissions and Engine Design and Operation

I. Introduction

Part II of this report discusses the combustion process in jet engines that produces nitrogen oxides and sulfur oxides and the factors involved in their production. In this chapter, we consider what can be done in the design and operation of jet aircraft engines to reduce emissions of nitrogen oxides. Information on the science and technology of the problem is given in Appendix J.

II. Combustion of Fuel in Jet Aircraft Engines

Today, jet aircraft engines burn fuels that are essentially *hydrocarbon* material, and this will probably continue for at least the next two decades. The products of the complete combustion of pure hydrocarbon fuels are carbon dioxide and water. Sometime in the future, hydrogen may be used as fuel for jet aircraft engines, producing water as the only product of complete combustion. For both fuels, the amounts of these exhaust products are proportional to the amount of fuel consumed.

If the combustion of a hydrocarbon fuel is not complete, small but

---

This chapter presents the findings of the Panel on Engines and Fuels: F. D. Rossini, *Chairman;* L. Brewer, A. Ferri, F. S. Johnson, and R. H. Miller. Consultants to the Panel: J. M. English, I. Glassman, S. S. Penner, and R. A. Rudey.

significant amounts of soot, carbon monoxide, and unburned hydrocarbons are produced. Fortunately, under cruise conditions, jet aircraft engines operate at very high efficiency, with essentially complete combustion, so that the amounts of these minor products are negligible.

However, since the combustion of the fuel takes place in air, which is about four fifths nitrogen and one fifth oxygen, there is opportunity for nitrogen oxides to be formed when the temperature becomes high enough. The possible consequences of introducing nitrogen oxides directly into the stratosphere have been discussed in previous chapters. In this chapter, we examine the formation of nitrogen oxides and the possible changes in design and operation of jet aircraft engines that can reduce their amounts.

III. Nitrogen Oxides from Jet Aircraft Engines

The nitrogen oxides formed in the effluent from jet aircraft engines are a mixture of nitric oxide and nitrogen dioxide called NO_x. As a measure of the amount of NO_x, an emission index is used, defined as grams of NO_x (calculated as though all the nitrogen oxides were in the form NO_2) per kilogram of fuel burned. The amount of NO_x emission depends on the peak temperature attained in the combustion process and on the amount of time the gas mixture of nitrogen and oxygen is at that temperature. Nitrogen oxides, NO_x, are formed in significant amounts only when the peak temperature exceeds about 1800 K. Thus, the amount of NO_x produced can be controlled within appropriate limits by holding the peak temperature down to this level. Figure 6 is a simple diagram of the conventional combustion chamber (combustor) in a jet aircraft engine, together with its temperature profile. With regard to the production of NO_x, several points are important: (1) in this combustor, only a part of the air is initially mixed with the fuel, thus giving a rich fuel mixture and a high local combustion temperature well above 1800 K; (2) additional air mixed peripherally into the combustion chamber lowers the temperature of the gaseous system to that which can be accommodated at the turbine inlet—a temperature presently between 1200 and 1400 K in cruise conditions, which is a limitation imposed by the material of construction and design of the turbine blades.

The efficiency of an engine is a function of the turbine inlet temperature not of the local peak temperature in the combustor. The formation of nitrogen oxides can be reduced by lowering the local peak temperature closer to the temperature desired at the turbine inlet, without sig-

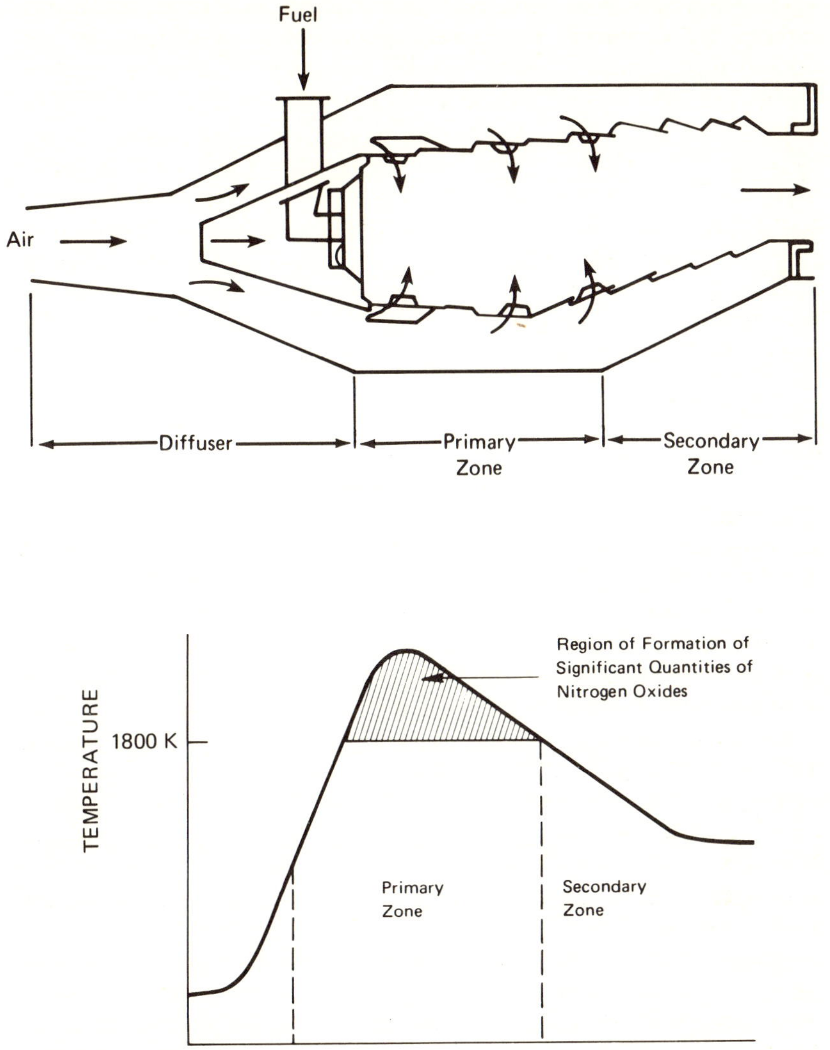

FIGURE 6 Schematic diagram of conventional combustor.

nificantly affecting the efficiency of the engine. Recent laboratory experiments (see Appendix J, Section IV.B.1) indicate that, in principle, local peak temperature can be reduced by providing for *prevaporization and premixing* of the fuel and air, thereby significantly reducing the nitrogen oxides. Figure 7 is a much simplified schematic illustration of how, for cruise conditions, a combustor can be arranged to do this.

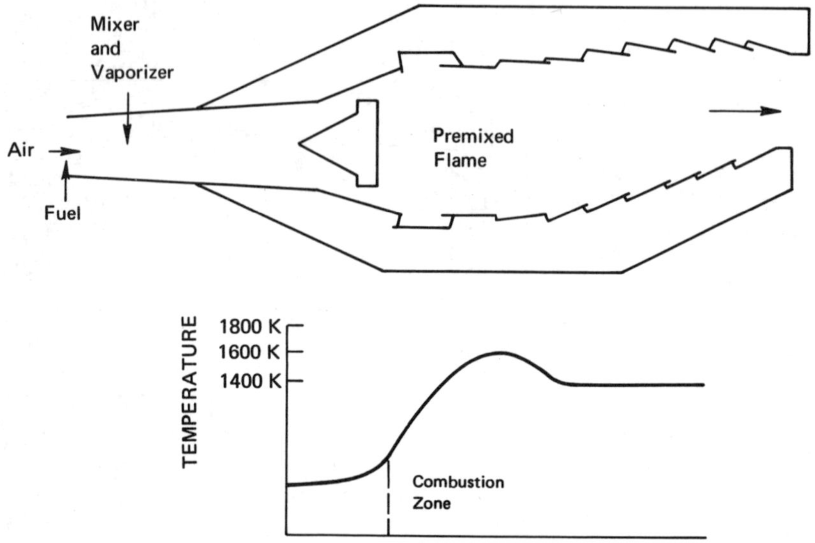

FIGURE 7 Diagram of combustor for premixing and vaporizing of fuel and air.

In this arrangement the fuel is premixed with a large fraction of the air stream. As indicated in the accompanying temperature profile, there is relatively little excess of temperature above that at the turbine inlet.

Should the fuel itself contain combined nitrogen, about half of it may end up in nitrogen oxides. Present-day jet aircraft fuel from natural petroleum is relatively low in nitrogen content, so that no significant amounts of NO_x are produced from this source. "Synthetic" petroleum from tar sands is also relatively low in nitrogen content, but "synthetic" petroleum from oil shale and from coal is relatively high in nitrogen content. Jet aircraft fuel from the latter sources would require treatment to lower the nitrogen content to acceptable limits, using well-established technology.

Quite aside from the problems introduced by aircraft engines operating in the stratosphere, one might ask why engines have not already been designed to minimize NO_x and reduce the burden on the lower atmosphere. To date the contribution of jet aircraft, in cruising flight, to the NO_x burden of the lower atmosphere has been considered an almost insignificant part of the total; consequently, there has been little or no effort to control this factor in aircraft engine design and operation. There has been more concern over potential aircraft emissions in the vicinity of airports because of the density of aircraft there. Regulatory action to control the latter output is being taken. But in the stratosphere even small amounts of NO_x could have important consequences, as described in Chapters 2, 3, and 8.

The major thrust and progress over the last 10 years have been in the area of improving economic viability of long-range air transportation by the introduction of new fan engines with high bypass ratios having lower fuel consumption and greater thrust for a given engine weight. A high bypass ratio means that a large fraction of the air entering the engine bypasses the combustors and turbine. The reduction in fuel consumption has been significant. A 25 percent reduction in fuel consumption, under comparable conditions at cruise, was initially achieved in going from pure jets of the early jet aircraft to bypass ratios (ratio of mass of air bypassing engine core to the mass of air going through the core engine) of about 1; another 20 percent has been achieved in going from these early fan engines to the high bypass ratios of 6 to 7 on the fan engines of the wide-body jet aircraft. This has led to a steady development of larger, longer-range aircraft, which, if operated with reasonable passenger loads, are far more economical than their predecessors. A welcome by-product of this development has been a reduction of jet engine noise for a given thrust (the jet velocity, and hence the noise, is lowered as bypass ratios are increased).

Current commercial supersonic aircraft, the Concorde and the Tupolev-144, have an NO_x emission index under cruise conditions of about 18 g/kg. Indices are in the range of 14–16 g/kg for the high bypass engine of the earlier Boeing 747, the DC-10, and the L-1011 and about 6 g/kg for the still earlier Boeing 707 and DC-8. The historical increase in the NO_x emission index, rather than decrease, is due to the increasing combustion peak temperatures. In present combustors, where stoichiometric temperatures are approached, the NO_x formation is a function of combustor peak temperatures, and these increase with compression ratio and flight Mach number. When these engines were designed, emissions of nitrogen oxides were not considered to be a problem, and the measures taken to eliminate smoke resulted in higher local peak temperatures in the combustor and the formation of greater amounts of nitrogen oxides.

The laboratory experiments previously referred to were carried out in an attempt to find a solution to this problem of reducing NO_x emissions. Tests were conducted in a simplified combustion test chamber with simulated supersonic cruise conditions. In one laboratory, an emission index of 0.4 g/kg was obtained, with combustion efficiency of 99.3 percent (see Appendix J, Section IV.B.1). In another laboratory, an emission index as low as 0.3 g/kg was obtained, with combustion efficiency still maintained above 99 percent (see Appendix J, Section IV.B.2).

The experimental data mentioned have been reviewed by experts in engines, jet propulsion, engine effluents, and physical chemistry. In their considered judgment, no fundamental technical reason can be identified that would inhibit reduction of the NO_x emission index by a factor of 10 to 20 times, possibly more, in operational jet aircraft engines. However, such reductions will require a major developmental effort and can be achieved over the next 10 to 15 years only if adequate resources and high priority are given to the task.

The Panel on Engines and Fuels CONCURS with this opinion. The Panel re-emphasizes that this can be accomplished only if a large effort, motivated by significant incentives, is mounted during the period. It recognizes that aircraft and engines must be designed to operate efficiently and reliably over a wide variation in operating conditions, and that several years of development effort will be required to ensure operational suitability of the new engine combustor configuration. The Panel also recognizes that, as material and turbine blade design and cooling technology develop, it is probable that the turbine inlet temperatures will increase, and hence the possibility of greater production of NO_x exists.

The added cost of bringing into being new jet aircraft engines embodying radical changes in combustor design and operation that will reduce NO_x emissions by factors ranging from 10 to 20 below today's supersonic aircraft might be as low as $100 million. This assumes that, over the next 10 to 15 years, engine manufacturers will, in course, be developing new jet aircraft engines and that the incorporation of changes in design and operation of the combustor to reduce nitrogen oxides will be about 10 percent of the overall development cost of a new engine, estimated to be of the order of $1 billion.

IV. Conclusions and Recommendations Regarding NO_x Emissions from Jet Engines

We RECOMMEND *that research and development on combustors and other engine components be increased immediately with a view to reducing the effect of engine emissions on stratospheric ozone. We* RECOMMEND *that the development program be aimed toward a reduction in the emission of oxides of nitrogen by a factor of 10 within the next decade, and that the research program be aimed at substantial further improvement in emissions to be realized in practical engines during the ensuing decade.* Within 10–15 years, new jet aircraft engines, using hydrocarbon fuel, can be designed, developed, tested, and put into use having NO_x emission indices reduced substantially below those of current supersonic aircraft (Concorde and Tupolev) provided that adequate resources and high priority are put on the task. This conclusion assumes that the engine turbine inlet will continue to be limited to a temperature below 1800 K. Should new material technologies or revised engine configurations remove this limitation in favor of higher temperatures and efficiencies, NO_x reduction would be more difficult. Given adequate resources and high priority, we believe that a factor of 10 reduction is a fully realizable target and that a factor of 20 has a high probability of success in achieving the goal. The time frame of 10 to 15 years seems reasonable in relation to the rate of scientific, technical, and industrial productivity on such problems. This schedule would dictate, however, an immediate start in research and development, if the new engines are to be available for use by 1985.

There may be an engine emission problem for the wide-body subsonic jets if the fleet size increases and new advanced larger bypass engines are built. Such engines could have larger compression ratios and, therefore, higher NO_x emission indices than present engines unless new burner technology is introduced and a retrofit program is adopted. The immediate solution to such a problem appears to be a reduction of flight altitude, by a few kilometers, and a corresponding reduction of flight speed proportional to the square root of the increase of air density, in order to operate at the best efficiency. An alternate possibility is reduction of inlet temperature. None of these options appears to be without serious economic cost.

Some believe that *hypersonic* aircraft may come into commercial service sometime in the future. Such aircraft would fly at Mach 8 to 10 at about 30–36 km (about 100,000–120,000 ft) in place of the Mach 2 to 3 at about 16–20 km (52,000–66,000 ft) of supersonic aircraft. These hypersonic aircraft would use hydrogen as a fuel, for which the main product of combustion is water and the minor product is nitrogen oxides. The Panel has not studied the problem of hypersonic transports, but such a study should be made on an appropriate time scale.

7

Climate and Aircraft Fuels

I. Introduction

In this chapter we consider the processes of combustion of fuel in jet engines that produce sulfur oxides and the factors involved in their production. We then consider the question of what can be done to reduce these emissions. The basic background information on the science and technology of the problem is given in Appendix J.

II. Combustion Products of Concern

As indicated earlier, the propulsion power for jet aircraft for at least the next two decades probably will be derived from the burning of hydrocarbon material—presently, natural petroleum; later, perhaps, "synthetic" petroleum. Natural petroleum contains small amounts of sulfur, which becomes oxidized along with the rest of the fuel during the combustion process. The sulfur oxides produced in this process contribute in some degree to the corrosion of engine parts, particularly in the jet turbine.

With the prospect of operating jet engines at much higher altitudes—well within the stratosphere—there may be new problems. The com-

---

This chapter presents the findings of the Panel on Engines and Fuels: F. D. Rossini, *Chairman;* L. Brewer, A. Ferri, F. S. Johnson, and R. H. Miller. Consultants to the Panel: J. M. English, I. Glassman, S. S. Penner, and R. A. Rudey.

bustion products known to produce aerosols might, over the years, create an inadvertent modification of the earth's climate. The potential climatological impact of increased aerosols in the stratosphere was considered in Chapter 4, along with the question of whether increased NO_x and the resulting reduction of ozone in the stratosphere could have a temperature effect that would alter the earth's climate.

III. Fuel Desulfurization

Fortunately, potential climate problems can be avoided by reducing to relatively low values the amount of sulfur in jet fuels. Process technologies exist in the petroleum industry for removal of sulfur, as well as nitrogen and trace metals, either from the petroleum feedstock or from a product stream. With this technique, these contaminants can be reduced simultaneously. Hydrogenation of complex molecules at a temperature of about 400°C and a pressure of about 70 atm releases sulfur, nitrogen, and the free metals.

Because of the increasing requirements for low sulfur in other petroleum products, such as automobile fuels, and the attractive economics of the operation, the tendency is toward increased use of hydrodesulfurization of a very large fraction of the petroleum. One large petroleum company, operating with high-sulfur crude oil, is desulfurizing the entire petroleum, with the elemental sulfur coming out on a large moving belt. Another large petroleum company, in its pilot plant operations, is routinely processing raw shale oil, reducing the sulfur content from 7000 to 16 ppm. The sulfur content in current jet aircraft fuels is about 500–1000 ppm; the specification limit at the present time is 3000 ppm. The lower amount is beneficial from the standpoint of reducing corrosion of the turbine blades. The possible decrease in lubricity of the fuel when the sulfur content is decreased should be investigated.

IV. Conclusions and Recommendations Regarding Sources of SO_x in the Stratosphere

We RECOMMEND *that it be recognized that control of the sulfur content of fuel for stratospheric airliners will be required in the future. The degree of sulfur content should become a matter of international regulation in the light of further research on the effect of stratospheric flight on climate.*

We CONCLUDE that the amount of sulfur oxide in effluents from jet aircraft engines using hydrocarbon fuel can be reduced by applying technology already available in the petroleum industry. If desirable, currently available techniques should permit lowering the amount of sulfur in jet fuel to an amount of the order of 10–20 ppm. Based on information provided by the petroleum industry, the Panel estimates that the cost of reducing sulfur in these fuels to a level of 10–20 ppm would be of the order of one cent per gallon. The Panel BELIEVES that a reduction in this cost for jet fuel will occur if part of the cost of desulfurization is allocated to other petroleum products that are desulfurized in the same process.

8

Economic Impact and Decision Analysis

I. Introduction: The Decision Problem

Any decision problem has the following basic structure: there is a set of alternative policies among which we must choose, and for each possible policy there is a set of predicted consequences. Some of these consequences are favorable, and we refer to them as benefits; others are unfavorable, and we think of them as costs. The task is to choose among the alternatives the one that has the largest aggregate of favorable net consequences.

The choice among different sets of consequences is a value judgment. How one evaluates and compares different consequences involves ethics as well as facts. One tries to go as far as possible in achieving a common measure; in practice, measuring benefits and costs in terms of money and aggregating them by addition is quite legitimate for many but not all purposes. At some point, however, there can be no basis for further rational argument; reasonable people may disagree.

The benefits derived from high-altitude commercial travel are reasonably expressed in dollars. Similarly, some costs can be evaluated at their current market price (e.g., a change in climate that might affect food supply). However, illness and death are not valued by the market,

---

This chapter presents the findings of the Panel on Decision Analysis—Monitoring and Surveillance: K. Arrow, *Chairman;* D. Hunten, H. Jacoby, H. Odishaw, R. Solow, J. Tukey, and R. Zeckhauser. Consultants to the Panel: R. d'Arge, J. M. English, M. Menter, S. Shavell, and M. Weinstein.

though they are often valued implicitly through political decisions as to what society is willing to invest in measures that reduce their incidence, e.g., expenditures on highway or automobile safety features. These implicit valuations differ greatly from case to case and hence are difficult to use as an objective basis for decision. For example, the disutility of being ill is not properly measured by the medical costs, especially if the illness is not cured or is cured only after prolonged suffering. Although the law courts and the government put values on workdays lost and on death, there does not appear to be any objective definition of costs of suffering from skin cancer, particularly since it is unlikely that any compensation will be paid. The reader who is willing to put a money value on the incidence of skin cancer can do so.

Therefore the costs of high-altitude travel are only partially reducible to money. We conclude that it is reasonable to use dollar values for agricultural and certain other effects of climate change but not for such effects on health as skin cancer. Even when dollar values are legitimate measures of benefits and costs, their distribution may be of concern. The benefits and costs may fall on different groups and on different nations and regions, and a simple total should be supplemented by some indication of distribution.

Decisions on the feasibility, economic viability, and desirability of commercial supersonic travel involve many issues other than climatic impact. Economic viability involves the study of cost–benefits from high-speed flight; at least in the free-world market, these benefits must be compared with the industrial development costs. Neither CIAP nor the present study bear on this or many other cost–benefit questions.

A. UNCERTAINTIES

In addition to stochastic uncertainties that can be handled by traditional statistical techniques, there is a second source of uncertainty—the uncertainties that arise whenever we attempt to develop models of large, complex systems with many variables. A classic example is the difficulty of making great improvements in long-range weather predictions. Even when a relation has been well established under laboratory conditions, its application to a large complex region such as the stratosphere is difficult. Because a calculation of a consequence of such a model is subject to uncertainty, we must regard it as a probability distribution rather than a single value. In this study, we have asked experts to give their estimates of the probability ranges of the parameters that determine the consequences of any given amount of stratospheric travel. The questionnaire used is reproduced in Appendix K. Methods of this kind have been used in connection with technological forecasting for some years. Although their validity cannot be regarded as established, they seem to be useful; and their results seem in accord with commonsense and the views of the experts.

The reader should bear in mind that in judging the consequences of one policy or another about stratospheric travel we are not fundamentally concerned with NO_x concentrations or ozone reduction; we are interested in human welfare. This can only be evaluated at the end of a long chain of calculations, and at virtually every stage there is uncertainty.

B. ALTERNATIVES

In view of the potential consequences for human welfare, what alternative approaches could be taken to stratospheric travel? They are (1) to prohibit all large-scale stratospheric travel; (2) to permit travel to the full extent demanded and at altitudes determined by considerations of speed and efficiency, the design of engines being determined by similar considerations with no consideration being given to the volume of nitrogen and sulfur oxides emitted; (3) to impose restrictions on the altitudes at which flights take place, since the effects of the effluents are so altitude-dependent; (4) to restrict the numbers of airplanes allowed to fly; (5) to impose a tax on the effluents of the airplanes, possibly varying according to altitude; and (6) to impose restrictions on NO_x emissions and the sulfur content of fuel.

The institution of emission controls will entail costs. In the case of NO_x this means that there must be at least some idea of the design costs in reducing NO_x emissions by considerable factors. The costs of desulfurizing the fuel are fairly low and reasonably well known. The costs of reducing NO_x emissions involve engine development and like all development costs are somewhat more uncertain. Nevertheless, even high estimates of development costs imply very low costs per passenger trip. Thus, suppose that the additional cost of development to reduce NO_x emissions to a prescribed factor is $150,000,000. Assume further that interest plus amortization on the SST's is 20 percent per annum and that there will be 500 airplanes. Then the cost per plane is $60,000 per year. The additional cost per passenger-trip is less than 30 cents. Clearly, gross changes in the assumptions would not significantly alter the conclusions.

The aeronautical experts of the study group did not believe that changes in engine design to reduce NO_x emissions would be likely to increase capital or operating costs of the engines. However, it is possible that an emphasis on NO_x elimination may involve costs in other dimensions of performance.

Altitude controls that would avoid the problems created by NO_x or other effluents would make supersonic travel impractical. Limits on the number of planes flying in the stratosphere, either as a total or by altitude, are conceivable but would cause additional problems. An industry would be developed that would have a strong interest in expanding. Also, since development involves a large fixed cost, the economic pressure to spread that cost over a large number of planes would be very strong. A scarcity situation would be created, so that, if supersonic travel is sufficiently advantageous, there will be long queues or high prices. It would appear likely that setting numerical quotas would be an unsatisfactory solution to the problems that may arise from stratospheric travel.

The taxation of effluents is an attractive policy in many contexts, such as water pollution from industrial or residential use or atmospheric contamination from heating and surface transportation sources. It perhaps should be explored further, but in our judgment it is not likely to be useful here. The problem is the great length of time required for development. The tax is designed to impose the social cost on the passengers using the service; the developer of the aircraft is not directly affected and hence the development, especially if it is government-financed, might not be deterred by a prospective tax on passengers. Suppose the tax is sufficiently high that demand is reduced to the point that the SST would not be worthwhile. Then the resources used for development would have been wasted. Further, the pressure to use the developed plane will be considerable.

In principle, there are still more sophisticated and complex kinds of policy options—those based on monitoring or feedback. The amount of stratospheric travel might be allowed to develop to a certain level, while regulations for subsequent development are developed on the basis of information derived from monitoring; if no adverse effects appear, a large expansion of travel might develop. However, the feedback is likely to require a long time scale. The natural variability of most of the quantities in question (e.g., ozone concentration) is such that this approach would be quite unsatisfactory as a means of avoiding unacceptable changes. Nevertheless, a continuing research program aimed at more intense observations and additional modeling may, in the long run, provide the basis for better regulation of future stratospheric travel.

The Panel has attempted to analyze the consequences of several alternative possible fleet sizes suggested by the Department of Transportation. Very roughly, these may correspond to what the demand could produce in the future. No claim is made for the accuracy of these fleet forecasts and certainly none for their desirability; they are intended only to suggest the magnitudes of the problems that could result.

II. The Scenarios and the Consequences

In order to understand the magnitude of any environmental effects of stratospheric flight it is necessary to have some idea of likely fleet sizes. We use as benchmarks the upperbound projections made in Monograph 2 of the Climatic Impact Assessment Program of the Department of Transportation for the years 1990 and 2005. These are probably overestimates of the amount of travel that will be demanded. The CIAP has also presented, too late for use in these calculations, an "expected" fleet for a number of years in the future; see DOT CIAP *Report of Findings on the Effects of Stratospheric Pollution by Aircraft,* Exhibit 1.5. The expected fleet for 1990 is about 75 percent of the upperbound fleet; but for the year 2005, the expected fleet is about 30 percent of the upperbound fleet. The figures presented in Appendix K could be adjusted in these ratios.

Two adjustments must be made to the CIAP figures to compare them with figures in this report. The large advanced SST's used in CIAP estimates are assumed to have a capacity of 256 and to fly for 4.5 hours a day. The formerly projected U.S. SST was planned to have a capacity of 298, and it is here assumed to fly for 5 hours a day. Therefore the CIAP numbers of large advanced SST's should be reduced in the proportion (256/298)/(4.5/5). Similarly, the CIAP upperbound projections of subsonic transports flying in the stratosphere were reduced in the proportion 5.4/7, since the CIAP projections assumed that they were flying at cruising altitudes only 5.4 hours per day.

We have then four primary models: 100 present SST's, 100 U.S. SST's, the CIAP upperbound fleet for 1990, and the CIAP upperbound fleet for 2005, both adjusted as indicated. We will also consider three all-subsonic models. For comparison, we consider a fleet of 1000 wide-bodied subsonics at an average height of 13.5 km (44,000 ft) with 1974 characteristics. Then we also consider the two CIAP upperbound fleets as modified by the assumption that all supersonic travel is replaced by an equal amount of subsonic travel at 13.5 km (44,000 ft). Descriptions of the models are presented in Section VII of Appendix K.

The Panel on Decision Analysis—Monitoring and Surveillance has attempted to compare the consequences for human welfare of the different possible models. As emphasized in this chapter, the consequences are subject to uncertainty. It is desirable also not only to have some kind of best estimate of consequences but also to have some idea of possible, even if unlikely, extreme consequences. These uncertainties, it must be emphasized, concern in chief measure the very models that are used to analyze the data. The uncertainties themselves cannot therefore be assessed by using these models but must be based upon judgment as to the limitations of the models and the data.

Members of the Committee's study group were asked for their estimates of the uncertainty in each link of the chain of consequences. Because different people are expert in different areas, the judgments at each stage are based on those of different experts, e.g., one will be more knowledgeable about the diffusion of nitrogen oxides, another will know more about the effect of a given chemical concentration on ozone reduction.

We asked each expert to express his uncertainty in the form of a probability distribution. These probabilities are subjective expressions of belief or credibility based on expert knowledge and acquaintance with alternative possible hypotheses. This method, it must be stated, has not been used systematically in the making of significant decisions in the past, so we have little relevant experience. But there is a strong argument that rational decision-making under uncertainty requires that the decision-maker postulate some probability distribution over the unknown elements.* Admittedly, the expert's ability to assess his own uncertainties will be imperfect, and therefore the probability distributions obtained should themselves have ranges about them. In particular, an estimate for a point at the extreme of the distribution (specifically, when an expert is asked to state a value for some parameter, which, in his judgment, will be exceeded with a probability of only 0.01) should not be taken too seriously. When all is said and done, the probability distribution is likely to provide an answer that is both more useful and more meaningful than any other.

The numbers derived in Appendix K by this method are in reasonable agreement with those derived in other parts of the report by more conventional methods. For estimates of the percentage increase in skin cancer per hundred supersonic airliners, Appendix K (Table K.8) arrives at figures between 60 percent and 70 percent larger than those

* The theoretical argument was developed by F. P. Ramsey, "Truth and Probability," in *The Foundations of Mathematics and Other Essays* (Harcourt Brace, New York, 1931), and later by L. J. Savage, *The Foundation of Statistics* (Wiley, New York, 1954). For an excellent exposition, see H. Raiffa, *Decision Analysis* (Addison-Wesley, Reading, Mass., 1972).

implied by Table 4 of Chapter 2. However, this is within the uncertainty factors mentioned in connection with the latter table. Moreover, the figures for erythemally effective exposure in this table are appropriate for estimating small increases in the incidence of skin cancer; for large increases, the figure suggested in Chapter 3 for incidence of skin cancer is 50 percent bigger and is in reasonable agreement with the numbers appearing in Appendix K, Table K.8.

A practical argument for using probability distributions in this study is that the uncertainties on the final consequences are compounded out of uncertainties in separate steps that have to be combined somehow. In this method, we simply use the ordinary law of probability theory (on the assumption that the uncertainties are independent) to arrive at a distribution of possibilities about the ultimate consequences in which we are interested. The details of this procedure are set forth in Appendix K.

The major possible consequences of flight in the stratosphere are an increase in skin cancer and in climatic change in the form of temperature and precipitation change. In the latter case, the effect on agricultural production and on other human activities is of major concern. The translation from temperature to further consequences will be discussed below, but this final transformation is based on limited information. The effects of nitrogen oxides on ultraviolet radiation at the earth's surface have consequences on the biosphere other than those on skin cancer. Again, only qualitative information is available.

In the absence of any attempt at NO_x emissions control, a fleet of a hundred present SST's will be expected to lead to an increase in the incidence of skin cancer of about 2 percent (Appendix K, Table K.8). If we assume roughly 250 new reported cases per 100,000 per year at present for the United States (see Figure 3 in Chapter 3), this will be about 10,000 additional cases per year in the United States. A present SST flying some 5 hours per day in the stratosphere might make 1400 flights a year and increase the incidence of all kinds of skin cancer by 100 cases per year.

These numbers are based on the median value estimates. The analysis also says that there is a 10 percent probability of as much as a 12 percent increase in skin cancer, or about 1 case for every two flights. There is a 1 percent possibility of as much as an 82 percent increase in skin cancer (or almost three cases per flight)—this is by no means impossible though most unlikely. For a fleet of 100 U.S. SST's the numbers are much higher, as shown in Table K.8, and they became extraordinarily high indeed for the CIAP upperbound fleets.

In the following discussion, in order to provide a convenient description and to avoid the appearance of greater certainty than is warranted, we will use the term "distinctly possible" for the value for which the analytic calculations show odds of 9 to 1 against exceeding. Similarly, we will use the term "conceivable" for the value for which the calculated odds are 100 to 1 against exceeding.

What would the results of NO_x emission control be? The relatively easy to achieve reduction by a factor of 3, to 6 g/kg, would reduce the median increase in skin-cancer incidence for the 1990 upperbound fleet to 7 percent, with 30 percent distinctly possible and 240 percent conceivable. This is beyond what is likely to be acceptable; in the year 2005, we find a median increase of 35 percent, a distinctly possible value of 200 percent, and a conceivable value of 1200 percent, even with the reduction in emissions. Because the CIAP upperbound fleet assumption is probably an overestimate of real market demands, these calculated numbers must also be regarded as overestimates. They can be divided by factors of 2 or 3 to approximate CIAP "expected" fleets, assuming linearity.

Reduction of nitrogen oxide emissions by a factor of 20 brings a reasonable degree of control with the DOT CIAP Monograph 2 upperbound fleet projected for 1990, although it is still conceivable that there may be a 70 percent increase in incidence. Indeed, for such a fleet even reduction by a factor of 100 is not without some danger, as can be seen in Table K.8 of Appendix K.

What are the risks in the absence of supersonic travel? Subsonic aircraft already fly at 12 km (about 40,000 ft) in the lower stratosphere. With the present fleet of 400 wide-body jets, risks of increased skin cancer can be regarded as negligible. However, if subsonic airliners had an average cruising height of 13.5 km (44,000 ft) with present NO_x emission rates, a fleet of 1000 such planes (there are about 400 now) would give rise to some risks. The median at the increased altitude is only about 4 percent, but a value of 260 percent is conceivable. This large range reflects the great uncertainty about injection coefficients in the bottom layer of the stratosphere. To meet the demand filled by the CIAP upperbound fleet using only subsonic travel would require 1000 wide-bodied planes in 1990. By the year 2005, the demand reflected by the upperbound fleet would pose considerable risks. A reduction in NO_x emission to 3 g/kg would reduce the risks to the 1000-plane level; further reduction by a factor of 20 below the present levels would essentially eliminate the problems from subsonic flight. It does appear, however, that continuance of the present NO_x emission level even for subsonic flight is not without considerable risk if altitudes are increased to an average of 13.5 km (44,000 ft).

The percentage figures given apply to people of pigmentation like United States whites living at latitudes typical of those of the United States. The average change in incidence for the northern hemisphere will be somewhat smaller because most of the white population in the rest of the hemisphere lives at higher latitudes than in the United States. The total number of cases in the world must be several times greater than in the United States alone.

The above gives some indications of the risks that can be avoided by insisting on lower NO_x standards. It might be asked how costly this will be in terms of development costs. It is estimated in Chapter 6, Section III, that to achieve a reduction of a factor of between 10 and 20 might cost of the order of $100 million. This one-time cost is to be compared with the annual costs of not reducing the emissions.

We therefore RECOMMEND *that nations seek the adoption by international regulation of whatever constraints may be necessary on aircraft emissions and fleet sizes in the stratosphere to prevent undesirable changes in the intensity of ultraviolet light at ground level and undesirable changes of climate. In particular, we* RECOMMEND *that national and international regulatory authorities adopt standards and policies designed to encourage the entry into stratospheric service of only those new types of aircraft, subsonic and supersonic, that are equipped with engines for which maximum effort has been made to improve emissions —and only if the resulting improvements are judged to be adequate.*

We also RECOMMEND *that national and international regulatory authorities be alerted to the existence of potentially serious problems arising from growth of future fleets of stratospheric airliners, both subsonic and supersonic. The most clearly established problem is a potential reduction of ozone in the stratosphere, leading to an increase in biologically harmful ultraviolet light at ground level.*

The stratospheric emission from present world fleets of subsonic commercial airliners is not considered to pose a significant risk. Production of 16 Concorde supersonic airliners having present emission indices might lead in the long run to several thousand additional cases of skin cancer per year in the world, of which perhaps a thousand would be in the United States. Whether or not one regards these numbers as significant, one must conclude that *a large further increase in numbers of supersonic and high-flying wide-body subsonic airliners with present emission indices does pose significant risks.*

The analysis in Appendix K (see Table K.9) includes four possible influences of stratospheric effluents on climate, especially surface temperature, as discussed in Chapter 4, Sections III and IV. There is a great deal of uncertainty about all climatic effects, even as to sign. The direct effects of water vapor and nitrogen oxides are relatively less uncertain than those due to ozone or sulfate particles. Even so, the range of uncertainty is large. For the most extreme case of the CIAP 2005 upperbound fleet, which is probably an overestimate of future travel demands, analysis of uncertainties similar to that for skin-cancer incidence shows median drops in temperature as great as 0.3°C, while distinctly possible drops range up to 1.5°C (see Appendix K, Table K.9).

As the discussion in Chapter 5 shows, it is not clear whether cooling the earth's surface will increase or decrease human welfare. Lowering the temperature seems to lead to an increase in the output of many major crops at the expense of some marginal areas, particularly those in the far north, which must inevitably decrease their output because of the shortening of the growing season. It also seems clear that rice production is adversely affected by cooling and by reduction in incident radiation, which could result from the presence of SO_x particles.

CIAP Monograph 6 shows that there are costs other than those in agriculture, particularly in the form of increased changes in marine resources and cost of living. However, the economic implications of temperature changes are not clearly understood.

It is clear from Table K.9 of Appendix K that reduction in NO_x emissions indices to 0.9 g/kg (a factor of 20 from the present level) and a simultaneous reduction of the sulfur index to 50 ppm (a factor of 20 from the present level) would essentially eliminate any climatic temperature changes caused by effluents from stratospheric aircraft.

What if only subsonic travel is permitted? A fleet of 1000 wide-bodied subsonic jets, even flying in the lower stratosphere, would produce no great risk of temperature change, nor, according to our estimates, would the 1990 CIAP upperbound fleet. However, if the fleets continue to grow, the prediction for the year 2005 (upperbound) would certainly present more risks. The median prediction is small—a temperature decrease of about 0.06°C—but a fall of 1°C is conceivable, and a drop of 0.15°C is distinctly possible.

The effects on human welfare of a decrease in temperature can be divided into two kinds, those on agricultural production and those on the costs and burdens of living in cities. As is clear from Chapter 5, there are a variety of effects on crops. Indeed, in the aggregate it appears that, for most grains, cooling would be beneficial rather than the reverse. The chief negative effect would be in the forced abandonment of far northern wheat lands, for example, certain parts of the Soviet Union and Canada. Rice is a different matter; the effects of both temperature and solar radiation itself work in the same direction. A drop of 1°C mean temperature accompanying a 1.2 percent fall in solar radiation would translate to a loss of about 17 percent of the world's rice crop, if adjustments were not made. Various adjustments such as new lands would mitigate this loss but only at a considerable cost. The estimates given in CIAP Monograph 6 (Tables 3.133 and 3.139) suggest that by the year 1990 the annual losses in rice due to a 1°C fall in temperature would be about $2 billion a year.*

Reasonably reliable figures are available for the effect of temperature changes on urban costs. These are measured in two ways: by looking at wage differences between cities of different temperatures to measure the willingness of individuals to take lower wages in warmer climates and by direct computation of the costs of lower temperature, primarily health costs and psychic costs arising out of dislike for the temperature changes. Both methods yield a figure of $7.3 billion, flow beginning in 1990 (revised as in the footnote). This figure is for the United States and should be expanded to the entire northern hemisphere. Indeed there is a latitude dependence in that the predicted temperature decreases will always be greater in the more northern latitudes. Hence, the Soviet Union, Scandinavia, and Canada should incur greater costs than the United States.

It might not be unreasonable to multiply the United States' urban costs for a 1°C fall in temperature by a factor of 5 to obtain total northern hemisphere urban costs of $36 billion per year per °C drop. Is this a big number? A 1°C fall in temperature is distinctly possible for the CIAP upperbound fleet for the year 2005 based on Appendix K. Assuming approximately 1.65 million trips a year, the cost would come to about $20,000 for each trip. Assuming about 180 passengers (60 percent load), the cost per passenger trip would be $100, a sizable, though not overwhelming, figure. Incidentally, for a linear system, these estimates of cost per trip and per passenger trip are independent of assumed fleet size; thus, for this calculation one would *not* scale down these costs by a factor of 2 or 3 in going from CIAP "upperbound" to CIAP "expected" fleet projections.

It has been estimated that the cost of reducing the sulfur by a factor of 20 or more is approximately 1 cent per gallon. These costs per plane trip come to about $440 or about $1.80 per passenger trip. It is obvious that a comparison of benefits and risks makes clear the case for desulfurization of fuel. This is in addition to the less precise, but prob-

* Costs given in DOT CIAP Monograph 6 are adjusted to reflect a flow beginning in the year 1990 rather than one discounted back to the year 1974; this amounts to about doubling the CIAP figures if an interest rate of 5 percent is assumed.

ably more significant, argument that substantial changes in temperature are likely to have unforeseen and adverse consequences, particularly in marginal regions.

There is an additional factor not brought out in Table K.9 of Appendix K. Although the effects of both sulfate aerosols and ozone reduction on temperature are uncertain, current models suggest that they tend to be in opposite directions. Hence, once steps are taken to prevent ozone reduction, the effects of unrestricted sulfur content might be greater than before the engines are modified to reduce NO_x. For this reason, the timing of the introduction of low sulfur fuel may be important, as indicated in the recommendation in Chapter 7.

III. The Need for International Cooperation and Regulation

As the stratosphere is a world resource and the effect of effluents is not restricted to a local region, knowledge of the state of the stratosphere and of such emissions as NO_x and SO_x is an international concern. Stratospheric monitoring, both scientific and regulatory (see Chapter 2) will require an observational program with stations at sites throughout the world. International cooperation is essential for the protection of man and his environment.

The International Council of Scientific Unions (ICSU) has mobilized the world scientific community for such broad research programs as the International Geophysical Year and the International Biological Program. Recently, ICSU established a Scientific Committee on Problems of the Environment (SCOPE), which has close ties to the United Nations Environment Program (UNEP). The UNEP has stated that "Monitoring systems should be developed first for pollutants liable to affect weather and climate"; such systems are expected to be part of the UNEP Earth Watch Program. Recognition of the United Nations' interests in the stratosphere was formalized at the Geneva Conference on the Human Environment (August 27, 1971). Although concerned with factors affecting climate, UNEP has not yet focused on aircraft engine emissions in the stratosphere.

Current commercial supersonic aircraft programs in the Soviet Union, France, and the United Kingdom lend urgency to concern about inadvertent modification of the stratosphere in the future. A similar concern exists in connection with the development of higher-flying wide-body subsonic airliners in the United States. All nations are involved, whether or not they contemplate operating stratospheric aircraft. The evidence so far available is sufficient to indicate that action should be taken to inform all governments and all relevant scientific and regulatory international organizations of the potential effects of stratospheric aircraft emissions. Awareness should lead to continued research, development, monitoring, and analysis and to the adoption of appropriate constraints through international and national regulations.

Domestic and international institutions have the authority to impose ameliorative constraints. Current U.S. federal statutes separately authorize the Administrator of the Environmental Protection Agency and the Secretary of Transportation to establish aircraft engine emission standards for environmental protection. Thus far, proposed rules have been limited to supersonic and subsonic aircraft flight in the vicinity of airports. A working group of the International Civil Aviation Organization (ICAO) is studying the need for international constraints governing aircraft engine emissions for flight under 1 km (3000 ft).

Domestic statutes and regulations can only govern operation of aircraft in the airspace of a particular country. Although such authority might be sufficient to impose desired standards on foreign aircraft seeking to enter that country's airspace, it would not extend to foreign aircraft that do not seek to enter that airspace, even though stratospheric operation of these aircraft abroad could affect the country's airspace and biosphere.

A multinational treaty would extend desired constraints, but would be binding only on parties to the treaty. Most nations are members of ICAO, and, as such, each nation under the governing Convention on International Civil Aviation (the so-called Chicago Convention) is obligated to accept ICAO standards unless, within a stated period, it files a "deviation" from the standard. Nations generally have adopted the ICAO standard as a minimum. When a deviation has been filed, it generally has been to advise the Council of the adoption domestically of a more stringent standard than that of ICAO.

The ICAO Council has approved the report of its Air Navigation Commission recommending action to establish stratospheric aircraft engine emission standards, but it concluded in regard to establishing a "permissible level" of such engine emissions that prime responsibility rested with the World Meteorological Organization (WMO), with ICAO providing technical guidance as needed. The ICAO assigned a "priority 1" to the determination of guidelines for methods of control (operational practices) of engine emissions; however, the WMO Executive Council, which met most recently in June 1974, has not yet discussed the problem of stratospheric engine emissions.

The United States National Environmental Policy Act of 1969 * declares "that it is the continuing policy of the Federal Government . . .

* Public Law 91-190; 83 Stat. 852.

to use all practicable means and measures . . . to create and maintain conditions under which man and nature can exist in productive harmony, and to fulfill the social, economic, and other requirements of present and future generations of Americans." Further, "it is the continuing responsibility of the Federal Government to use all practicable means . . . to the end that the Nation may

(1) fulfill the responsibilities of each generation as trustee of the environment for succeeding generations:

.

.

.

(3) attain the widest range of beneficial uses of the environment without degradation, risk to health or safety, or other undesirable and unintended consequences:"

Further, the Act directs all agencies of the United States Government to ". . . lend appropriate support to initiatives, resolutions and programs designed to maximize international cooperation in anticipating and preventing a decline in the quality of mankind's world environment."

We RECOMMEND *that international organizations concerned with atmospheric science, biological science, medical science, and aerospace science devote increased attention to studies concerned with the stratosphere and with possible inadvertent modification thereof. We further* RECOMMEND *that, to prepare for formal international negotiations looking toward regulation of stratospheric flight, international organizations concerned with the environment and with aerospace flight form international commissions of government-appointed scientists and engineers to review and agree on the state of knowledge concerning the effect of engine emissions on the stratosphere and the means available for modifying them. These scientific understandings should include the effect of stratospheric modification on the intensity of ultraviolet light at the ground, on climate, and on agriculture. They should be updated at frequent regular intervals.*

APPENDIXES

Appendix A

Stratospheric Ozone and Its Photochemistry

I. The Natural Stratosphere

The natural stratosphere is the subject of CIAP Monograph 1. The properties treated in detail there include temperature, composition, radiation, chemistry, and atmospheric motions. There are extensive tables of quantitative data, a number of illustrative figures, and detailed discussion of these topics. Some of these results are presented here as an introduction to the problem of stratospheric ozone and its photochemistry.

A. TEMPERATURE

Starting at the surface of the earth, one defines a series of layers of the atmosphere: troposphere, stratosphere, and mesosphere. These layers are defined in terms of their temperature structure, as illustrated in Figure A.1 (Newell, 1969). Averaged over reasonably long periods of time, the temperature of the troposphere decreases smoothly with increase in altitude. At an elevation that varies systematically with latitude and season, the temperature ceases to decrease with elevation; this property defines the tropopause, which lies between 8 and 16 km. The stratosphere is the region above the tropopause. In the stratosphere, the

This Appendix presents the findings of the Panel on Atmospheric Physics and Chemistry: F. Johnson, *Chairman;* J. Friend, *Vice Chairman;* J. Chamberlain, D. Hunten, H. Johnston, F. Kaufman, H. Landsberg, and H. Panofsky. Consultants to the Panel: R. Hudson, C. Leith, E. Reiter, and G. Robinson.

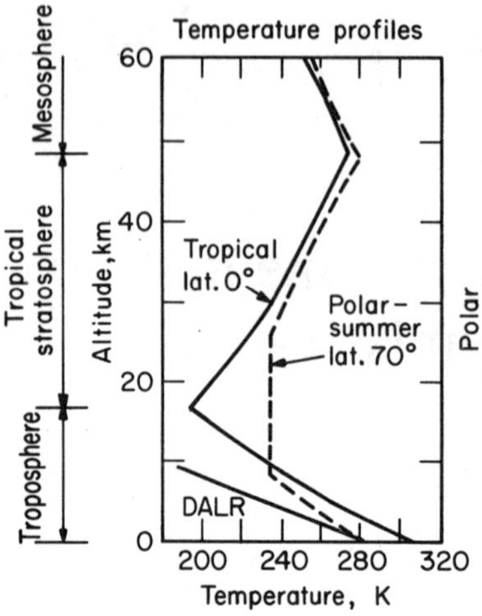

FIGURE A.1 Sample temperature profiles at tropical and polar zones. DALR, dry adiabatic lapse rate. Troposphere, stratosphere, stratopause, and mesosphere defined in terms of vertical temperature profiles.

temperature typically is constant or increases with an increase in altitude. This increase, however, is reversed at about 45 to 50 km—the stratopause. The region above the stratosphere is named the mesosphere. These features are indicated by two temperature profiles in Figure A.1.

If a sample of dry air is lifted adiabatically in the atmosphere, it will expand and cool, cooling occurring at a rate of 10 K per kilometer. This "dry adiabatic lapse rate" is labeled DALR in Figure A.1. Although air in the troposphere is well stirred, the temperature decreases with height more slowly than the dry adiabatic lapse rate. Principally this difference depends on water present in the atmosphere. Condensation of water in the atmosphere releases the latent heat of vaporization and thereby warms the air. The increase in temperature in the stratosphere is largely caused by ozone, O_3. Ozone strongly absorbs ultraviolet radiation between 200 and 300 nm and weakly absorbs solar radiation from 300 to 330 nm and from 450 to 700 nm; it also has important infrared absorption near 10^4 nm. This act of absorption and subsequent chemical reactions inject heat into the stratosphere. An increase of temperature with height is called an inversion. If a parcel of air is lifted in a region of temperature inversion, it cools at the adiabatic lapse rate and is thus colder and heavier than surrounding air. Thus it tends to resist being lifted. A temperature inversion confers great stability to a mass of air against vertical mixing.

The two temperature profiles in Figure A.1 show substantial differences between polar and tropical regions. One could get an indication of how temperature changes with latitude by a series of such profiles, but

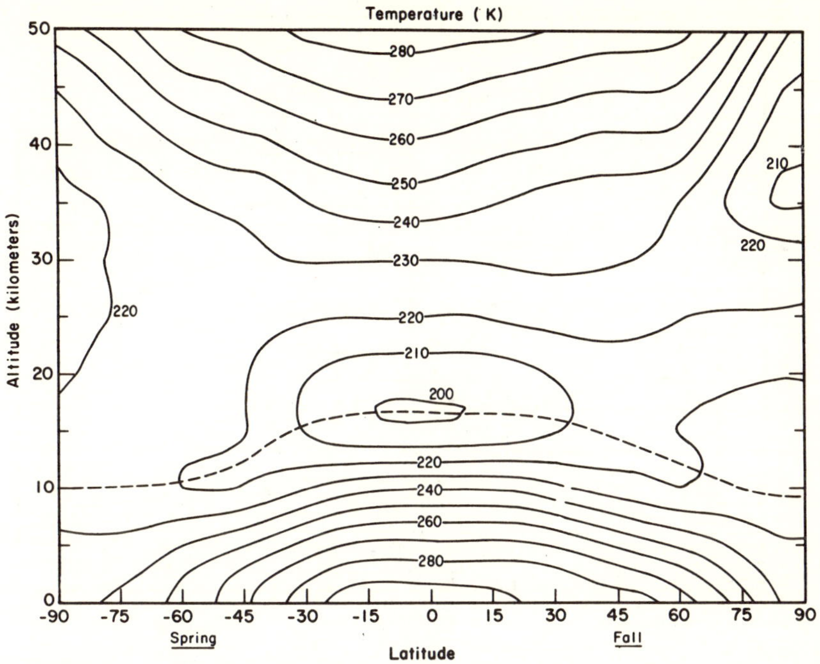

FIGURE A.2 Average temperature contours for March 22: —90°, north pole; 0°, equator; + 90°, south pole. Dotted line is approximate location of average tropopause. Zonal average.

considerably more information can be given by contour lines of zonally averaged temperatures, Figures A.2 and A.3 (Crutcher, 1969; Sissenwine, 1969). One axis of such a figure goes from the north pole (−90°) through the equator (0°) to the south pole (90°), and the other axis on a greatly exaggerated scale gives the vertical height from the surface of the earth to the top of the stratosphere. At each point, temperatures are averaged around the globe at a fixed latitude (zonal average). Contour lines of constant temperature then show how temperature varies with latitude and elevation. Such maps can be prepared from average temperature tables for various seasons of the year.

B. COMPOSITION

The concentration of air decreases with height as indicated by Figure A.4. At the base of the stratosphere, the air density is about 10 percent of that at the earth's surface, and at the top of the stratosphere the air density has decreased to 0.1 percent of that at the surface.

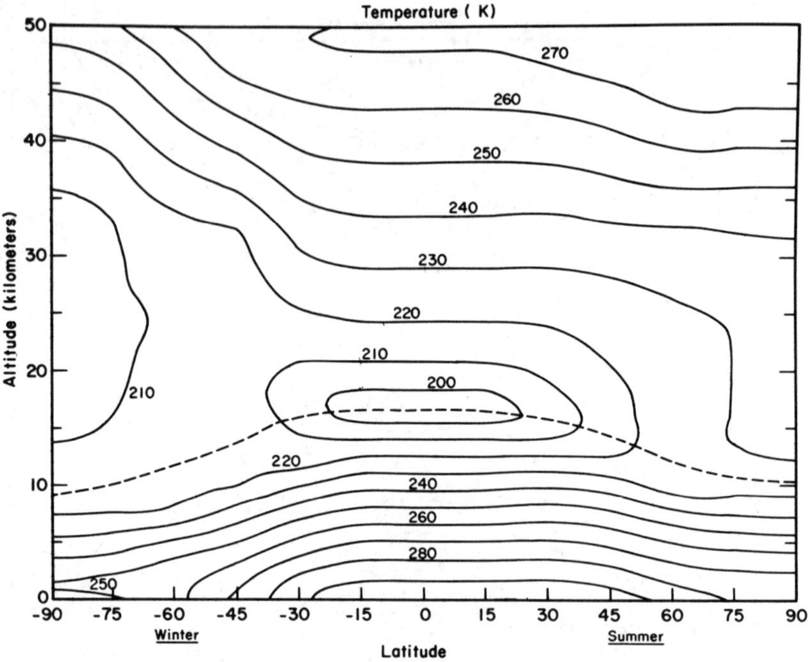

FIGURE A.3 Same as Figure 2, January 15.

Atmospheric ozone has been extensively observed by ground-based optical methods (Dütsch, 1969), by balloons (Hering and Borden, 1964), by rockets (Krueger, 1973), and by satellites (Lovill, 1972). Stratospheric ozone is unevenly distributed in space. High spatial resolution soundings by balloons (Hering and Borden, 1964) reveal multiple layers of ozone up to about 25 km, especially in temperate and polar regions; tropical zones usually show one well-defined maximum concentration of ozone. An average ozone profile for the temperate zone is included in Figure A.4. The maximum ozone concentration occurs on an average at 20 to 25 km.

The overhead vertical column of ozone is expressed in units of molecules cm^{-2} or in terms of the thickness of a sample of pure ozone at standard temperature and pressure (STP). These thicknesses are usually given in units of centimeters or in Dobson units (10^{-3} cm STP). The global average ozone column is about 0.3 cm or 300 Dobson units. The distribution of ozone in the northern hemisphere in May 1969 as seen from an orbiting satellite is given by Figure A.5 (Lovill, 1972).

Stratospheric ozone is variable in time. The day-to-day variations at

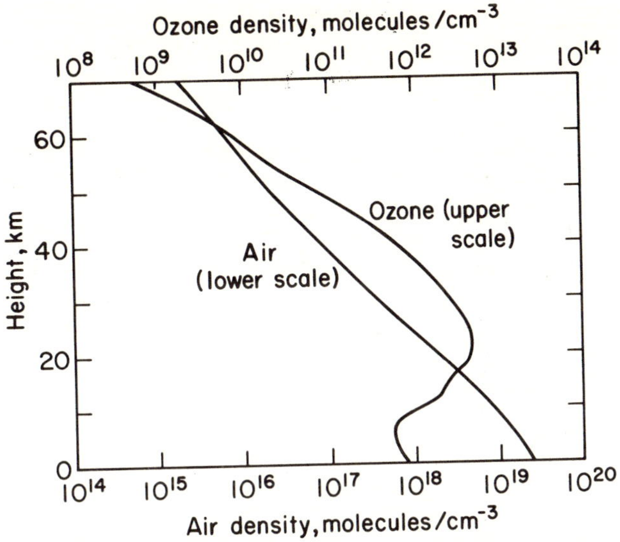

FIGURE A.4 Midlatitude vertical distributions of air density (lower scale) and ozone density (upper scale) in the troposphere, stratosphere, and lower mesosphere. The air density distribution is from the U.S. Standard Atmosphere (1962), and the ozone density distribution is from Krueger and Minzner (1973). Note that the air density scale is 10^6 times the ozone density scale such that the ratio of the two curves at any height subjectively gives the ozone mixing ratio in ppm (v).

Aspendale, Australia, for four months are illustrated by Figure A.6, and the variation of monthly average ozone for 10 years at the same station is also given (Canadian Department of Transport, 1960–1970). The systematic natural variation of ozone column with season and with latitude is given by Figure A.7. Especially in tropical zones, there is a "quasi-biennial" cyclic variation of total ozone (Angell and Korshover, 1973). Christie (1973) made an estimate of total global ozone—on the basis of a sparse data base before 1960, as he pointed out—and found some evidence for an 11-year cyclic variation of ozone (Figure A.8). The increasing trend in ozone during the 1960's can be seen in this figure; this trend was detected by Komhyr *et al.* (1971), and it has been widely discussed.

By averaging data over approximately a 10-year period, Dütsch (1969) has prepared average distributions of ozone as a function of latitude and season. A global perspective of ozone can be obtained from

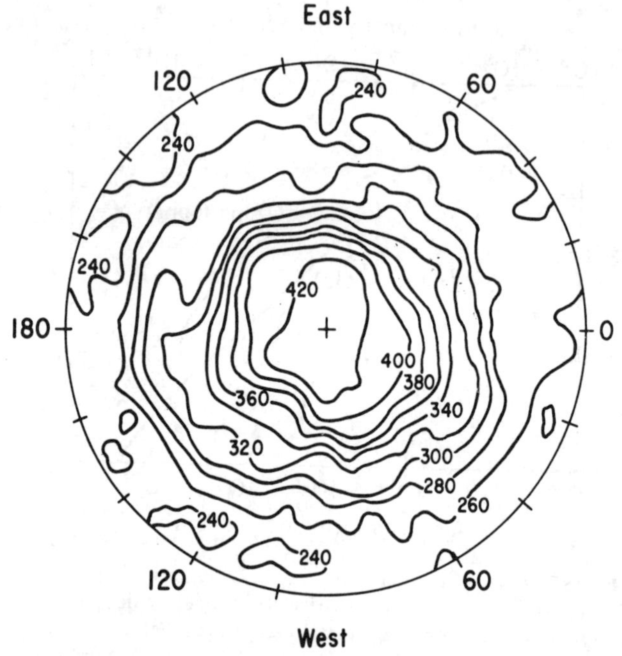

FIGURE A.5 Total ozone in the northern hemisphere as measured from Nimbus IV satellite, May 1969. Contour lines are in units of milliatmosphere-centimeters STP (Dobson units). (From Lovill, 1972).

a zonal average contour map. Zonal average contours of ozone concentration are given in Figure A.9 for March 22. The high concentration of ozone in the springtime polar stratosphere is noteworthy, and the height of maximum ozone concentration can be seen to vary from about 17 km in polar regions to about 20 km in midlatitudes to about 25 km in tropical zones. The ratio of ozone concentration to total gas concentration is called the mole fraction or the mixing ratio by volume (a closely related quantity is the ratio of mass of reference material per unit volume to the mass of air in the same volume or mixing ratio by mass). The ozone distribution in Figure A.9 is presented in terms of mole fractions (parts per million, ppm) in Figure A.10. The distribution of mole fractions is quite different in appearance from that of concentrations. The maximum mixing ratio is above the tropics at about 35 km.

Concentration units are most convenient for quantitative work with

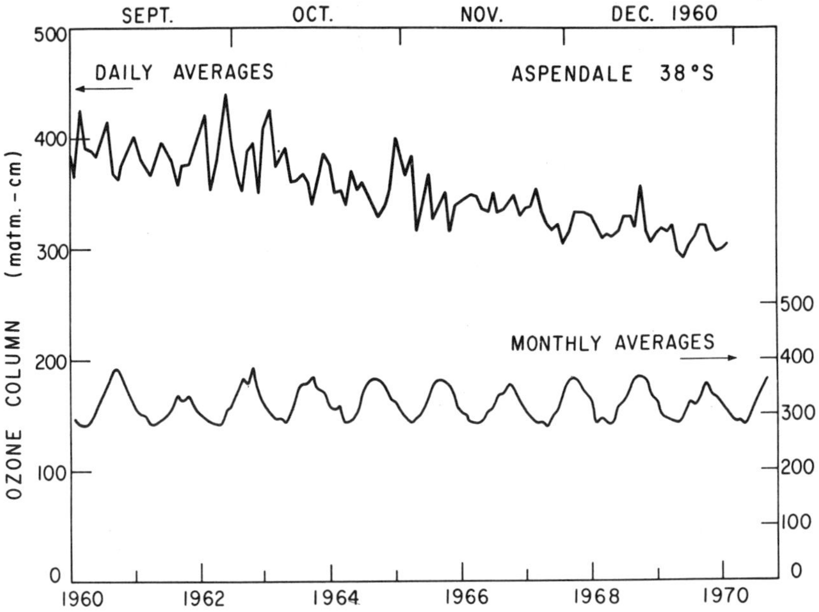

FIGURE A.6 Total ozone column at Aspendale, Australia, showing daily fluctuations and seasonal variations.

chemical reaction rates and absorption of solar radiation. In the vertical dimension of the atmosphere, where air density decreases with altitude, the key to the sign and magnitude of diffusion processes (molecular or turbulent) is not the gradient of concentration but the gradient of mole fraction.

There are numerous minor species in the stratosphere. Observed profiles for carbon dioxide, water, methane, hydrogen, carbon monoxide, nitrous oxide, nitric acid, nitrogen dioxide, and nitric oxide are given in Figure A.11 (Hard, 1974). The oxides of nitrogen have been observed in the stratosphere only recently. The range of observations of nitric oxide (NO) and nitrogen dioxide (NO_2) are given in Figures A.12 and A.13. The range of observations of stratospheric nitric acid is given by Figure A.14.

In addition to the observed minor species, there are numerous atoms, free radicals, and unstable trace species in the stratosphere. These species are inferred from chemical measurements in the laboratory; and their concentration can be calculated from the composition of the stratosphere, from the intensity of solar radiation above the stratosphere, and from the molecular properties of both major and minor species.

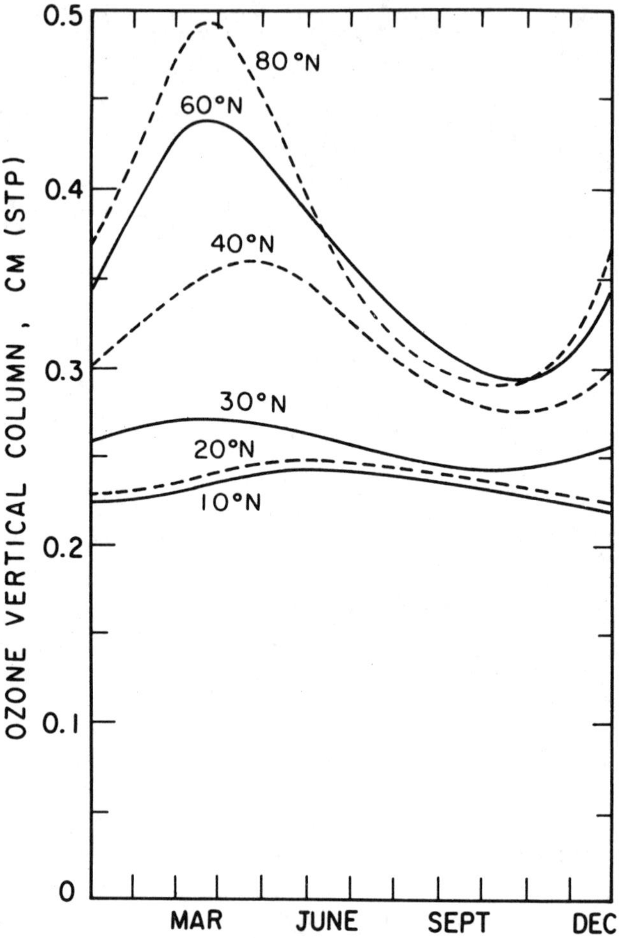

FIGURE A.7 Variation of integrated vertical ozone column with season at tropical, temperate, and polar zones.

Contours of calculated O(^{3}P) concentrations for March 22 are shown in Figure A.15 (Johnston and Whitten, 1973). These figures are based on the observed average ozone distribution; and they give zonal averages for any single time, or they may be interpreted as 24-h averages at a given longitude. A two-dimensional distribution of calculated O(^{1}D) concentrations is given in Figure A.16. Much of the strong chemical reactivity in the stratosphere is a consequence of atomic oxygen there, both O(^{3}P) and O(^{1}D). An interesting aspect of Figures A.15 and

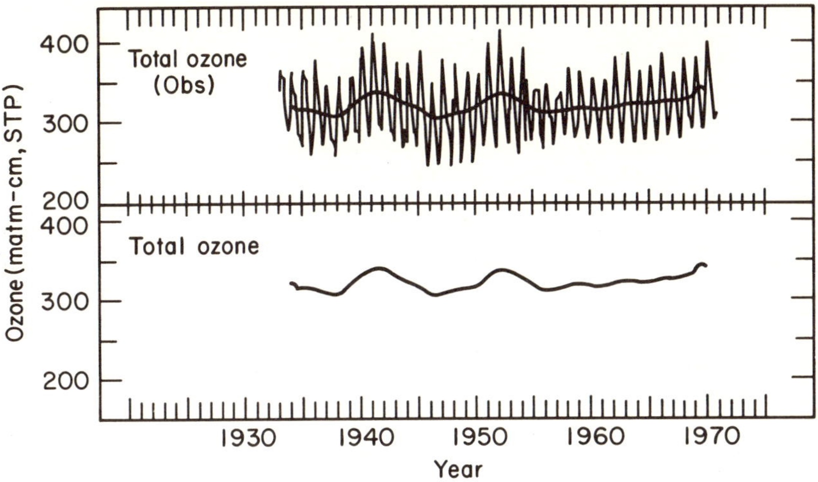

FIGURE A.8 Long-term variations in total ozone in the northern hemisphere. The sharp seasonal variations (compare Figure A.6) are shown in the top panel. An 11-year running mean is inscribed in the top panel and repeated in the bottom panel. A possible 11-year cycle is indicated before 1960. An increasing trend is indicated for 1960–1970, but the 1970 high value is still less than high values before 1960. It is generally recognized that the data base before 1960 is too sparse to offer firm support to these suggested long-term trends.

A.16 is the almost uniform stratification of oxygen atoms and the rapid increase with height.

Dynamic photochemical models of the stratosphere are used to calculate the formation of nitric oxide from nitrous oxide and singlet oxygen atoms in the stratosphere. These models predict the distributions of nitric oxide, nitrogen dioxide, and nitric acid, subject to assumptions about boundary values for these substances. These predictions can be compared with observed distributions of the nitrogen oxides as a partial validation of the models. Some calculated profiles for the active oxides of nitrogen, $NO_x = NO + NO_2$, are given in Figure A.17, and these results may be compared with Figures A.12 and A.13. The calculated partition of $NO_y = NO + NO_2 + HNO_3$ among its three components is shown in Figure A.18.

C. AIR MOTIONS AND STRATOSPHERIC TRANSPORT

Air motions consist of components that span the scale from molecular mean-free paths up to the circumference of the earth. Motions in the

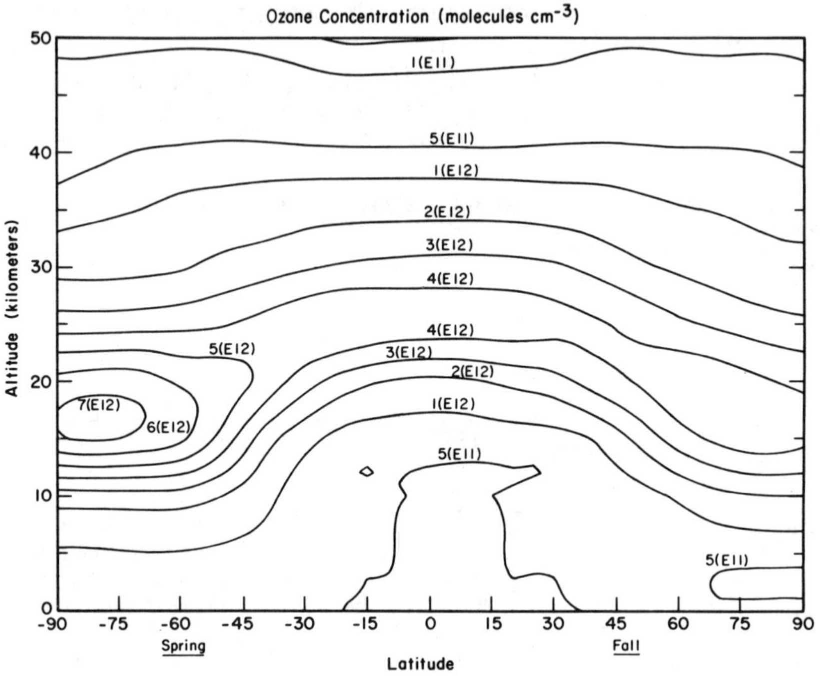

FIGURE A.9 Average (March 22) ozone concentration in units of molecules cm^{-3} expressed as zonal-average contour lines (7E12 means 7×10^{12}). The variation of the ozone maximum with latitude can be read from the figure. The large concentration in the lower, spring, polar stratosphere is consistent with Figure A.7.

stratosphere are treated in CIAP Monograph 1, Chapters 2, 6, and 7 and in Monograph 3, Chapters 2, 3, and 4. This section will discuss briefly a few aspects of this subject.

The response of the atmosphere and the terrestrial environmental system to a significant release of various kinds of effluent from a fleet of supersonic aircraft is in large part determined by the nature of the dispersion of the effluent away from the source region. Thus, subsequent predictions of responses to this dispersed effluent are limited by the accuracy with which such dispersion can be quantitatively predicted.

The net dispersion away from the source region is dependent on the cumulative effects of various types of atmospheric motion. Briefly stated, these motions might be classified as (1) mean meridional circulations; (2) quasi-stationary waves; (3) transient disturbances related to rem-

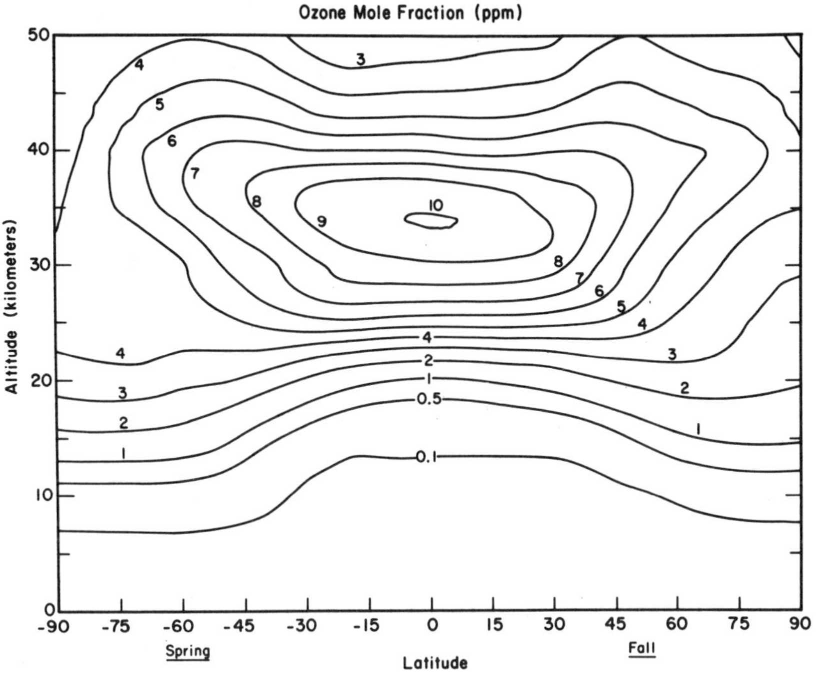

FIGURE A.10 Average (March 22) ozone mole fractions or mixing ratio by volume (ppm).

nants of tropospheric disturbances; (4) gravity waves; and (5) small-scale turbulence. These various types of motion do not necessarily operate independently of one another. For example, the stratospheric mean meridional circulations (the circulation remaining after averaging all motions around a latitude circle) appear to be mainly induced by the action of the quasi-stationary waves. The interactions between various types of stratospheric motion influence the ultimate nature of effluent dispersion. Therefore, an accurate description of the dispersion process should take these factors into account.

1. *Stratospheric Residence Time and Box Models*

It has been found from observations of radioactive debris that material in the lower stratosphere has a "residence time" of a few years. This concept, superficially simple, rapidly breaks down if pushed very far.

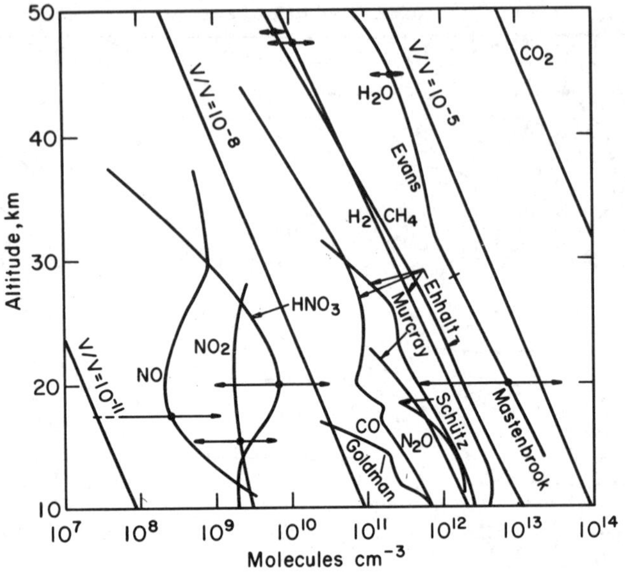

FIGURE A.11 Stratospheric trace-gas measurements: NO, NO_2, HNO_3, average profiles estimated from Figures A.12–A.14; CO based on Ehhalt and Heidt (1974), Farmer *et al.* (1974), Goldman *et al.* (1973), Seiler and Warneck (1972); N_2O based on Ehhalt and Heidt (1974), Farmer *et al.* (1974), Harries *et al.* (1974), Murcray *et al.* (1973b), Schutz *et al.* (1970); H_2 based on Ehhalt and Heidt (1974), Ehhalt (1974), Ehhalt *et al.* (1974), Scholtz *et al.* (1970); CH_4 based on Ackerman and Muller (1973), Cumming and Lowe (1973), Ehhalt and Heidt (1974), Ehhalt *et al.* (1972), Farmer *et al.* (1974); H_2O based on Mastenbrook (1971), Murcray *et al.* (1973b), Brousaides and Morrissey (1967), Evans (1974). Figures A.11–A.14 are adapted from an informal letter report by T. M. Hard (1974).

As discussed below, a unique residence time is generally not found, and it is also a function of height above the tropopause.

Use of the term "stratospheric residence time" almost implies a two-layer model of the atmosphere, the stratosphere, and the troposphere; an equivalent is the transfer between connected boxes. A substance widely distributed and inert in the stratosphere and rapidly rained out of the troposphere is a candidate for the concept of stratospheric residence time. However, the residence time so found is not constant. It is expected to be different for every different initial distribution of inert tracer. With

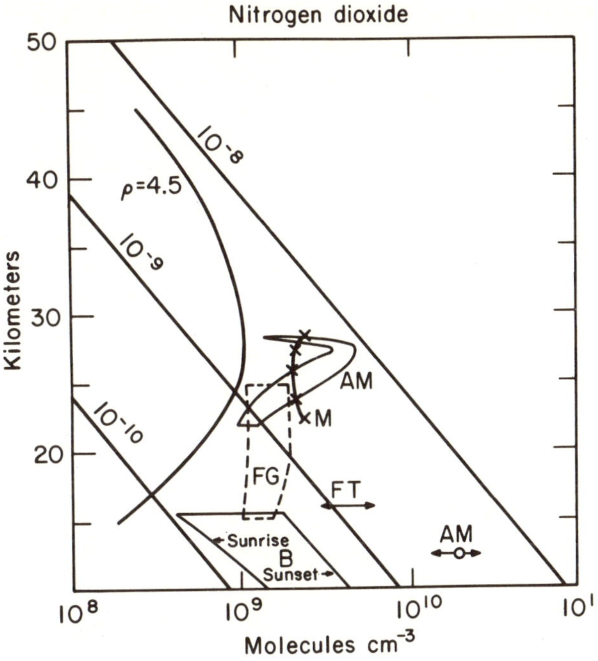

FIGURE A.12 Vertical profiles of observed NO_2. The curve $\rho=4.5$ is the amount of NO_2 that would destroy ozone 3.5-fold faster than the elementary reaction $O+O_3 \rightarrow O_2+O_2$ destroys ozone, which is the amount of NO_2 required to destroy 70 percent of the ozone formed from solar radiation. The various observations were made by A, Ackerman and Muller (1972, 1973); B, Brewer *et al.* (1973, 1974); FG, Fontanella *et al.* (1974); FT, Farmer *et al.* (1974); M, Murcray *et al.* (1974).

this model, it is expected that the number of molecules N of the substance will follow the rate equation:

$$dN/dt = -(1/\tau)N. \qquad (1)$$

In this case, τ may be defined as a "residence time." The integrated form of this equation is

$$\ln N = \ln N_0 - (t/\tau). \qquad (2)$$

The negative slope of a plot of ln N against time is the reciprocal of the residence time, τ.

If the substance is inert in the stratosphere and troposphere and is only slowly removed from the troposphere, there will also be transport

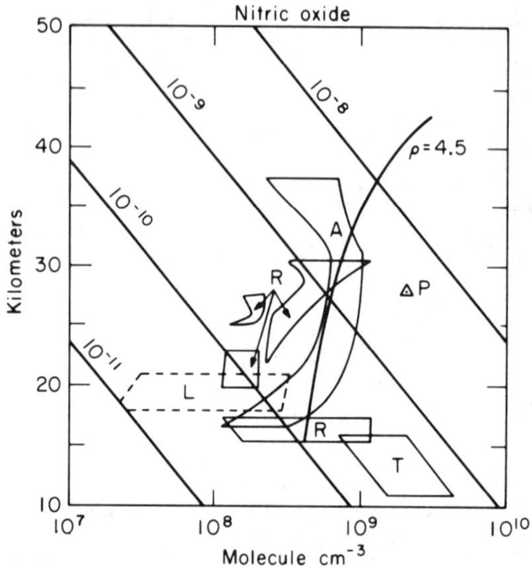

FIGURE A.13 Vertical profiles of observed NO. The curve $\rho=4.5$ is the amount of NO, as such, that would destroy ozone 3.5-fold faster than the elementary reaction $O+O_3 \rightarrow O_2 + O_2$. The observations were A, Ackerman *et al.* (1973); L, Loewenstein *et al.* (1974); P, Patel (1974); R, Ridley *et al.* (1974); T, Toth *et al.* (1974).

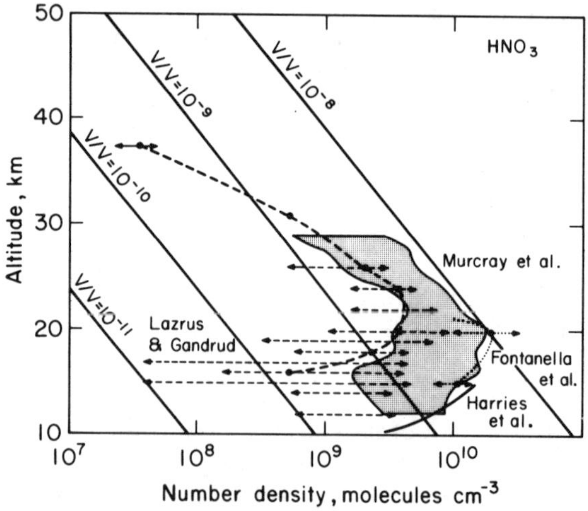

FIGURE A.14 Vertical profiles of observed HNO_3: Fontanella *et al.* (1974); Harries *et al.* (1974); Lazrus and Gandrud (1974); Murcray *et al.* (1973a, 1973b).

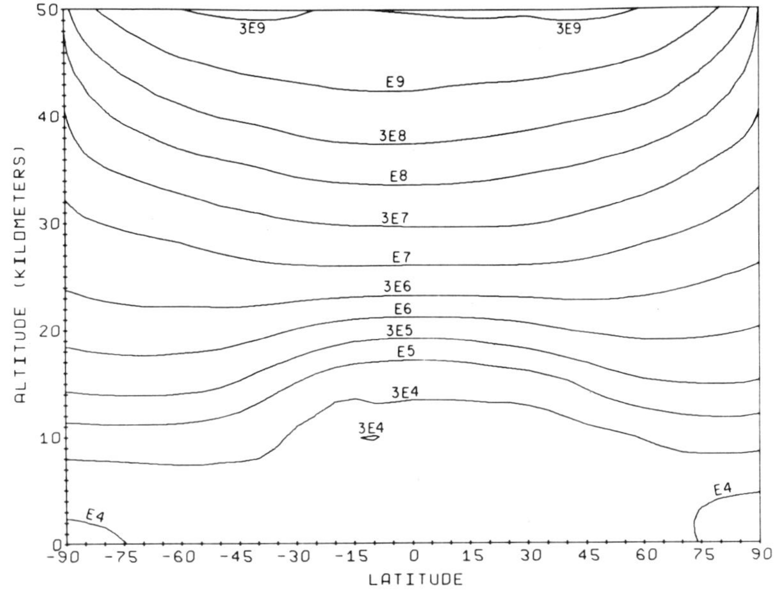

FIGURE A.15 Zonal average contour maps of triplet oxygen atom concentration, molecules cm^{-3} (March 22). Note the strong stratification of these concentrations and the rapid decrease with decreasing elevation.

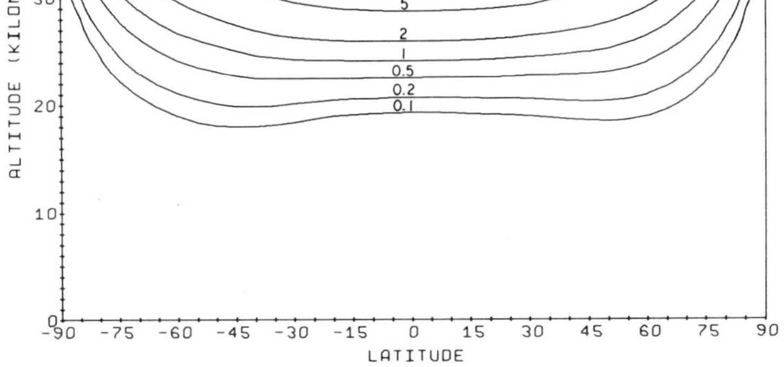

FIGURE A.16 Zonal average contour maps of singlet oxygen atom, O(^{1}D), concentration (March 22).

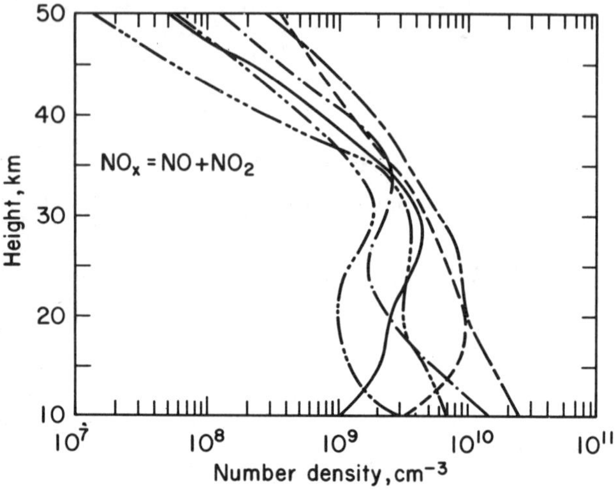

FIGURE A.17 Profiles of active nitrogen oxides (NO_x=NO +NO_2) as calculated by various models. —— – – – ——, Crutzen (1973a, 1973b); ——– – ——, Shimazaki and Ogawa (1974); —— • ——, Whitten and Turco (1974); —— – ——, Hesstvedt (1972); - - - - -, McElroy *et al.* (1974); ——, Chang (1973).

from the troposphere to the stratosphere, and the transport rate equation might be

$$dX_s/dt = -(1/\tau_s)(X_S - X_T), \tag{3}$$

where X is the mixing ratio and the subscripts S and T refer to stratosphere and troposphere, respectively. This equation may be rewritten

$$d \ln X_s/dt = -(1/\tau_s)(1 - \overline{X_T/X_S}). \tag{4}$$

For small finite differences in time, this expression may be solved for the residence time:

$$\tau_S = \frac{(t_2 - t_1)(1 - \overline{X_T/X_S})}{\ln (X_{S1}/X_{S2})}, \tag{5}$$

where $\overline{X_T/X_S}$ is the average value of this ratio over the time interval t_1 to t_2. The radioactive tracers from nuclear bomb tests appear to give a situation fairly close to the model discussed here. The radioactive tritium and carbon are subject to backtransport from the troposphere to the stratosphere, and data based on Eq. (2) tend to give apparent residence times that are too long. Radioactive strontium-90 and other

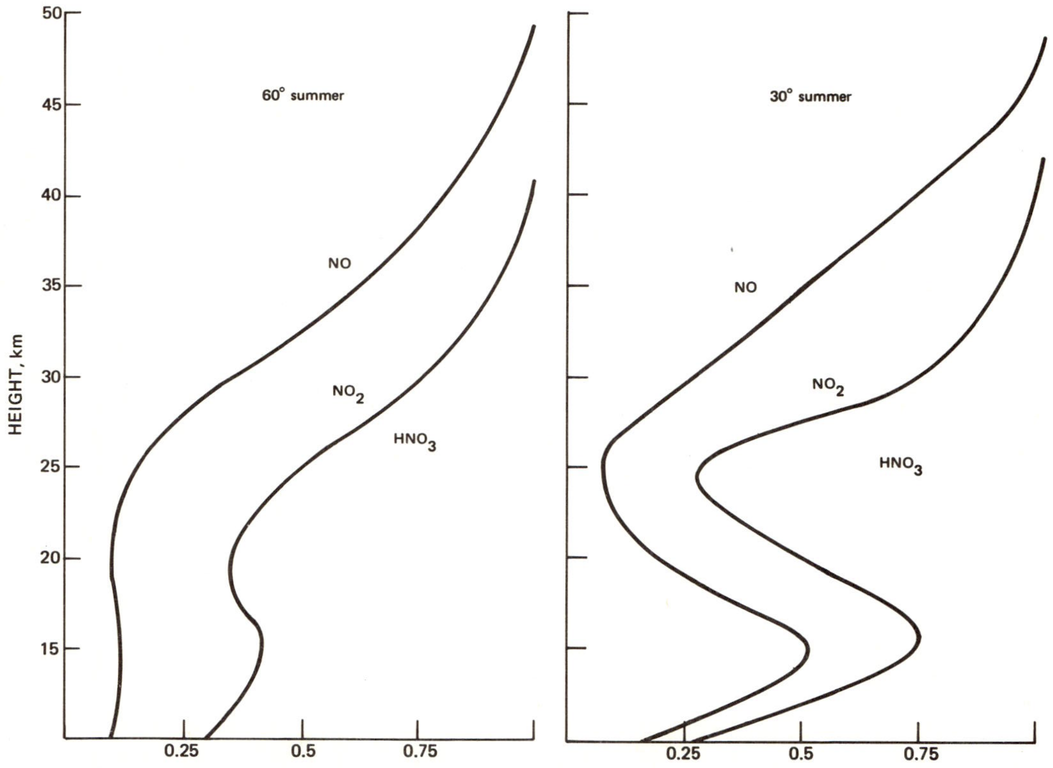

FIGURE A.18 Partition of NO_y (=$NO+NO_2+HNO_3$) among its various components as a function of elevation and latitude.

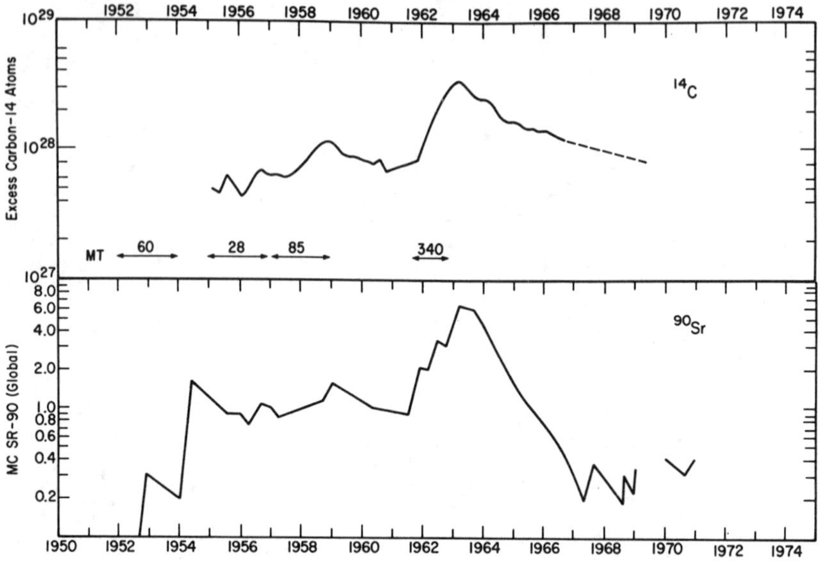

FIGURE A.19 Total stratospheric inventory of excess carbon-14 and strontium-90 from 1953 to 1971. Periods and yields (megatons, MT) of nuclear explosives are indicated in the top panel. Note the much more rapid removal, 1963–1966, of ^{90}Sr relative to ^{14}C after the end of the large-scale tests.

heavy metals (Telegadas and List, 1969) are lodged on particles that settle out at a nonnegligible rate (Feeley and Seitz, 1970; Hunten, 1975), and thus simple use of Eq. (2) for these tracers gives a residence time smaller than that appropriate for gases.

The last nuclear explosion of the massive 1961–1962 series occurred in December 1962, and there was extensive sampling of the atmosphere up to 33 km by balloons and by airplanes to measure the distribution of excess carbon-14, strontium-90, and other radioactive bomb debris. These data have been published in a series of Health and Safety Laboratory (HASL) reports (for example, Telegadas, 1971), and some of the data have appeared in the scientific literature. The total stratospheric burdens of excess carbon-14 (Telegadas, 1971) and strontium-90 (Telegadas, 1967) are plotted against time according to Eq. (3) for the period 1952–1970 in Figure A.19. The quantities of nuclear explosions in units of megaton (MT) equivalents of TNT are noted on the figure. The large-magnitude test series ended in December 1962, and there were no more tests until 1967. Thus ^{14}C and ^{90}Sr can be compared for a period of four years, 1963–1966 inclusive. It is apparent that the

TABLE A.1 Excess Carbon-14

Date	Number of Atoms of ^{14}C in Units of 10^{26} N_S	N_T	Mole fraction of ^{14}C in units of 10^{-16} X_S	X_T	X_S/X_T	Mean Residence Time (1/*e*) in Years Based on One-Year Time Interval τ_1	τ_2	τ_3
1/63	356	183	27.4	1.83	15.0	3.0	2.7	2.1
4/63	349	263	26.8	2.63	10.2	2.3	2.0	1.9
7/63	301	235	23.1	2.35	9.8	2.0	1.7	2.0
10/63	272	244	20.9	2.44	8.6	2.1	1.8	2.3
1/64	255	249	19.6	2.49	7.9	2.5	2.0	2.8
4/64	225	263	17.3	2.63	6.6	2.7	2.2	3.6
7/64	183	258	14.1	2.58	5.5	5.2	4.2	4.7
10/64	170	245	13.1	2.45	5.3	6.0	4.7	4.7
1/65	170	249	13.1	2.49	5.3	6.6	4.9	
4/65	156	257	12.0	2.57	4.7	5.7	4.3	
7/65	151	275	11.6	2.75	4.2	6.0	4.6	
10/65	144	270	11.1	2.70	4.1			
1/66	146	256	11.2	2.56	4.4			
4/66	131	258	10.1	2.58	3.9			
7/66	128	254	9.8	2.54	3.9			

residence time as inferred simply from Eq. (2) is over twice as long for gaseous ^{14}C as for the particulate ^{90}Sr.

Table A.1 includes stratospheric inventory N_S of ^{14}C, tropospheric inventory N_T, stratospheric inventory mole fraction X_S, and the tropospheric inventory fraction X_T. From the data in Table A.1, mean residence times were calculated in three different ways, where t_1 was each successive time and t_2 was one year later. The first method, τ_1, is based on Eq. (2). The second method, τ_2, uses Eq. (5) and includes the correction for transport of ^{14}C from the troposphere to stratosphere. The third method smoothed the data with a four-quarter running average, and then the residence time τ_3 was calculated from Eq. (5). These residence times increased from about 2 years to about 4 years between 1963 and 1965. Thus the concept of residence time clearly has serious limitations in this context.

In the examples given above, a tracer was injected into the stratosphere, and then with no further injections of material the transfer of the tracer out of the stratosphere was characterized in terms of a varying residence time. Another use of the term applies to a material that is continuously produced in the stratosphere and removed by air transport to the troposphere, where it is washed out. The defining equation is

$$dN/dt = P - (N/\tau), \quad (6)$$

where P is the rate of production of the material and the other terms are as in Eq. (1). Over a long period of time, a steady state is obtained between production and removal, and then the total inventory N is constant so that dN/dt is zero. In this case, the residence time can be evaluated

$$\tau = N/P, \quad (7)$$

if one can evaluate the global stratospheric inventory N and the global production rate P.

A possible example for this approach is "odd nitrogen," $NO_y = NO + NO_2 + HNO_3$. The global inventory of nitric acid can be deduced from an extrapolation of the observations by Lazrus and Gandrud (1974) and the value is 2.2×10^{34} molecules. However, nitric acid is only a fraction f of the total NO_y, as indicated in Figure A.18. Thus, the total inventory is $2.2 \times 10^{34}/f$. The global source rate of NO_y production has been evaluated by Isaksen (1973), and the value is 1.4×10^{34} molecules yr^{-1}. Thus the stratospheric residence time of NO_y is

$$\tau = \frac{2.2 \times 10^{34}}{1.4 \times 10^{34} f} = \frac{1.5}{f} \text{ years.}$$

From Figure A.18 it can be seen that HNO_3 is about one half of the total NO_y, so that the residence time is about 3 years; that is, the fraction $1/e$ of odd nitrogen is transported out of the stratosphere in 3 years. However, another substance with a different stratospheric distribution would be expected to have a different residence time.

2. *One-, Two-, and Three-Dimensional Models*

(a) *One-Dimensional Models* A one-dimensional model can be considered to be the limit in which many narrow boxes are used to make up the atmosphere. An empirically derived profile for the vertical eddy-diffusion coefficient is used to describe vertical motion. Horizontal spreading is assumed to be much faster than vertical spreading and to fill either the whole globe or a hemisphere. There are two advantages: (1) Essentially all atmospheric chemical reactions can be, and often are, included. (2) The description of transport is deliberately kept simple enough to permit the eddy-diffusion coefficients to be determined empirically from tracers such as CH_4. They can then be verified on other tracers such as N_2O and radioactive debris.

The one-dimensional models can therefore be used today as an approximate representation of nature with a detailed and complete set of chemical reactions. A particularly simple one-dimensional model, discussed in Section I.3, involves the use of injection coefficients.

(b) *Two-Dimensional Models* These use an array of eddy-diffusion coefficients, small for vertical transport and large for north–south transport. It is necessary to include a coupling term or else to tilt the coordinate axes slightly to account for the vertical component of motion normally associated with meridional motion. Some models also include meridional circulation. As a minimum, therefore, three eddy-diffusion parameters are required at each height and latitude.

The difficulty is that available observations of tracers (or even those likely to be obtained in the foreseeable future) are simply inadequate to determine such a grid of parameters. When Gudiksen *et al.* (1968) applied the method to radioactivity, they were forced to scale the horizontal coefficients down by a factor of 8 from the values deduced by Reed and German (1965) on the basis of heat transport. Still further changes have been made for the ozone problem (Hesstvedt, 1974a, 1974b). Thus, even though a model may be made to represent worldwide ozone data very well, its utility for prediction must remain in doubt unless it can be validated. It may, in fact, be correct, but that is no help unless the correctness can somehow be ascertained.

Reasonably complete chemical reactions can be included in two-dimensional models.

(c) *Three-Dimensional Models* A number of models exist that solve the equations of motion of the atmosphere in a large computer. The use of eddy-diffusion coefficients cannot be abandoned entirely, as they must be used to represent transport on a scale smaller than the computational grid. However, most of the transport that occurs in the atmosphere appears to be simulated by the model; and from the spread of an inert tracer in the model, values can be generated for the eddy-diffusion coefficients to be used in one- and two-dimensional models. The validity of coefficients determined in this way, as well as those determined from atmospheric observations, must be established for the problem at hand. Generally, the agreement between model-generated coefficients and those determined from atmospheric observation is good, with some differences attributable to averaging with fairly coarse grid-point spacing and average slope of the tropopause.

Currently, the only numerical models that are capable of a valid physical description of most of the various kinds of stratospheric motion systems in an explicit way are the so-called three-dimensional general circulation models (Cunnold *et al.,* 1974; Dickinson, 1973; London and Park, 1973; Mahlman, 1973a, 1973b). However, preparation and execution of such a model is not sufficient to allow immediate utilization with confidence. Like one- and two-dimensional models, these models also must be tested for their ability to simulate the movement and behavior of known trace substances before they can be reliably extrapolated to a new situation. Current general circulation models including trace-substance transport have been validated by such comparisons. Unfortunately, such models are quite expensive to run and thus cannot be used extensively for simulating atmospheric experiments under a variety of different conditions. Because of their large size, they cannot at present include complete chemistry relating to atmospheric ozone.

3. *Injection Coefficient Models*

Many of the difficulties associated with the notion of residence time disappear if the concept of an "injection coefficient" is used instead (McElroy *et al.,* 1974; Hunten, 1974). This quantity is particularly useful to describe the steady-state concentration of a contaminant, such as NO_x, from sources within the stratosphere. The treatment is one dimensional, and the eddy-diffusion coefficient profile is a key input.

Once a contaminant is injected at some height in the stratosphere, its subsequent behavior must be traced. In most cases of interest here, it must be transported upward to the region where it perturbs the ozone chemistry. In most cases, it is not lost *in situ* but must be transported downward through the sluggish lower stratosphere into the troposphere, where it is rapidly lost. Calculations made for an impulsive injection of a tracer at 17-km altitude indicate that it reaches its maximum concentration at 25 km 1.6 years later, after which losses to the troposphere cause decreasing concentrations. The concentration at 25 km falls to half its peak value 3 years later (i.e., 4.6 years after injection) and to one tenth its peak value 7 years after injection.

The one-dimensional models are well enough understood at this time to permit operational predictions. Only vertical motions are considered, through an eddy-diffusion coefficient K, whose value at each height is determined by analysis of naturally occurring tracers. The principal ones are methane and nitrous oxide, which originate in the ground and are transported upward by the same processes that cause downward motion of contaminants from a high-altitude source. Use is also made of the detailed studies of the horizontal and vertical motion of radioactive debris from nuclear tests. These studies have shown that material injected at midlatitudes spreads rapidly around a latitude circle, and more slowly northward and southward. Most of the material remains in the original hemisphere. This is well illustrated by Figure A.20, which shows the distribution of excess carbon-14 from nuclear bomb tests a few months after the termination of testing in 1962. It is therefore assumed that material injected anywhere in a hemisphere is rapidly spread over that hemisphere and subsequently moves only up and down by eddy diffusion, and the one-dimensional transport treatment is applied to the hemispheric average.

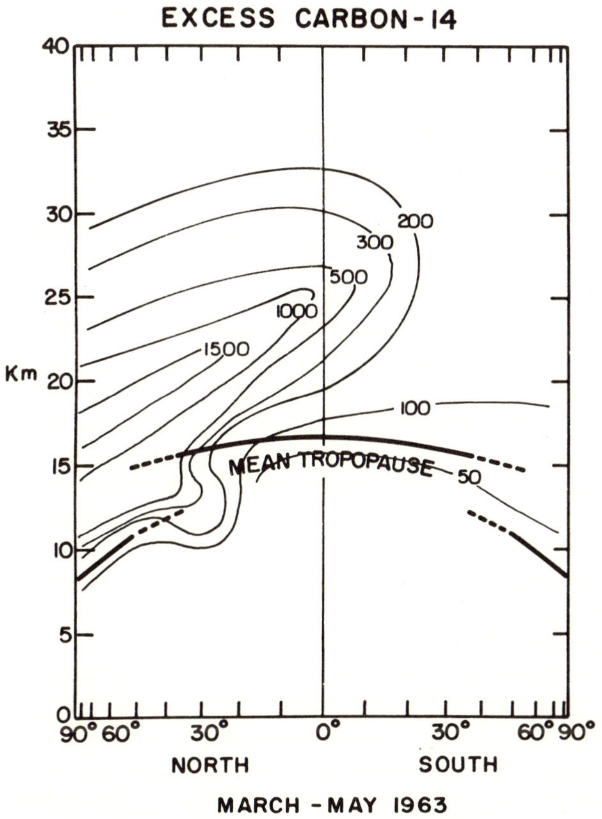

FIGURE A.20 Contour map of zonal-average, excess carbon-14 mixing ratio in units of 10^5 atoms of carbon-14 per gram of air. Note the bulge in the vicinity of the tropopause gap and the jet stream.

The principal distinction among one-dimensional models is their choice of the profile of eddy coefficient *K*. Until recently, there has been little information on which to base a choice, and a wide variety of assumptions has been made (CIAP, Monograph 3). In the last year or two, the methane mixing ratios measured by Ehhalt *et al.* (1972, 1974) and Ehhalt (1974) have become available. They have been analyzed separately by Wofsy and McElroy (1973) and Hunten (1974), with very similar conclusions: at the tropopause there is a large, essentially discontinuous drop in *K* from over 10^5 cm^2 sec^{-1} to about 10^3 cm^2 sec^{-1}, followed by a rapid rise to 4×10^4 cm^2 sec^{-1} or even more at 40 km.

The methane-based *K*'s give long residence times for contaminants in the lowest stratosphere, longer by a factor of 2 to 3 than have been generally accepted. This previous opinion has been based on data from radioactive particulates, whose residence time is shortened by gravitational setting (Feely and Seitz, 1970; Hunten, 1975). Data from carbon-14, which exists as CO_2, confirm the *K* profile obtained from methane (Johnston *et al.,* 1974). For convenience, this profile is taken to have a form that facilitates analytic solution. Such a form is an exponential variation with height; the corresponding scale height is H_K. The most satisfactory set of values is considered to be $K(13\text{ km}) = 10^3$ cm^2/sec, $K(40\text{ km}) = 3.5 \times 10^4$ cm^2/sec, $H_K = -9.43$ km (Hunten, 1974).

Two general classes of one-dimensional model have been studied. In the first, transport equations are written for each species, with transformations by chemical reactions. Diurnal variations are often explicitly included. Several of these are described in CIAP Monograph 3. A great simplification, and little loss of accuracy, can be obtained by use of grouped constituents, notably odd nitrogen, NO_y, already mentioned. Diurnal variations are often represented by suitable averages, which may be different for different substances.

For small perturbations of the natural atmosphere, the transport and chemistry of odd nitrogen can be treated as separable. The transport is exactly like that of an inert tracer, and for exponential variations of *K* an analytic solution of the eddy equation can be written (Hunten, 1974). It is convenient to describe the result by means of an injection coefficient, which is more appropriate to the problem at hand than a residence time (McElroy *et al.,* 1974). The molecular mixing ratio *X* at and above a source is constant in the steady state, and is given by

$$X = \alpha Q, \tag{8}$$

where α is the injection coefficient and *Q* is the source strength. If *Q* is in units of molecules cm^{-2} sec^{-1}, α has units cm^2 sec. Figure A.21 shows

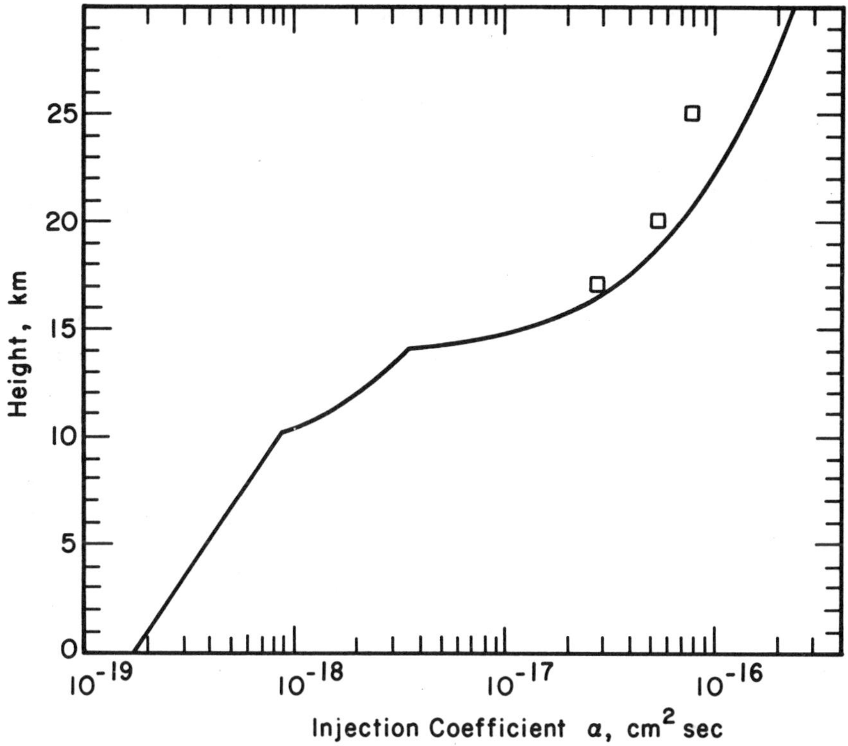

FIGURE A.21 Injection coefficients as a function of height. Squares (McElroy *et al.*, 1974) and solid line (Hunten, 1974) are for methane-based eddy coefficients and different tropopause heights.

injection coefficients as a function of height. The sluggishness of the lower stratosphere is graphically illustrated by the steep increase of α just above the tropopause. The curve is for the analytic model and includes a boundary layer, 4 km deep, to represent the variable height of the actual tropopause. The points (from McElroy *et al.*, 1974) are for a tropopause at 16 km. The injection coefficient may be conveniently expressed as

$$\alpha = C(\exp h_q - D), \tag{9}$$

where C and D are constants that depend on the eddy coefficient and the boundary conditions. The nondimensional height is

$$h = (z - z_t)/H_p, \tag{10}$$

where h_q is the value of h at the source, z_t is the height of the tropopause, and H_p is a scale height—that of the product of K and the atmospheric density. The constants are (compare Figure A.21):

Heights, km	z_t, km	H_p, km	C	D
0–10	10	6.2	0.087	0
10–14	10	6.2	0.29	0.70
14 up	14	18.1	16.7	0.979

The units of C are 10^{-17} cm² sec.

This model represents a collapse of the stratosphere to latitude 30°, along surfaces parallel to the tropopause. An adjustment should therefore be made to the height of any source at a different latitude. For example, at 40° latitude the tropopause is about 2 km lower than at 30°; an aircraft at 16.5 km should therefore be translated to 18.5 km, where $\alpha = 5.06 \times 10^{-17}$ cm² sec.

Multiplication of α by the source strength Q gives the mixing ratio at and above the source. If there are several sources at different heights, their contributions can be added. The distribution below the source can also be computed [it has the same shape as Eq. (9)]. But for ozone reduction, the important chemistry occurs well above the heights at which SST's will fly.

Fleet emissions Q' are commonly expressed in 10^9 g/year, or kT/yr. Conversion to molecules cm^{-2} sec^{-1} is done by the expression

$$Q = (10^9\, N_A/\mu A Y)Q' = 1.62 \times 10^5\, Q', \qquad (11)$$

where the second form holds for NO_2 with molecular weight $\mu = 46$. N_A is Avogadro's number, $A = 2.55 \times 10^{18}$ cm², the area of a hemisphere, and $Y = 3.16 \times 10^7$ sec, the length of a year.

A fleet of 100 present SST's would have the following effects:

Molecule	Source Q' (kT/yr)	Q (cm^{-2}sec^{-1})	X	X(natural)	δ (%)
NO_2	87.5	1.42×10^7	0.72×10^{-9}	10×10^{-9}	1.0
H_2O	6100	2.4×10^9	0.12×10^{-6}	4×10^{-6}	

Conversion from Q to X, the mixing ratio of contaminant, uses Eq. (8) with $\alpha = 5.06 \times 10^{-17}$ cm² sec. As the table shows, the resulting amounts, while smaller than the natural ones, are not so much smaller that they can be ignored, particularly for the much larger fleets that may fly in the future.

Catalytic destruction of ozone by odd nitrogen (NO_x) and odd hydrogen (HO_x) is discussed in some detail below and in CIAP Monograph

3. It is found that the NO_x effect is much more serious, and that all investigators agree closely on the effect of a given mixing ratio X. The published results of McElroy *et al.* (1974) are particularly suitable; a curve fit to their tabulated points gives

$$\delta = 1.405X - 0.0105X^2. \qquad (12)$$

Here X is the additional NO_x mixing ratio in ppb (10^{-9}), and δ is the percentage reduction of ozone. For the example in the table, $\delta = 1.0\%$. Equation (12) is valid only to the limit of the computations, $X = 24$ ppb.

Table A.2 shows results for a series of illustrative fleets. The injection coefficients were calculated by Eq. (9), with a 2-km height adjustment as discussed above, to allow for the difference in latitude between the aircraft (40°) and the model (30°). Combination of Eq. (8) with Eq. (11) gives, for one hemisphere,

$$X(\text{ppb}) = 0.00162Q', \qquad (13)$$

when Q' is in kT/yr of NO_2. Ozone reduction is obtained from Eq. (12), and the percentage increase of erythemally effective radiation from

$$\delta(\text{EER}) = 2\delta(O_3). \qquad (14)$$

Too simple must consider turbidity + zenith angle

For small changes, we assume that skin-cancer incidence is proportional to the flux of erythemally effective radiation.

Large changes are beyond the range of both Eqs. (12) and (14); the last line of Table A.2 therefore gives only a lower bound for δ and no estimate for δ(EER).

II. Ozone Chemistry in the Natural Stratosphere

The properties and numerical constants pertaining to chemical reactions of stratospheric interest are being assembled and critically reviewed by a division of the National Bureau of Standards. A table characterizing more than 250 reactions has been prepared, and these data are included in CIAP Monograph 1, Chapter 5. Upon detailed examination of these 250 reactions, it is found that not all are important to atmospheric ozone chemistry. It requires seven chemical reactions to give a minimum statement of ozone chemistry. It requires 30 to 50 reactions to carry out a quantitative treatment of stratospheric ozone chemistry, and these reactions are listed in Table A.3. Most current chemical models of the stratosphere usually use about 50 or 60 reactions, which give good balance between errors of omission and the error of carrying along excess unimportant details.

TABLE A.2 Decreases of Stratospheric Ozone and Increase of Erythemally Effective uv Radiation for Illustrative Fleets of Aircraft [a]

Fleet	Height (km)	Fuel Flow (10^9 kg/yr)	NO_x (10^6 kg/yr)	α (10^{-17} cm^2 s)	$X(NO_x)$ (10^{-9})	O_3 δ (%)	EER δ (%)
1. 400 Wide-body present subsonic	10.5	10.4	156	0.23	0.058	0.082	0.16
2. 100 Wide-body projected subsonic	13.5	2.6	39	1.79	0.113	0.16	0.32
3. 100 Present SST	16.5	3.49	62.8	5.06	0.52	0.72	1.44
4. 100 Large SST	19.5	9.13	164	8.92	2.37	3.27	6.5
5. CIAP 1990 upperbound fleet	16.5	1.62	29.2	5.06	0.24	11	22
	19.5	30.6	551	8.92	7.96		
6. CIAP 2005 upperbound fleet	16.5	8.09	146	5.06	1.20	>30	Large
	19.5	155	2790	8.92	40.3		

[a] The first four lines are intended for orientation and are not meant to indicate expected traffic, except for Line 1, which represents long-distance traffic for 1974. Lines 5 and 6 are taken, in simplified form, from the CIAP upperbound fleets. The injection coefficient α, multiplied by the source strength and some conversion factors, gives the increase in NO_x mixing ratio (X) at the ozone layer after equilibrium has been reached. The last two columns are estimates of the corresponding percentage reduction of ozone and percentage increase of erythemally effective radiation. Confidence limits are estimated to be a factor of 2 for flight at 19.5 km, a factor of 3 for flight at 16.5 km, and a factor of 10 either way for subsonic aircraft.

Without going into technical detail, some basic features of ozone chemistry are given here.

A. OZONE PRODUCTION

Mostly above 25 km, oxygen molecules are dissociated by radiation below 242 nm, and the oxygen atoms so produced rapidly add to molecular oxygen to form ozone (Chapman, 1930).

(1) $O_2 + h\mu$ (below 242 nm)$\rightarrow O + O$
(2) $O + O_2 + M \rightarrow O_3 + M$ (twice)

NET: $3\ O_2 + h\mu \rightarrow 2\ O_3$

The instantaneous zonal average (or 24-h local average) rate of formation of ozone is given by Figure A.22 (Johnston and Whitten, 1973).

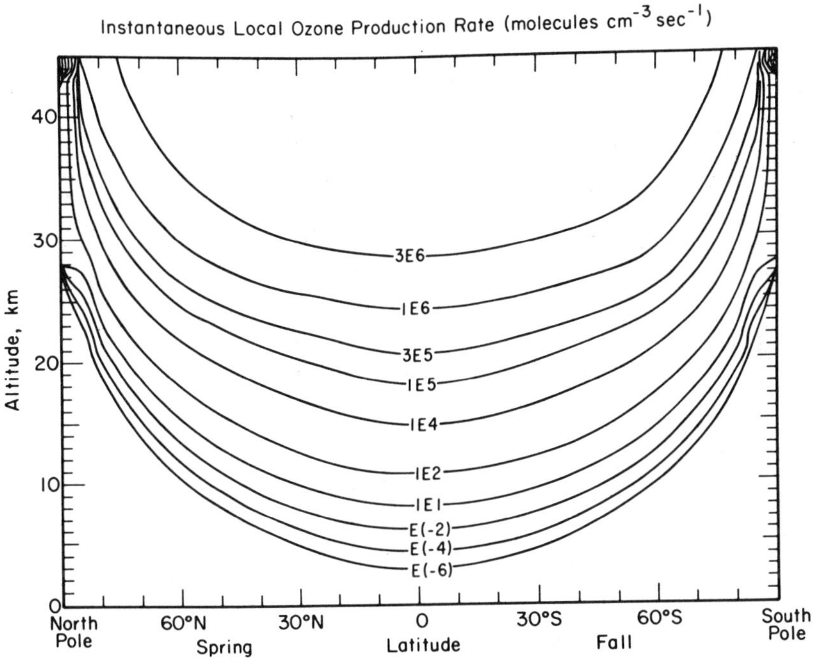

FIGURE A.22 Zonal average rate of formation of ozone from the photolysis of oxygen:

$O_2 + h\nu$ (below 242 nm)$\rightarrow O + O$
$O + O_2 + M \rightarrow O_3 + M$ (twice)

NET: $3\ O_2 + h\nu \rightarrow 2\ O_3$

TABLE A.3 Photochemical Reactions in the Stratosphere Classified in Various Ways

a. Classic Chapman, pure air reactions
- (1) $O_2+h\nu$ (below 242 nm) $\rightarrow O+O$
- (2) $O+O_2+M \rightarrow O_3+M$
- (3) $O_3+h\nu$ (uv and visible) $\rightarrow O_2+O$
- (4) $O_3+O \rightarrow O_2+O_2$

b. Nitrogen oxide (NO_x) catalyzed destruction of ozone
- (5) $O_3+NO \rightarrow O_2+NO_2$
- (6) $NO_2+O \rightarrow NO+O_2$
- (7) $NO_2+h\nu$ (below 400 nm) $\rightarrow NO+O$

c. Pure air reactions, considering excited oxygen atoms
- (3a) $O_3+h\nu$ (below 310 nm) $\rightarrow O_2+O(^1D)$
- (3b) $O_3+h\nu$ (above 310 nm) $\rightarrow O_2+O(^3P)$
- (4a) $O_3+O(^1D) \rightarrow O_2+O_2$
- (4b) $O_3+O(^3P) \rightarrow O_2+O_2$
- (8) $O(^1D)+M \rightarrow O(^3P)+M$

d. Destruction of ozone by free radicals (HO_x) based on water
- (9) $HO+O_3 \rightarrow HOO+O_2$
- (10) $HOO+O_3 \rightarrow HO+O_2+O_2$
- (11) $HO+O \rightarrow H+O_2$
- (12) $HOO+O \rightarrow HO+O_2$
- (13) $H+O_3 \rightarrow HO+O_2$
- (14) $H+O_2+M \rightarrow HOO+M$

e. Sources and sinks of NO_x
- (15a) $O(^1D)+N_2O \rightarrow 2NO$
- (15b) $O(^1D)+N_2O \rightarrow N_2+O_2$
- (16) $N_2O+h\nu \rightarrow N_2+O(^1D)$
- (17) $NO+h\nu$ (below 192 nm) $\rightarrow N+O$
- (18) $N+NO \rightarrow N_2+O$
- (19) $N+O_2 \rightarrow NO+O$
- (20) $N+O_3 \rightarrow NO+O_2$
- (21) N_2+cosmic rays $\rightarrow N+N$
- (22) $HO+NO_2 \xrightarrow{M} HNO_3$
- (23) $HNO_3+h\nu \rightarrow HO+NO_2$
- (24) $HO+HNO_3 \rightarrow H_2O+NO_3$

f. Sources and sinks of HO_x radicals
- (25a) $O(^1D)+H_2O \rightarrow 2HO$
- (25b) $O(^1D)+H_2 \rightarrow H+HO$
- (25c) $O(^1D)+CH_4 \rightarrow HO+CH_3$
- (26) $HO+CH_4 \rightarrow H_2O+CH_3$
- (27) $CH_3+O_2 \rightarrow CO_3OO+M$
- (28) $CH_3OO+NO \rightarrow CH_3O+NO_2$
- (29) $CH_3O+O_2 \rightarrow H_2CO+HOO$
- (30) $HO+HOO \rightarrow H_2O+O_2$
- (31) $HOO+HOO \rightarrow H_2O_2+O_2$
- (32) $H_2CO+h\nu \rightarrow H+HCO$
- (33) $HCO+O_2 \rightarrow CO+HOO$

TABLE A.3—Continued

(34) $H_2O_2 + h\nu \rightarrow HO + HO$
(35) $HO + H_2O_2 \rightarrow H_2O + HOO$
(36) $O + H_2O_2 \rightarrow HO + HOO$

g. Coupling of NO_x and HO_x systems
(37) $HOO + NO \rightarrow HO + NO_2$
(22) $HO + NO_2 \xrightarrow{M} HNO_3$
(23) $HNO_3 + h\nu \rightarrow HO + NO_2$
(24) $HO + HNO_3 \rightarrow H_2O + NO_3$
(38a) $NO_3 + NO \rightarrow 2NO_2$
(38b) $NO_3 + h\nu(<580\ \text{nm}) \rightarrow NO_2 + O$

h. NO_x reactions during the night
(39) $NO + O_3 \rightarrow NO_2 + O_3$ X O_2
(40) $NO_2 + O_3 \rightarrow NO_3 + O_2$
(41) $NO_3 + NO_2 \xrightarrow{M} N_2O_5$ — on aerosol particles
(42) (N_2O_5 photolyzed during the day)
(43) $NO_3 + NO_2 \rightarrow NO_2 + O_2 + NO$

i. Reaction of chlorine species
(44) $Cl_2 + h\nu \rightarrow Cl + Cl$
(45) $HO + HCl \rightarrow H_2O + Cl$
(46) $O(^1D) + CF_2Cl_2 \rightarrow ClO + CF_2Cl$
(47) $CF_2Cl_2 + h\nu$ (below 210 nm) $\rightarrow CF_2Cl + Cl$
(48) $Cl + CH_4 \rightarrow HCl + CH_3$
(49) $Cl + O_3 \rightarrow ClO + O_2$
(50) $ClO + O \rightarrow Cl + O_2$
(51) $ClO + NO \rightarrow Cl + NO_2$

j. Ionic reactions (see CIAP Monograph 1, Chapter 5)

k. Other reactions of N_2O_5, NO, HNO_2, CO, H_2CO, and H_2O_2 and of the various atomic and free-radical species are important for certain special problems. However, these reactions do not materially change the predictions concerning ozone made by the 51-step mechanism given here.

The rates of ozone formation in Figure A.22 are based on a three-dimensional model of the average observed ozone and temperature distributions and a planar wave of incoming solar radiation. Thus the figure includes all angles between sunlight and the earth (day and night). The globally integrated rate of formation of ozone is 5.0×10^{31} molecules sec^{-1}.

If the local concentration of ozone (Figure A.9) is divided, point by point, by the local rate of formation of ozone (Figure A.22), then one has calculated the photochemical "ozone replacement time." The ozone replacement times calculated this way are given by Figure A.23 for March 22. The ozone replacement times cover a wide range; the con-

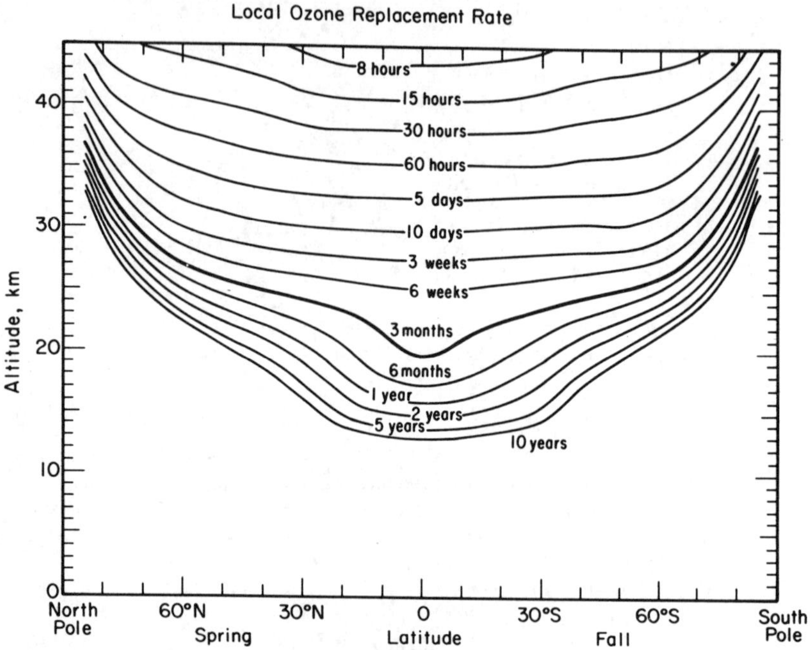

FIGURE A.23 Zonal average ozone replacement time (March 22). This is given by the local ozone concentration (Figure A.9) divided point by point by the ozone formation rate (Figure A.22).

tours shown span periods from 8 hours to 10 years. The ozone replacement times are to be regarded as a measure of the instantaneous rate of formation of ozone in much the same way that the rate of radioactive decay of uranium is characterized by a half-life of 4.5 billion years (Johnston, 1974).

The ozone replacement times shown in Figure A.23 can be compared with the residence times of inert tracers, roughly 2 years. Near the equator, the photochemical replacement times are short compared with stratospheric residence times, except for the levels close to the tropopause, where the times are comparable. Below 20 km at middle and high latitudes, the ozone replacement times are much longer than residence times. Above the 3-month contour in Figure A.23, one is in the "ozone formation region," where the ozone distribution is largely determined by photochemistry. Below the 10-year contours, ozone is detached from its source, and its distribution is largely determined by air motions. In the region where replacement times are between 3 months and 10 years, the ozone distribution is a complicated function of photochemical formation, air transport, and photochemical destruction.

The dominant role of air transport in determining the ozone distribution is illustrated by Figure A.24, where the 4-month and 10-year ozone replacement contours are superimposed on contours of ozone concentration for January 15. The large quantities and high concentrations of ozone in the detached region arrived there mostly by air motions.

There are additional, very minor sources of ozone in the atmosphere. Any process that turns NO to NO_2 forms ozone, because NO_2 is rapidly photolyzed to make NO and an oxygen atom, which forms ozone by Reaction (2). Molecular oxygen converts NO to NO_2 ($2NO+O_2 \rightarrow 2NO_2$), but this rate is vanishingly slow if NO is less than one part per million in the atmosphere. The "smog" reaction based on the oxides of nitrogen and methane forms some ozone in the stratosphere, but the amount is small (Crutzen, 1972a, 1972b; Chameides and Walker, 1973). The crossover point, above which NO_x destroys ozone and be-

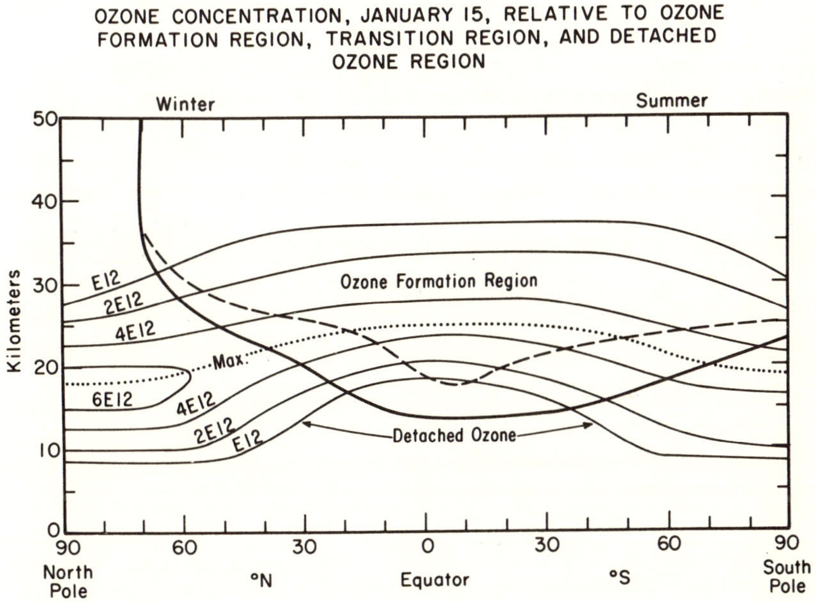

FIGURE A.24 Illustration of the role of horizontal air motions in moving ozone from the ozone formation region to the polar regions. This figure is a partial superposition of Figure A.9 and contours similar to those given in Figure A.23, but for January 15.

low which NO_x forms ozone through methane oxidations, is 13 km at midlatitudes (Johnston and Quitevis, 1974).

B. OZONE SINKS

Ozone transported into the troposphere is destroyed by contact with the surface of the earth. Given tropospheric ozone concentrations, the distribution of land and water, and measured efficiencies of ozone decomposition by various surfaces, one can calculate the global rate of ozone destruction by this process. Various estimates (in units of 10^{29} molecules sec^{-1}) are 5.4 to 8.6 (Aldaz, 1969); 6 (Brewer and Wilson, 1968); 1.8 to 3.0 (Fabian and Junge, 1970); 1.3 to 2.4 (Tiefenau and Fabian, 1972); 0.024 to 2.4 (Bauer, 1973). The surface sink strength ($\sim 2 \times 10^{29}$) is the order of magnitude of 1 percent of the stratospheric source strength (5.0×10^{31}).

There are numerous chemical and photochemical reactions that destroy ozone in the stratosphere (Johnston, 1968). These are conveniently classified as the O_x system, the HO_x system, the NO_x system, the ClO_x system, and the MO_x system, where M is a metal.

1. *The Pure Oxygen* (O_x) *System* (*Chapman, 1930*).

Ozone is weakly photolyzed to produce stable triplet oxygen atoms, $O(^3P)$, by visible light (450–650 nm) and by ultraviolet radiation above 310 nm. Ozone strongly absorbs ultraviolet radiation below 310 nm to produce excited singlet oxygen atoms, $O(^1D)$. A small fraction of singlet atoms react with water, nitrous oxide, methane, hydrogen, and ozone; but most singlet oxygen atoms are rapidly quenched by any molecule M to form triplet atoms. In either case, the photolysis of ozone almost always promptly leads to $O(^3P)$. The photolysis of ozone does not constitute ozone destruction, since it is rapidly followed by reformation of ozone from Reaction (2).

$$(3)\quad O_3 + h\nu \rightarrow O_2 + O$$
$$(2)\quad O + O_2 + M \rightarrow O_3 + M$$

NET: No chemical reaction

There is no net chemical reaction, but biologically damaging ultraviolet radiation is absorbed and heat is released in the atmosphere. The heat released following the photolysis of ozone sets up the steep temperature inversion between 16 and 50 km, which stabilizes the stratosphere (Figure A.1).

A very small fraction of $O(^3P)$ reacts with ozone to complete a couplet of reactions that do destroy ozone:

$$
\begin{array}{ll}
(3) & O_3 + h\nu \rightarrow O_2 + O \\
(4) & O + O_3 \rightarrow O_2 + O_2 \\
\hline
\text{NET:} & 2\,O_3 + h\nu \rightarrow 3\,O_2
\end{array}
$$

Reaction (4) is the "rate determining step" in this process. The zonally averaged instantaneous rate of Reaction (4) is given by Figure A.25 for January 15. The globally integrated rate of ozone destruction by this process is 0.86×10^{31} molecules sec^{-1} or about 17 percent of the gross formation rate (Johnston and Whitten, 1973).

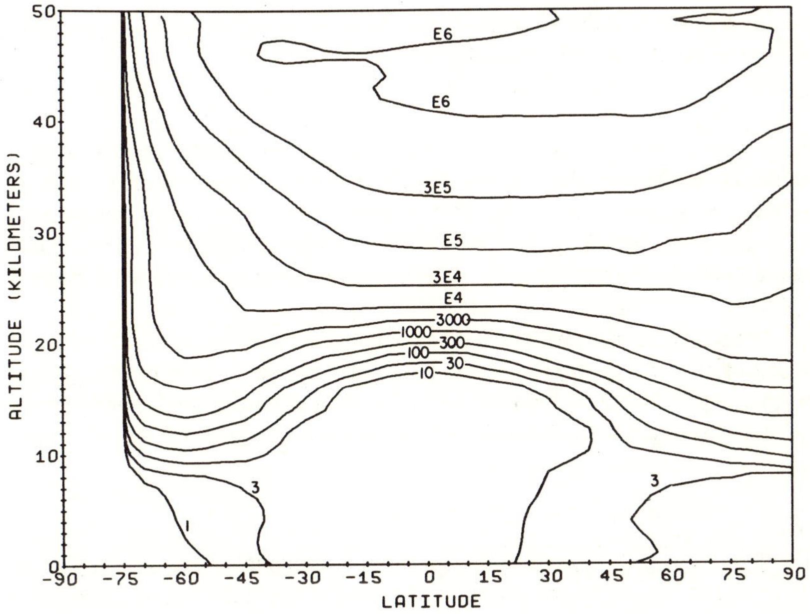

FIGURE A.25 Zonal average rate of destruction of ozone by the pure oxygen reactions:

$$
\begin{array}{ll}
 & O_3 + h\nu \rightarrow O_2 + O \\
 & O_3 + O \rightarrow O_2 + O_2 \\
\hline
\text{NET:} & 2\,O_3 + h\nu \rightarrow 3\,O_2
\end{array}
$$

2. HO_x *Free Radicals Based on Water (Bates and Nicolet, 1950; Nicolet, 1970; 1971)*

The photolysis of ozone by ultraviolet radiation below 310 nm produces excited singlet oxygen atoms, which react rapidly with water to form hydroxyl radicals:

(3a) $O_3 + h\nu$ (below 310 nm) $\rightarrow O_2 + O(^1D)$
(25) $O(^1D) + H_2O \rightarrow 2HO$

Singlet oxygen atoms also react with hydrogen and with methane to form hydroxyl radicals. The hydroxyl radicals react with ozone to produce hydroperoxyl radicals, HOO. Both HO and HOO destroy ozone, by direct interaction and indirectly by reaction with oxygen atoms:

(9) $HO + O_3 \rightarrow HOO + O_2$
(10) $HOO + O_3 \rightarrow HO + O_2 + O_2$
(11) $HO + O \rightarrow H + O_2$
(12) $HOO + O \rightarrow HO + O_2$
(13) $H + O_3 \rightarrow HO + O_2$

There are more than a score of reactions of HO_x radicals in the stratosphere. With the three-dimensional global average distributions of ozone, water, and the oxides of nitrogen, the global rate of destruction of ozone by the HO_x system was found to be 0.57×10^{31} molecules sec^{-1}, which is about 11 percent of the ozone source strength (Johnston, 1974).

The fairly easily identifiable sinks of ozone are the earth's surface, 1 percent; the O_x system, 17 percent; and the HO_x system, 11 percent (January 15, 0 to 45 km). Thus about 70 percent of the ozone formed is unaccounted for by these sinks.

3. *Chlorine and Its* ClO_x *Derivatives*

In the stratosphere, chlorine, Cl_2, is photodissociated to chlorine atoms Cl, HCl is converted to atomic chlorine by hydroxyl radicals, and the atoms react with ozone in a catalytic cycle (Hoshizaki *et al.*, 1973).

(49) $Cl + O_3 \rightarrow ClO + O_2$
(50) $ClO + O \rightarrow Cl + O_2$

NET: $O_3 + O \rightarrow O_2 + O_2$

There are over a score of reactions involving Cl, ClO, ClOO, HCl with O_x, NO_x, and methane (Stolarski and Cicerone, 1974; Wofsy and McElroy, 1974; Crutzen, 1974a, 1974b). So far, attempts to measure the concentration of HCl or ClO_x in the stratosphere have been unsuccessful; and from the limit of sensitivity of the analytical methods used, it appears that ClO_x is not important as a natural sink for ozone. Artificial sources of chlorine compounds in the stratosphere include organic chlorides produced at the earth's surface, especially chlorofluorocarbons, e.g., Freons (Molina and Rowland, 1974), and eventually the Space Shuttle (Hammond and Maugh, 1974).

4. *Metallic Oxides from Meteorites (the* MO_x *system)*

In the laboratory, ozone is catalytically decomposed on several metallic oxides (Johnston, 1968). The materials Ag_2O, MnO_2, MgO, and CuO are known to be especially powerful catalysts for ozone destruction. Small quantities of some of these materials are deposited in the atmosphere by meteorites. The role of metal oxide particles as a natural sink for ozone is difficult to evaluate. The size distribution, the effective surface area, and the collision efficiency for ozone destruction are not known for stratospheric MO_x particles. There is no known correlation of ozone with meteoritic showers, and MO_x is probably not a major sink. The role of finely divided aluminum oxide as a catalyst for ozone destruction is of interest with respect to the Space Shuttle, which produces HCl and Al_2O_3 as major exhaust products.

5. *The Oxides of Nitrogen (the* NO_x *system)*

Nitric oxide and nitrogen dioxide destroy ozone directly and indirectly (via oxygen atoms) through the catalytic cycle (Crutzen, 1970; Johnston, 1971):

$$(5)\quad NO+O_3 \rightarrow NO_2+O_2$$

$$(6)\quad NO_2+O \rightarrow NO+O_2$$

$$\text{NET:}\quad O_3+O \rightarrow O_2+O_2$$

Reaction (5) is usually followed by the photolysis of nitrogen dioxide:

$$(7)\quad NO_2+h\nu \text{ (below 400 nm)} \rightarrow NO+O$$

which reforms ozone by way of Reaction (2). Thus, Reaction (6) is the rate-determining step in this catalytic cycle. The rate constant for

Reaction (6) is 10^3 to 10^4 greater than that for Reaction (4) at typical stratospheric temperatures, and thus trace amounts of NO_2 are highly effective in destroying ozone.

The first measurements of NO_2 and NO in the stratosphere have been made in the last two years (Figures A.12 and A.13). There is a great variability in these measurements, and there appears to be a real variability in latitude and in season. Although NO_x is a very efficient catalyst for ozone destruction, it is not clear whether the quantity and distribution are sufficient to account for the 70 percent discrepancy between the ozone source and the recognized O_x, HO_x, and surface sinks. The O_x sink accounts for about 20 percent (including seasons other than the January 15 case given above) of the ozone source; it is easy to calculate the concentrations of NO_2, as such, that would destroy ozone 3.5 times faster than the O_x sink. The line labeled $\rho = 4.5$ in Figure A.12 is the amount of NO_2 required to destroy ozone 3.5 times faster than the $O+O_3$ reaction (Johnston, 1971). In the lower stratosphere, the observed NO_2 is distinctly larger than the required amount, but in the rapid ozone formation region 30 to 40 km there is an absence of data

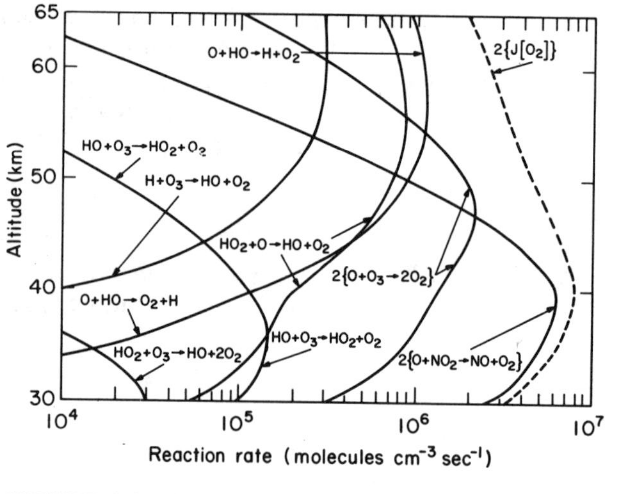

FIGURE A.26 Rates of formation of ozone, destruction of ozone by NO_x mechanism,

$$NO + O_3 \rightarrow NO_2 + O_2$$
$$NO_2 + O \rightarrow NO + O_2$$

$$\text{NET: } O_3 + O \rightarrow O_2 + O_2$$

and destruction of ozone by several HO_x reactions (McElroy *et al.*, 1974).

on NO_2 concentrations. NO_2 and NO are rapidly interconverted by sunlight and ozone. The amount of NO required to destroy ozone 3.5 times faster than the O_x sink is given by the line labeled $\rho=4.5$ in Figure A.13. This line is in the middle of observed values of NO, except that the situation is uncertain at high elevations.

The comparisons of how much NO_2 or NO is needed to destroy ozone 3.5 times faster than the O_x sink (Figures A.12 and A.13) show that, within the uncertainty of the observed NO and NO_2 distributions, the NO_x system is capable of destroying 70 percent of the ozone produced between 0 and 45 km. However, the amount of NO_x in the natural stratosphere is so uncertain that there is some room for ClO_x or MO_x or other processes as a natural sink. The present indication, however, is that the NO_x system is the most important sink for ozone in the stratosphere, as illustrated (McElroy *et al.,* 1974) by the relative role of ozone destruction by the O_x, HO_x, and NO_x reactions (Figure A.26).

C. SOURCES, SINKS, AND INVENTORY OF NO_x

Since NO_x ($NO+NO_2$) appears to be the most important sink for stratospheric ozone, its sources, sinks, and inventory require special consideration.

Nitrous oxide (N_2O) is formed in the soil by bacteria, is inert (similar to CO_2) in the troposphere, is slowly transported into the stratosphere, and is destroyed there in three ways, one of which produces nitric oxide.

$$\text{(15a)} \quad N_2O+O(^1D)\rightarrow NO+NO$$

This mechanism is the principal source of NO_x in the stratosphere (Crutzen, 1971; McElroy and McConnell, 1971; Nicolet and Vergison, 1971). Isaksen (1973) evaluated this process for the entire globe. Isaksen's results at 45° latitude in the summer for the production of nitric oxide [Reaction (15a)] is given as a function of height by Figure A.27, and the elevation of maximum rate is 22 km. Most production of NO occurs between 18 and 26 km. If larger eddy coefficients are used in the model (McConnell and McElroy, 1973), the height of maximum production is considerably higher—as high as 35 km. However, the increased rate of vertical mixing and loss of NO_x to the troposphere means that the densities of NO_x remain approximately the same. The integrated vertical column for NO production as a function of latitude is given by Figure A.28, and the global production rate of NO from this mechanism is $(4 \text{ to } 5)\times 10^{26}$ molecules sec^{-1}.

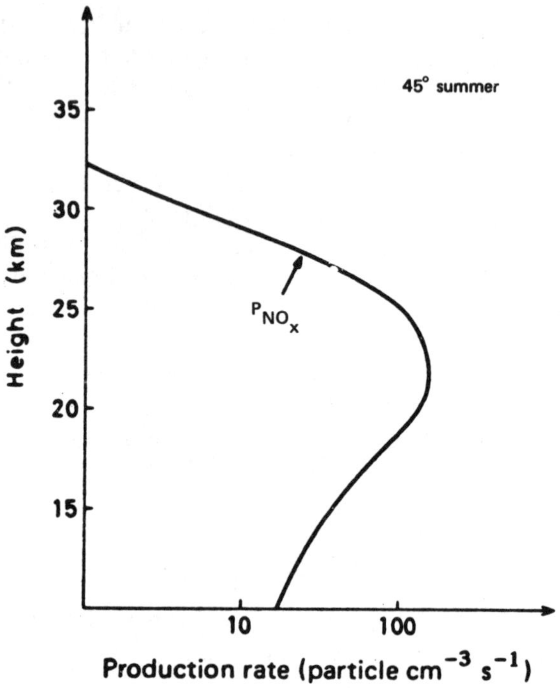

FIGURE A.27 Profile of rate of production of nitric oxide from nitrous oxide, $N_2O+O(^1D) \rightarrow NO+NO$. The maximum rate of input of natural NO is indicated to be 20 to 25 km (Isaksen, 1973).

The possible role of ammonia as a source of stratospheric NO_x has been examined, but this source appears to be small (McElroy *et al.*, 1974). Cosmic rays produce a small amount of nitric oxide in the polar regions of the stratosphere; and the rate of production is a strong function of the 11-year solar cycle. The annual production of nitric oxide at the minimum phase of the solar cycle is given by Figure A.29; the rate of NO production at solar cycle maximum is about twice this minimum amount (Warneck, 1972; Ruderman and Chamberlain, 1973; Brasseur and Nicolet, 1973). The cosmic-ray source is much smaller than the nitrous oxide source, but it is of great interest in calibrating stratospheric models because the cosmic-ray source of NO is localized and periodic. A periodically variable signal can often be picked out of a very noisy background.

There are two principal sinks for NO_x. In the upper stratosphere NO

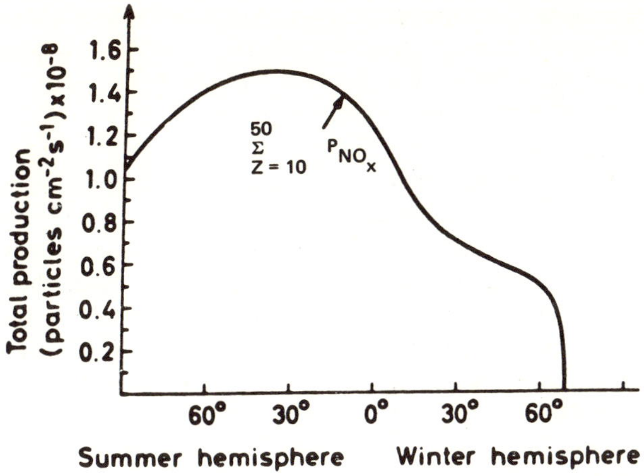

FIGURE A.28 Global distribution of total production of NO_x above the tropopause at winter solstice.

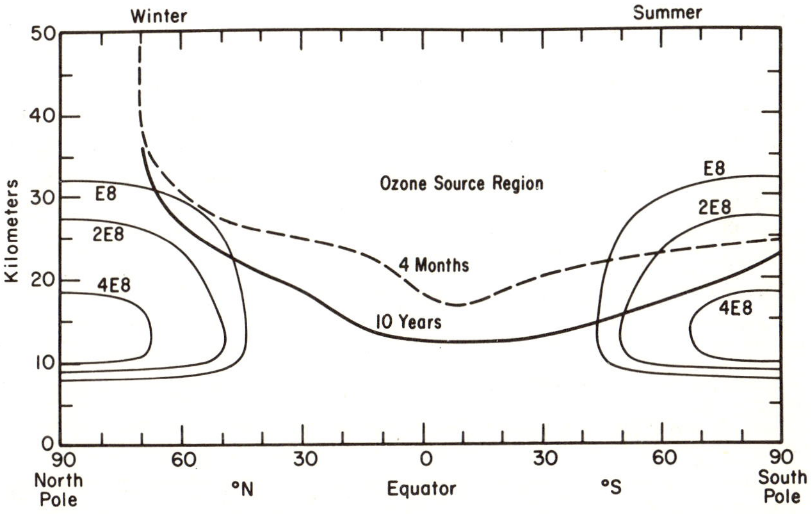

FIGURE A.29 Formation of nitric oxide from cosmic rays. The rate of production of atmospheric ions and nitric oxide from cosmic rays varies by about a factor of 2 as a function of the 11-year solar cycle. This figure gives the magnitude of NO produced per year at the minimum of the 11-year cycle, or it can be interpreted as the modulation amplitude of the periodic signal.

is photodissociated to nitrogen and oxygen atoms. The nitrogen atom has three principal paths: it may react with O_3 or O_2 to regenerate NO or it may react with NO to form N_2 and O, thus destroying NO_x (Brasseur and Cieslik, 1973). Although this process is very slow, it seems to protect the stratosphere from the large amounts of NO_x otherwise diffusing down from the ionosphere. In the lower stratosphere, most of the "odd nitrogen" is in the form of nitric acid (Figure A.18). Nitric acid exists as the molecule HNO_3 in the dry stratosphere, but it dissolves in water and is removed by rain in the troposphere. From the observed gradient of the mixing ratio of nitric acid with elevation, Lazrus and Gandrud (1974) calculated the global flux of nitric acid from the stratosphere to the troposphere: $(4.2 \pm 2.1) \times 10^{26}$ molecules sec^{-1}. This quantity is in excellent agreement with Isaksen's (1973) calculation of the global source of NO_x.

The inventory of the oxides of nitrogen in the stratosphere is more uncertain than the sources and sinks. The active oxides of nitrogen, NO and NO_2, are reversibly converted to nitric acid by the reactions

$$(22)\quad HO + NO_2 \xrightarrow{M} HNO_3$$
$$(23)\quad HNO_3 + h\nu \rightarrow HO + NO_2$$
$$(24)\quad HO + HNO_3 \rightarrow H_2O + NO_3$$

Also some active NO_x is reversibly converted to dinitrogen pentoxide, N_2O_5. Lazrus and Gandrud (1974) observed nitric acid over a wide range of latitude and elevation. With some extrapolation, their data indicate 2.2×10^{34} molecules of nitric acid for the global stratosphere, which is a mole fraction (mixing ratio by volume) of 1.7×10^{-9}. The total NO_y (NO, NO_2, and HNO_3, $2N_2O_5$ being insignificant by comparison) is about twice the amount of HNO_3. Thus the indicated inventory of NO_y is about 4.4×10^{34}, and the average mole fraction is about 3.4 ppb. There is a substantial uncertainty in this estimate.

III. The Ozone Problem in a Perturbed Stratosphere

CIAP Monograph 2 considers the properties of current (Concorde, Tupolev) supersonic transports (SST) and predicts the upper limit on the number and size of future SST's in 1990 and 2005. The properties of the formerly proposed U.S. SST can be obtained from previously published statements of fuel utilization (SCEP, 1970) and from current estimates of the properties of the exhaust gases (CIAP, Monograph 2). Monograph 2 also discusses the possibility of redesigning SST engines to reduce the quantity of nitric oxide emitted. Consequences of these perturbations are discussed here.

A. COMPARISON OF ARTIFICIAL AND NATURAL SOURCE STRENGTHS OF NO_x

The source strength *S* in grams per year of a component of a supersonic aircraft exhaust is given by

$$S = 8766NnFf(\mathrm{EI}), \tag{12}$$

where *N* is the number of aircraft, *n* is the number of engines per aircraft, *F* is the mass of fuel flow per engine in kg h^{-1}, *f* is the fraction of the day that each aircraft is at the cruise height, the emission index (EI) is the number of grams of exhaust component per kilogram of fuel burned, and 8766 is the average number of hours per year. The fuel flow *F* and emission index (EI) for the Concorde are given by Figure A.30 as a function of flight time. The average value of the fuel flow is near 5000 kg h^{-1} for each engine, and a value of 19,000 kg h^{-1} is used

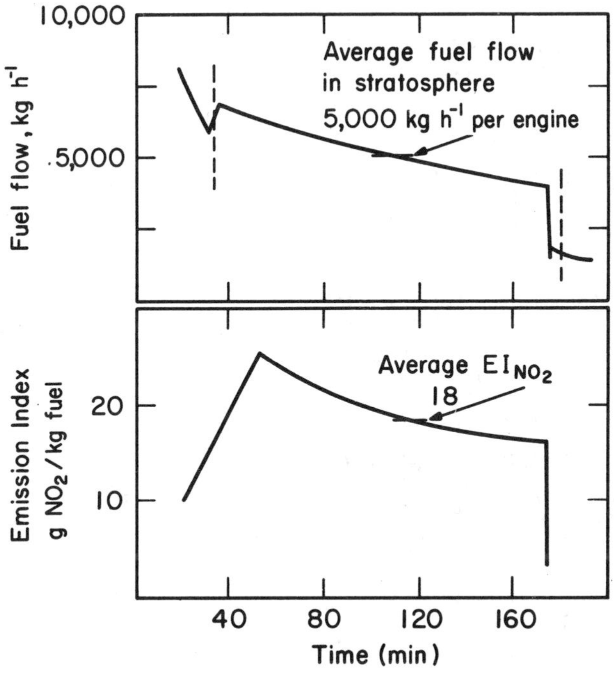

FIGURE A.30 Upper panel, fuel flow as a function of time from takeoff, per engine; lower panel, emission index of oxides of nitrogen (expressed as grams of NO_2 per kilogram of fuel) as a function of time from takeoff (both for Concorde-Olympus 593 Mk 602).

here for Concorde/Tupolev SST's. For future large SST's, a total fuel flow of 50,000 kg h^{-1} is assumed. For the formerly proposed SST, the fuel flow per hour would have been about 60,000 kg h^{-1}, and model calculations for large SST's at 20 km typically use this figure. All known SST's have four engines. The projected upper limits on rates of fuel consumption by supersonic (15–21 km) aircraft in various altitude bands is given by Figure A.31 (Grobecker, 1974). About 10^{11} kg yr^{-1} is the rate of fuel consumption by SST's to be expected in the period 1995–2000 if the upper limit on usage is attained, and the rate might rise an order of magnitude higher by 2020. The NO_x EI expressed in terms of grams of NO_2 per kilogram of fuel is given for the Concorde in Figure A.30, and the average value is about 18. This value is presumably appropriate for future SST's if design modifications to reduce NO_x emissions are not made. The value associated with the formerly proposed SST is 22.

The fraction of the day at cruise height is taken to be 5/24. The annual source of NO_2 is then 0.87×10^9 g yr^{-1} per Concorde or Tupolev and 2.3×10^9 g yr^{-1} per large SST. For 500 large SST's, the NO_2 source strength is 1.15×10^{12} g yr^{-1} or 1.15 MT yr^{-1}. These source strengths are 5×10^{26} molecules sec^{-1} for 500 large SST's and 2×10^{26} molecules sec^{-1} for 500 Concordes and/or Tupolevs. For an SST larger than the

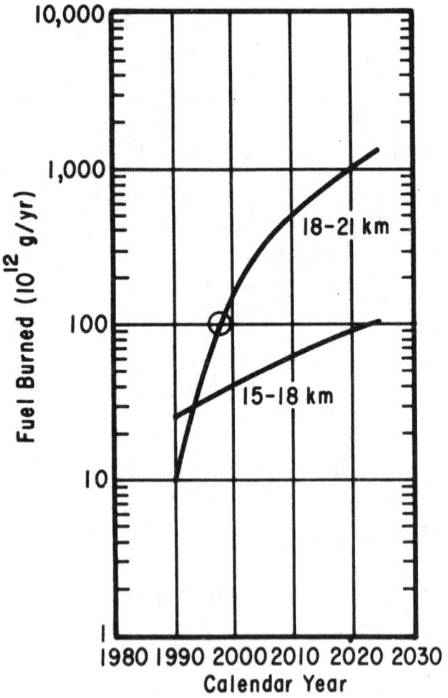

FIGURE A.31 Fuel burned in various altitude bands by upper limit fleet of supersonic aircraft (exclusive of HST) as projected by Grobecker (1974) to the year 2025. The NO_x produced is the product of the fuel burned and emission index. For comparison with subsequent figures, the NO_x produced (at EI=18, Figure A.30) at 18–21 km would be 1990, 0.18×10^{12} g yr^{-1}; 1997, 1.8×10^{12} g yr^{-1}; 2020, 18×10^{12} g yr^{-1}.

TABLE A.4 Comparison of Natural and Artificial Global Average Sources of Stratospheric Oxides of Nitrogen (in Units of 10^{26} molecules sec^{-1})

Natural source	4 to 5	Isaksen (1973)
Natural sink	4 ± 2	Lazrus and Ganrud (1974)
500 1971 U.S. SST's (EI=22)	6.5	
500 large SST's	5	
500 Concorde/Tupolev SST's	2.2	
Economic SST [a]	7.5	

[a] A model that burns 10^{11} kg of fuel per year, a figure projected as possible for 1997–2000.

Concorde, of which several hundred copies of the aircraft must be produced to achieve economic viability, the order of magnitude of fuel consumption must be about 10^{11} kg yr^{-1} (compare Figure A.31). Thus the annual rate of insertion of NO_x in the stratosphere for an emission index of 18 g kg^{-1} would be 1.8×10^{12} g yr^{-1}. This point is circled in Figure A.31, and an entry corresponding to it is included in Table A.4 as the Economic SST. It can be seen that 500 SST's, with current emission indices, would more or less double the natural global source of NO_x, which itself plays a dominant role in the ozone balance.

B. RESULTS OF MODEL CALCULATIONS OF NO_x PERTURBATION OF STRATOSPHERIC OZONE

There have been at least 12 investigators who have calculated the reduction of ozone by nitrogen oxide catalysts from supersonic aircraft exhausts. The authors and a reference code are as follows:

1. Johnston (1971)
2. Crutzen (1972a, 1972b, 1974b)
3. Hesstvedt (1972, 1973, 1974a, 1974b)
4. Chang *et al.* (1973); Chang and Deuwer (1973b), Chang and Johnston (1974)
5. Stewart (1973); Stewart and Hoffert (1973)
6. McElroy *et al.* (1974)
7. Whitten and Turco (1973, 1974)
8. Shimazaki and Ogawa (1974)
9. Vupputuri (1974)
10. Widhopf (1974); Widhopf and Taylor (1974)
11. Cunnold *et al.* (1974)
12. Hunten (1974)

identical to Fig 14, p 27 CIAP

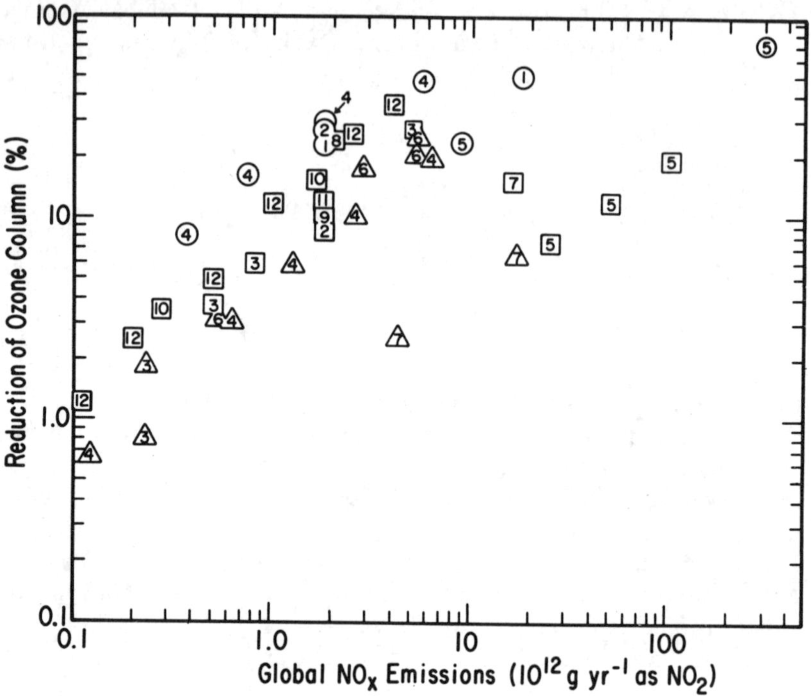

FIGURE A.32 Calculated reduction of vertical ozone column in terms of uniform global rate of NO_x addition at 20 km. One-dimensional models except for reference codes 3, 10, 11. The codes are identified in the text.

In subsequent figures, the numbers enclosed in a circle denote early work, which omits nitric acid chemistry. If an investigator has two or more papers on the subject, the number for the latest is enclosed in a square, and for earlier work the reference number is enclosed in a triangle. For studies that reported ozone reduction as a function of the mass of NO_x inserted per year at 20 km, the *global average results* are presented in Figure A.32. For a given rate of NO_x injection the models show up to a factor of 10 spread in computed ozone reduction. The different types of model are discussed below.

1. *Box Model* (*Johnston, 1971*)

The effluents from 2 years' operation of 500 formerly proposed U.S. SST's were uniformly distributed in a spherical shell or "box" surrounding the earth. The shells were assumed to be 1, 4, 7, 10, 13, 16, 19, and

22 km thick. These distributions were added to an NO_x background (derived to give the average, observed midlatitude ozone profile). The steady-state ozone columns before and after the SST perturbation were calculated. The location of the boxes is given in Figure A.33, and the percentage reduction of ozone for each box is given by the numbers on the figure. For the 1-km spread of nitrogen oxides, between 20 and 21 km, the NO_x perturbation (9×10^{10} molecules cm^{-3}) rapidly re-

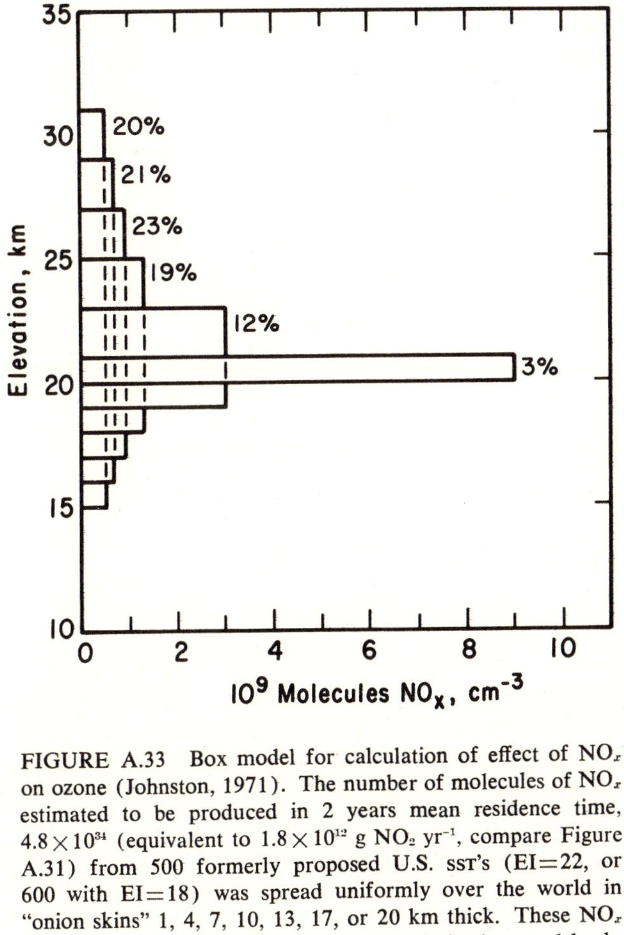

FIGURE A.33 Box model for calculation of effect of NO_x on ozone (Johnston, 1971). The number of molecules of NO_x estimated to be produced in 2 years mean residence time, 4.8×10^{34} (equivalent to 1.8×10^{12} g NO_2 yr^{-1}, compare Figure A.31) from 500 formerly proposed U.S. SST's (EI=22, or 600 with EI=18) was spread uniformly over the world in "onion skins" 1, 4, 7, 10, 13, 17, or 20 km thick. These NO_x distributions were added to a previously derived natural background that gave approximately the correct ozone profile. The percentage reduction of ozone for each assumed spread of artificial NO_x is indicated on the figure.

duced ozone to a very low value in the box, but it contacted only a small portion of the ozone profile. The reduction of the ozone column was only 3 percent. For wider assumed spreads of NO_x, the reduction of ozone in the box was less, but the total effect on the ozone column was greater. A 4-km spread gave a 12 percent reduction of the ozone column; a 7-km spread gave 19 percent; these and other values can be read from Figure A.33. The amount of reduction of the ozone column was relatively insensitive to spread between 7 and 22 km. A spread of 10 km was taken as standard, which gave a worldwide average ozone reduction of 23 percent. For a high traffic corridor, this model gave an ozone reduction twice as great as the worldwide average, that is, an ozone reduction by about a factor of 2. This model was also used for a wide

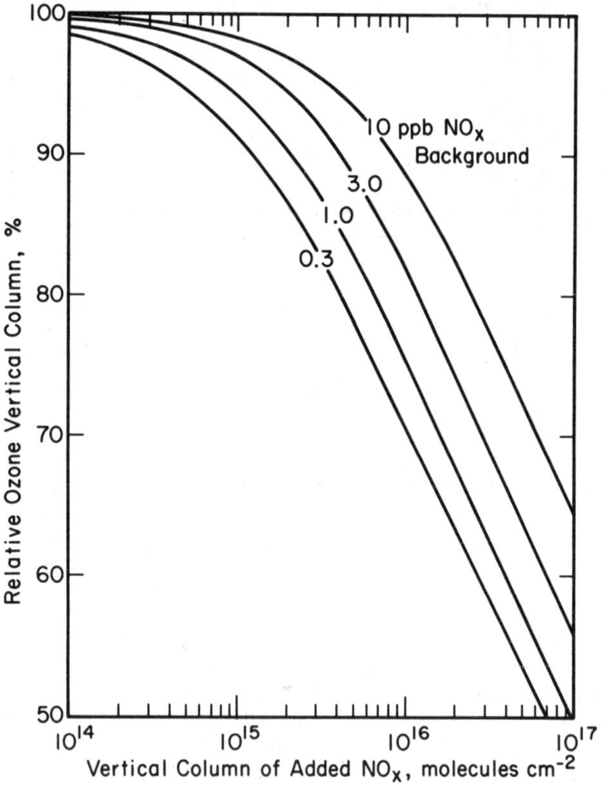

FIGURE A.34 Box model results for wide range of NO_x backgrounds, 0.3 to 10 ppb, and for wide range of columns of artificial NO_x (Johnston, 1971).

range of assumed NO_x backgrounds and a wide range of artificial NO_x columns, and the results for a 10-km vertical spread of exhaust gases are given in Figure A.34. The range of calculated global-average ozone reduction was between 10 and 30 percent with up to a factor of 2 reduction at the corridor of maximum traffic.

2. *Models with One-Dimensional Transport*

Seven groups (Crutzen, Chang, Stewart, McElroy, Whitten, Shimazaki, and Hunten) have reported results of ozone reduction by nitrogen oxides from SST exhausts, using one-dimensional models for vertical transport of substances in the atmosphere. All recent studies used O_x, HO_x, NO_x, and HNO_x chemical reactions, and in general rate constants were obtained from the recommended list by the National Bureau of Standards (CIAP Monograph 1, Chapter 5). Some studies used more extensive sets of minor reactions than others. The complex manner with which the nitric acid formation rate depends on total gas concentration and temperature was approximated by simple empirical expressions by some authors, and large differences in the calculated partitioning between NO_2 and HNO_3 were caused by these approximations.

To carry out computations with the one-dimensional transport models, one must have a vertical eddy-diffusion profile, and the profiles used by the seven modelers are given by Figure A.35. These K profiles were derived from considerations of (a) heat-flux data; (b) radioactive fallout from nuclear bomb tests, primarily involving particulate tracers such as strontium-90 and tungsten-185; (c) vertical profiles of ozone; (d) vertical profiles of methane; and (e) other considerations. These profiles show large differences from one to the other, and these differences lead to widely different predictions about transport of trace species in the atmosphere. Some of the more extreme profiles have been simply postulated and have no empirical basis. They may have been intended as inputs to sensitivity studies, rather than as representations of actual atmospheric behavior.

The large spread in the calculated reduction of ozone by the different authors is expected to arise primarily from differences in chemistry or from differences in the K profiles. The tests of the computations outlined below (Chang and Johnston, 1974; Chang, 1974) provide a sharp separation of the effects of chemistry and the effects of K profile in causing the differences in predictions among the models.

For the chemical kinetic systems under consideration, the total NO_y (i.e., $NO+NO_2+HNO_3$) is approximately a conserved quantity with the oxidation of N_2O as the principal source. Therefore, total NO_y

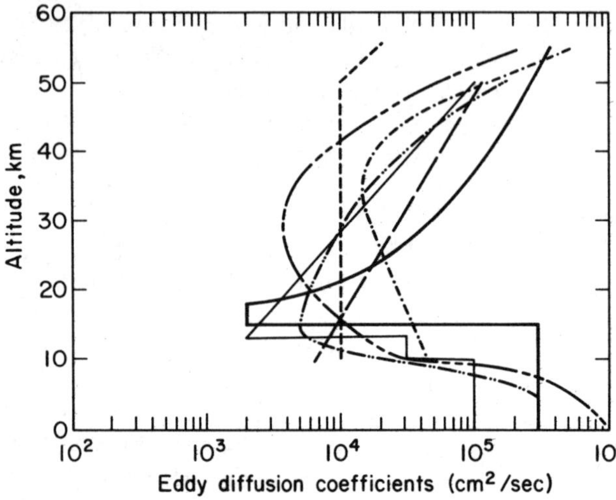

FIGURE A.35 Vertical profiles of eddy-diffusion coefficient as used in various one-dimensional models. ———, McElroy *et al.* (1974); ———, Hunten, (1974); – – –, Crutzen (1974); – • – • – • –, Whitten and Turco (1973); — • • — • •, Stewart (1973); — — —, Shimazaki and Ogawa (1974); — – – – – – –, Chang *et al.* (1974).

behaves very much as an inert tracer in the models. Since the parameterized transport terms in the one-dimensional models are linear operators in the mathematical sense, we can assume a linear relation between net NO_y sources and the steady-state total NO_y column concentration,

$$[NO_y]=\beta_i[S_{NO_y}], \tag{13}$$

where β_i is model dependent and represents the linear transport effect and S_{NO_y} means source of NO_y. Consequently, the relations between natural and perturbed NO_y are

$$[NO_y]_{natural}=\beta_i[S_{NO_y}]_{natural},$$

and

$$[NO_y]_{perturbed}=\beta_i([S_{NO_y}]_{natural}+[S_{NO_y}]_{SST}).$$

Therefore,

$$\Delta[NO_y]=[NO_y]_{perturbed}-[NO_y]_{natural}$$
$$=\beta_i[S_{NO_y}]_{SST}. \tag{14}$$

If we take the ratio $\Delta[NO_y]/[NO_y]_{natural}$ then β_i cancels. In other words, the quantity $\Delta[NO_y]/[NO_y]_{natural}$ to a large extent is independent of the individual differences in transport parameterization. The calculated *global average* percentage reductions of ozone are plotted in Figure A.36 against calculated percentage increase in nitrogen oxides for all models for which these quantities are available, including the lines from the box model of Figure A.34 with three different assumed NO_x backgrounds. The agreement between models in Figure A.36 is about a factor of 10 better than that for the same models in Figure A.32. This indicates that

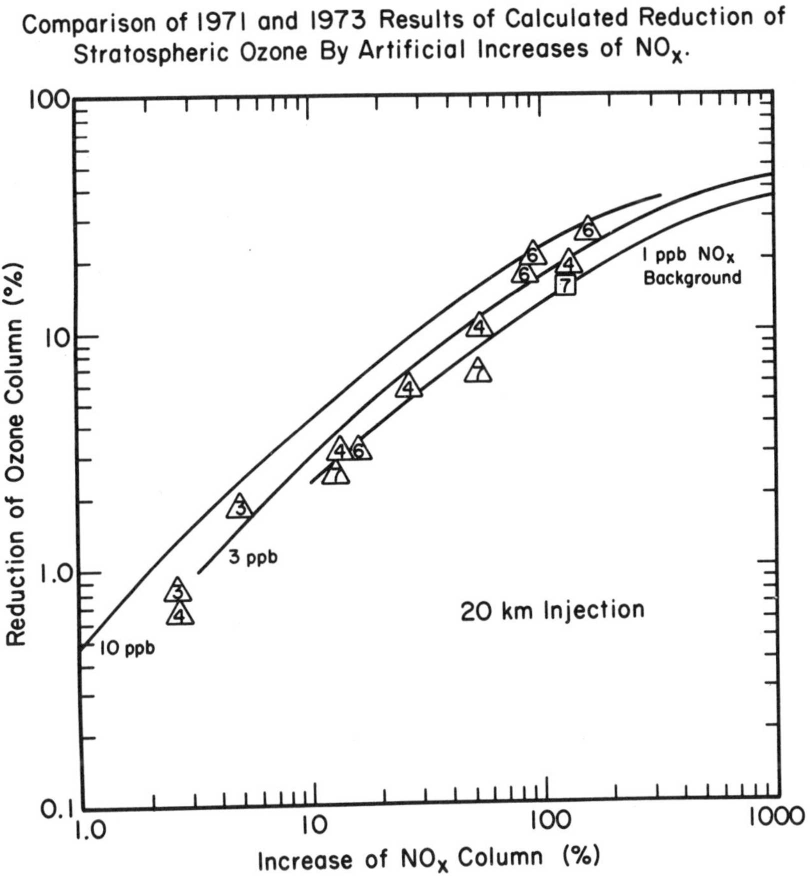

FIGURE A.36 Calculated percentage reduction of vertical ozone column as a function of percentage increase of nitrogen oxide background, $(-\Delta O_3/O_3)\times 100$ versus $(+\Delta NO_x/NO_x)\times 100$. 20-km injection height. The lines are from Figure A.34.

the spread in results in Figure A.32 is not primarily due to differences in chemistry.

In order to test for the sensitivity of calculated ozone reduction to the assumed *K* profiles, Chang (1974) used his set of chemical reactions and his computational procedures, and he carried out time-dependent calculations of the SST perturbation for six different vertical eddy-diffusion profiles (he did not at that time have access to Hunten's model). Thus all factors in the problem except the *K* profiles were the same, and the role of just this parameter was examined. For a uniform global input of 2.5×10^{12} g of NO_2 yr^{-1}, Chang calculated the percentage ozone reduction as a function of time for six models (Figure A.37). For this perturbation, Whitten's *K* model gives a 5 percent ozone reduction, Chang's gives an 11 percent ozone reduction, McElroy's gives a 19 percent ozone reduction, and the others fall between Chang's and McElroy's. With this NO_x input, Hunten's model gives a 26 percent

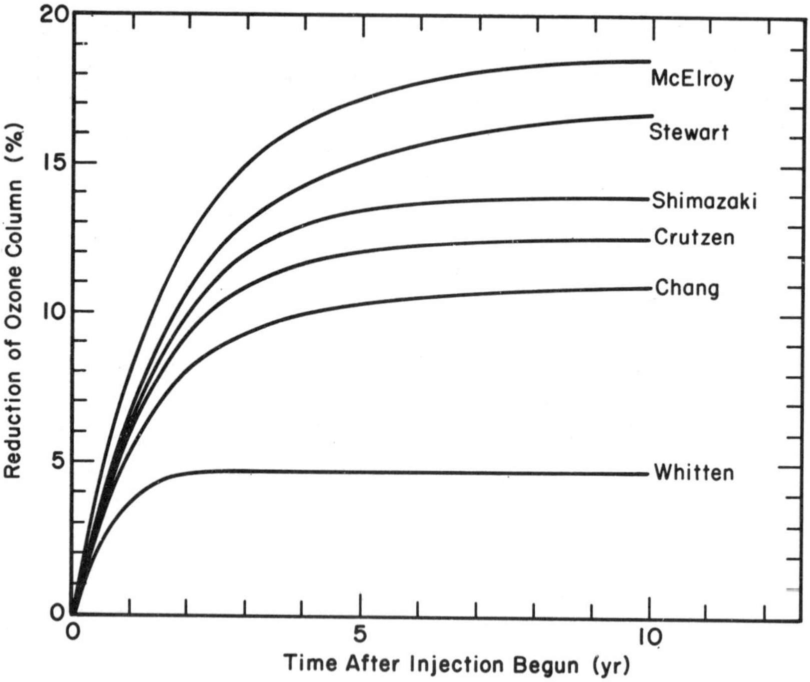

FIGURE A.37 Examples of time-dependent approach to steady state as determined from six models with different vertical eddy-diffusion coefficients (compare Figure A.35), the same chemical model, and the same boundary conditions (Chang, 1974).

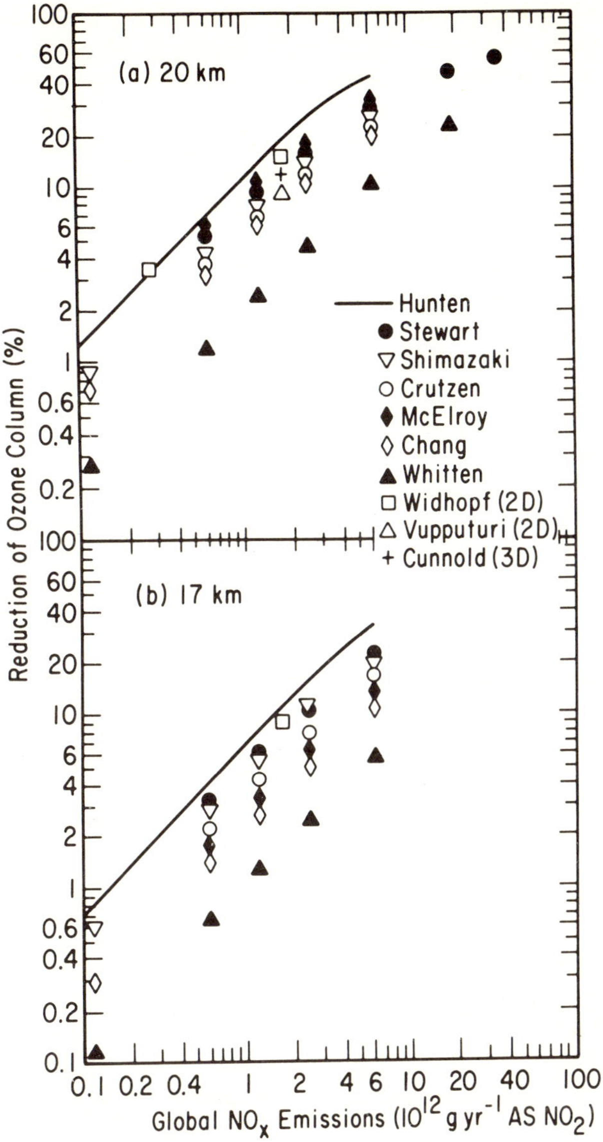

FIGURE A.38 (a) Calculated reduction of vertical ozone column in terms of uniform global rate of NO_x addition at 20 km. The points were calculated by Chang (1974) for six different *K* profiles. The line represents the ozone reduction predicted by Hunten (1974). Points are included for Widhopf (1974) and Cunnold *et al.* (1974). (b) Same except for NO_x insertion at 17 km.

reduction of ozone. Stewart's reported ozone reduction was low because he only carried out his calculations for 18 months of SST operation, whereas Figure A.37 shows that his model requires at least 10 years to reach a steady state. From Figure A.37, it can be seen that much of the spread of predictions in Figure A.32 is due to differences in the *K* profiles among the one-dimensional models.

Figure A.37 represents the ozone depletions as calculated by six models for one rate of NO_x insertion and at one elevation. Chang (1974) also carried out these calculations for a wide range of NO_x insertions and at two elevations, 20 and 17 km [Figure A.38, (a) and (b)]. The range of NO_x insertion corresponds to the SST fleet sizes indicated by Figure A.31 (Grobecker, 1974). The lines of Figure A.38, (a) and (b), correspond to Hunten's model for worldwide average insertion of NO_x. For flight operations mainly in one hemisphere (north or south), curves and data in Figure A.38 should be moved a factor of 2 to the left to approximate the confinement of exhaust gases to the hemisphere of release, or the NO_x values should be doubled above their true values to obtain the expected ozone decreases for the hemisphere concerned.

Johnston *et al.* (1974) proposed a test of these models for general accuracy of the *K* profile. None of these *K* profiles were derived from the data for carbon-14 from the nuclear bomb tests of 1961–1962 (Figure A.19). There are detailed maps of observed carbon-14 every three months for several years after the end of atmospheric testing in December 1962 (for example, Figure A.20). The quantity plotted in Figure A.20 is proportional to the mixing ratio of excess carbon-14. Between the equator and north pole the maximum mixing ratio is approximately parallel to the sloping tropopause. The elevation of maximum mixing ratio is between 17 and 25 km, and the bomb-produced carbon-14 was spread all over the northern hemisphere. This distribution of excess carbon-14 is highly appropriate for validation of *K* profiles to be used in models of SST perturbations of the stratosphere. Johnston *et al.* (1974) converted the mixing ratios of Figure A.20 to concentrations, averaged at fixed heights above the sloping tropopause, and obtained the hemisphere average carbon-14 distribution, presented in terms of a vertical profile at 30° N. Similar results were obtained for each quarter throughout 1963 and 1964. The change of excess carbon-14 concentration with time is given by Figure A.39. The distribution in January 1963 was taken as the initial condition, and each *K* profile (also including Brasseur and Nicolet's "*K* max." and "*K* min.") was used to calculate the distribution of carbon-14 at each subsequent period (for example, January 1964 in Figure A.40). In gen-

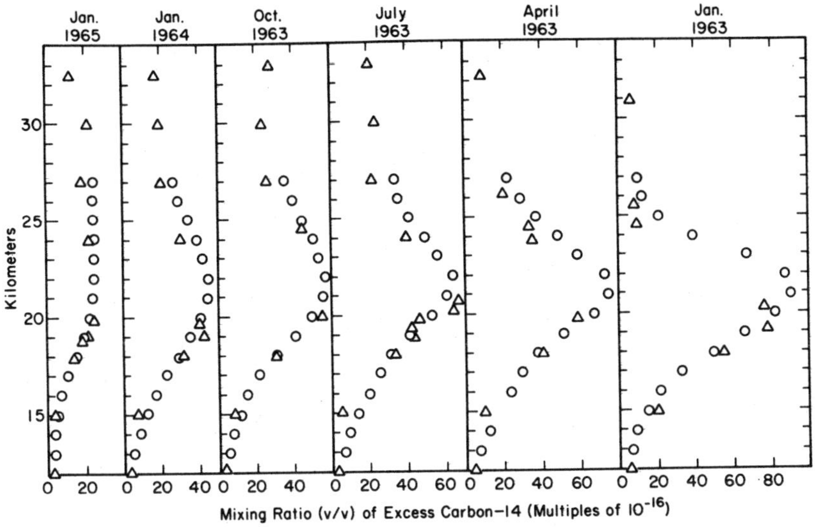

FIGURE A.39 Comparison of northern hemispherical average mixing ratios and locally observed mixing ratios of excess carbon-14 at 30° N. o, hemispherical average; Δ, balloon soundings at 30° N.

eral, these one-dimensional models sweep carbon-14 out of the stratosphere faster than was actually observed, in some cases much faster. Some models maintain roughly the observed integrated column of carbon-14, but the distribution is distorted in the direction of downward displacement of the peak of maximum concentration. Only Hunten's *K* profile gives a quantity and distribution that agrees well with that observed in January 1964.

A comparison of Figures A.38(a) and A.38(b) shows that there is a strong increase in ozone reduction with an increase in altitude of injection of contaminants. At a given rate of NO_x insertion, the reduction of ozone is about twice as great for input at 20 km as for that at 17 km. The Concorde or Tupolev SST's burn fuel at only about one third of the rate of a large SST (either the formerly proposed U.S. SST or a future "economic" design), and the smaller SST's cruise at about 17 km instead of at 20 km. Thus the large SST would be expected to reduce ozone about five times as much per aircraft as the Concorde, and the calculations for 20 km are representative of possible advanced supersonic transports.

The predictions of ozone reduction by seven one-dimensional models are summarized in Figure A.38. In view of the success of Hunten's

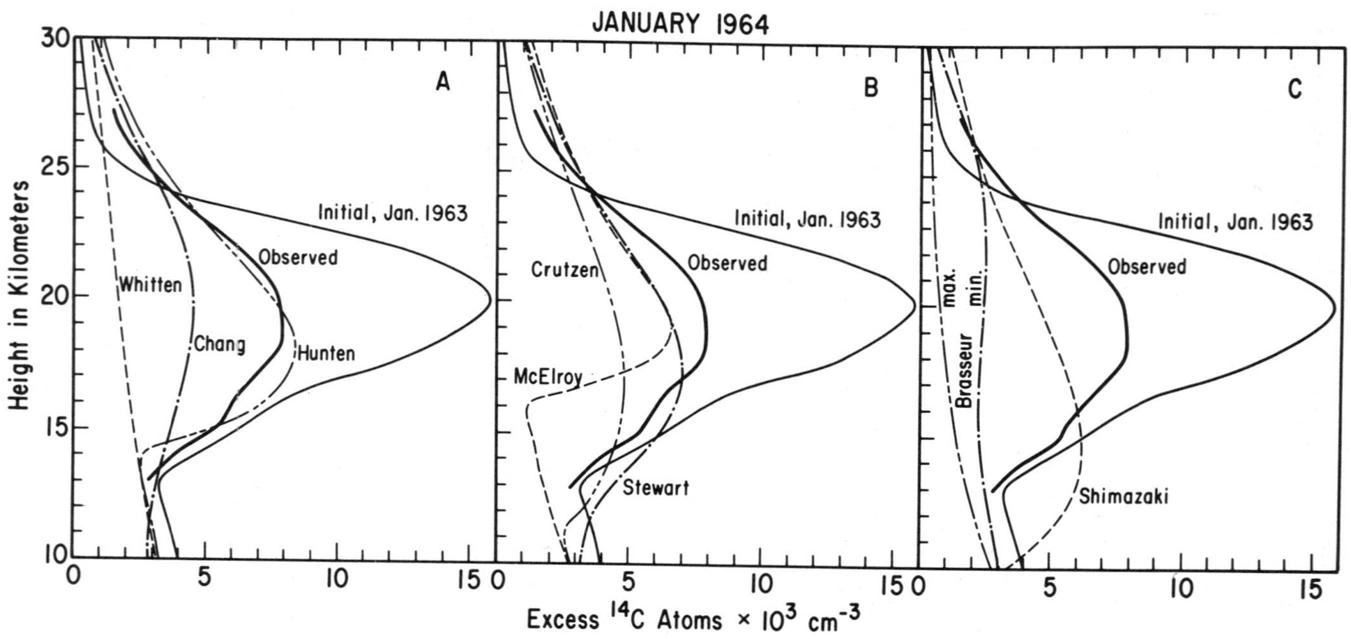

FIGURE A.40 Comparison of observed and predicted carbon-14 profiles for January 1964 for nine models of *K* for the same initial condition, that observed in January 1963. All three panels show the initial observed average profile of carbon-14 on January 1963 and the observed average profile one year later on January 1964. On each of the three panels are given profiles as calculated by the *K* profiles of Figure A.35, where the observed distribution of January 1963 was taken as the initial condition for the calculation.

model in predicting the quantity and distribution of carbon-14 in the stratosphere during the years following the atmospheric nuclear bomb tests of 1961–1962 (Figure A.40), it is regarded as the best one-dimensional model for predicting the perturbation of stratospheric ozone by supersonic transports.

3. *Models with Two-Dimensional Transport*

Partial results are available (September 1974) from three models that incorporate two dimensions of transport:

3. Hesstvedt (1974b)
9. Vupputuri (1974)
10. Widhopf (1974)

Hesstvedt's latest results are available only as global average effects; the detailed results as a function of latitude are not available for this review. The global average reduction of ozone as a function of artificial NO_x is given in Figures A.38(a) and A.38(b).

Vupputuri (1974) studied the natural atmosphere with two-dimensional transport, extensive O-N-H- chemistry, and several feedback mechanisms. He studied the perturbed atmosphere by doubling NO_x (including N_2O). He found that when the nitrogen oxides were doubled, the reduction of the vertical ozone column was between 10 and 19 percent, varying with latitude. (The one-dimensional models indicate an ozone reduction of 18 ± 4 percent for a doubling of stratospheric NO_x, Figure A.36). Vupputuri inserted 1.8×10^{12} g of NO_2 yr^{-1} between 40° and 50° N at 19–22 km, and he found ozone reductions between 7 and 16 percent. The global average reduction was 9.4 percent, 10.8 percent in the northern hemisphere and 7.9 percent in the southern hemisphere.

Widhopf (1974) reported a time-dependent meridional model for the troposphere and stratosphere. Large-scale eddy, as well as mean meridional advective, transport was included in the model. The chemical model included O_x, HO_x, NO_x, and HNO_x species, and it was the same as that used by Chang (1974). Steady-state distributions of O_3, N_2O, NO_2, NO, and HNO_3, calculated for the fall season in the northern hemisphere, show fairly good agreement with available atmospheric observations of these species. Agreement between the calculated and observed meridional variation of the total ozone was good. Three calculations of SST perturbations were reported, two at 20 km, one at

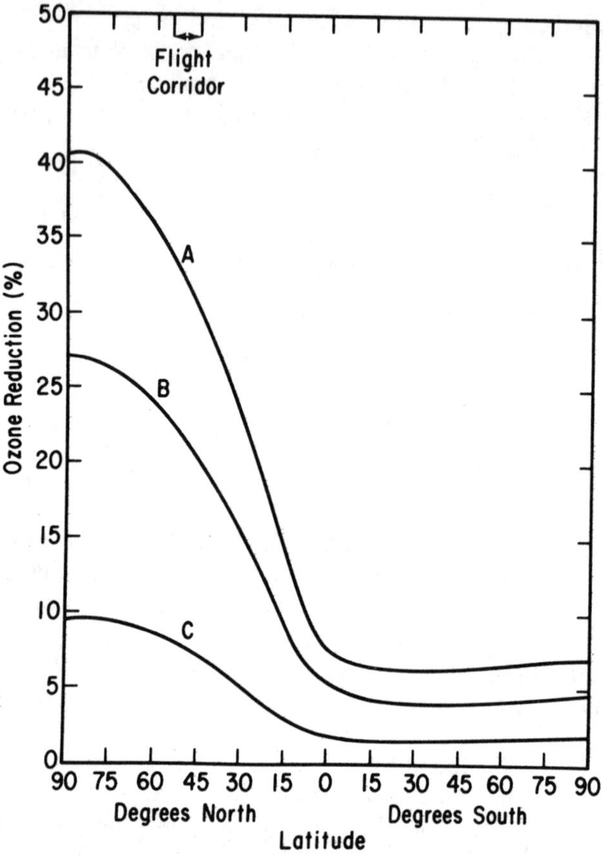

FIGURE A.41 Meridional variation of the reduction of the ozone column (between 10 and 50 km) for various NO_x insertions between 45° and 55° N: (A) 1.65×10^{12} g NO_x yr^{-1} at 20 km; (B) 1.65×10^{12} g NO_x yr^{-1} at 17 km; (C) 0.275×10^{12} g NO_x yr^{-1} at 20 km (Widhopf, 1974).

17 km, and all between 45° and 55° N. The results are given by Figure A.41 and are summarized here:

	A	B	C
Height, km	20	17	20
Added NO_2, 10^{12} g yr^{-1}	1.65	1.65	0.275
Ozone reductions (%)			
Global av.	15	9	3.4
Local max.	40	27	10
Northern hemisphere	23	15	5.3
Southern hemisphere	6	4	1.6

This calculation with two-dimensional transport gives somewhat greater ozone reduction for NO_x inserted at 20 km than that given by most one-dimensional models. However, the differences for insertion at 17 km are somewhat greater. The one-dimensional models give a factor of 2 less ozone reduction for 17-km injection than for 20-km injection. This two-dimensional model gives two thirds as great an effect at 17 km as at 20 km. Although these results are interesting, it must be remembered that they are highly dependent on the particular selection of an array of eddy-diffusion coefficients and that it is difficult to arrive at verifiable values for these coefficients.

4. *Models with Three-Dimensional Transport*

Mahlman (1973a, 1973b) developed a three-dimensional numerical model that executes dispersion experiments for inert atmospheric tracers. The model utilizes winds from a global general-circulation model that contains realistic topography, active water vapor, and seasonally varying radiation. The horizontal grid spacing is about 265 km. There are 11 vertical levels, the highest at about 31 km. A model experiment was carried out that simulated the continuous release of an inert tracer at 30° N latitude and at about 20-km elevation; and the tracer was removed by rainout near the earth's surface. The numerical integrations were continued for 11 months (not enough to attain a steady-state distribution).

The case of NO_y ($NO+NO_2+HNO_3$) from 500 formerly proposed U.S. SST's was treated by Mahlman. The resulting zonally averaged NO_y distribution after 11 months is shown by Figure A.42. The maximum enhancement of the NO_y concentration is 7.5×10^9 molecules cm^{-3} or about 5×10^{-9} mixing ratio. This model did not include any chemistry, and thus it is useful in this context primarily as a check on the NO_y distributions predicted by simpler models; a weak broad corridor effect and very little interhemispheric transport are indicated. The poor vertical resolution of the model near the tropopause is evident from the figure.

Cunnold *et al.* (1974) have presented initial results from a quasi-geostrophic, numerical, three-dimensional model. First, they calculated NO_x ($NO+NO_2$) distributions in the ambient and the perturbed atmosphere with a two-dimensional parameterized transport model in which ozone concentrations and temperatures were fixed at observed values but included annual variations. Integrations were carried out for 25 model years until the annual cycle of NO_y (NO_x+HNO_3) became steady. The calculated mixing ratios by volume for NO_x are shown in Figure A.43; at midlatitudes these NO_x values fall in the middle of observed NO and NO_2 values (Figures A.12 and A.13). The NO_x distri-

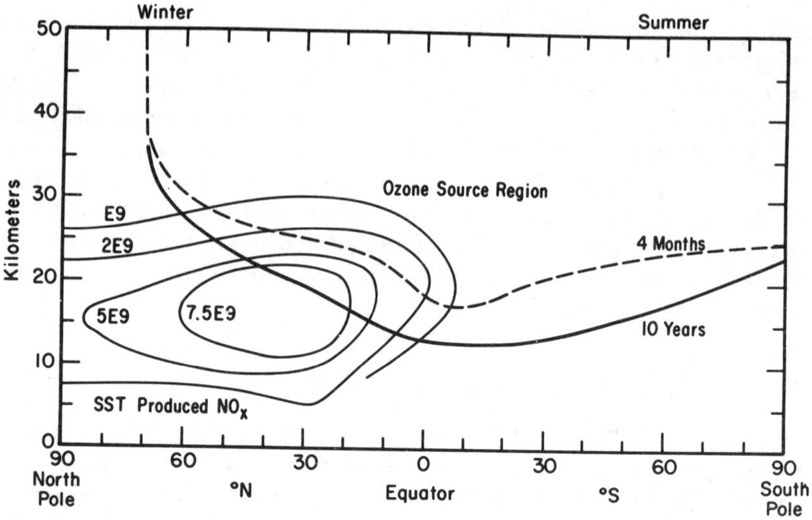

FIGURE A.42 Results of a three-dimensional global circulation model. Zonal average NO_x distribution after 11 months of operation of 500 formerly proposed U.S. SST's at 20 km and 30° latitude (Mahlman, 1973a, 1973b).

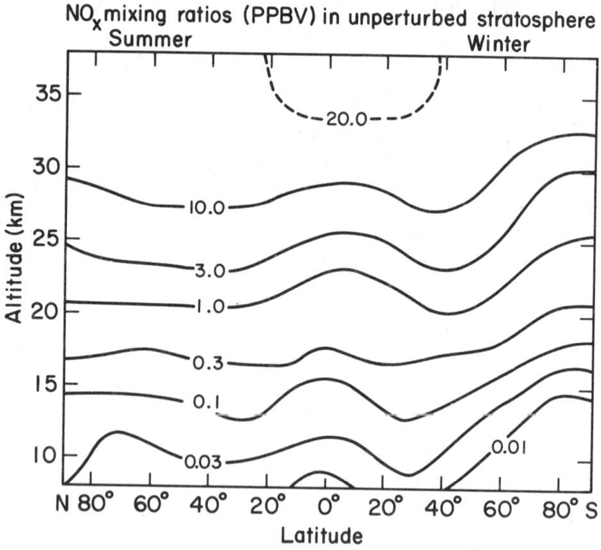

FIGURE A.43 Background distribution of NO_x as derived by MIT two-dimensional model (Cunnold *et al.*, 1974).

bution from the SST-perturbed stratosphere was treated by the same two-dimensional model. The perturbed result corresponded to an injection of NO_x of 1.8 MT yr^{-1} at 20 km between 40° and 50° N. The percentage increases in the NO_x mixing ratios due to this additional source of NO_x are shown in Figure A.44.

The predicted effect of artificial NO_x on ozone was then carried out with the full three-dimensional model, using the NO_x distributions from the two-dimensional model. The ratio of NO to NO_2 was calculated as a function of ozone and sunlight. Some results are given by Figure A.45. Run 17 is the control run for the unperturbed stratosphere, and run 18 is the simulation for the perturbed stratosphere. The latitudinal distribution of total ozone as a function of season for run 17 compares very well with observations (Dütsch, 1971). The global average reduction of ozone is 12 percent, and this point is entered as [11] in Figure A.32. The average decrease in the northern hemisphere is 16 percent, and that in the southern hemisphere is 8 percent. The largest local effect was a 25 percent decrease of the ozone column. The maximum reduction of ozone above the flight corridor was a factor of 2 greater than the worldwide average ozone reduction. The relatively large interhemi-

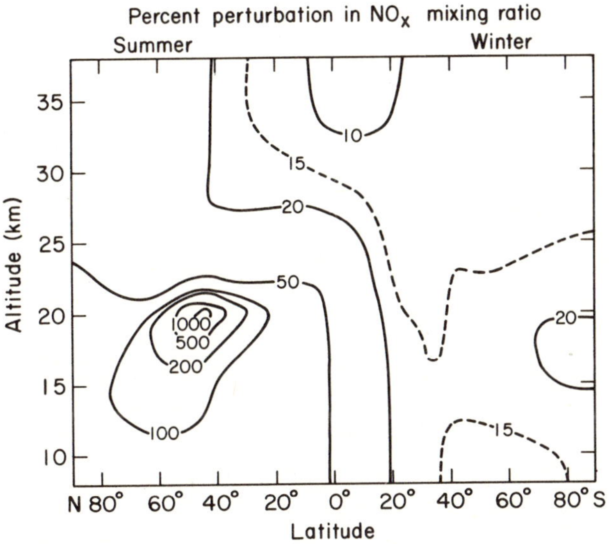

FIGURE A.44 Percent perturbation of atmospheric NO_x from injection of 1.8×10^{12} g NO_2 yr^{-1} at 40° to 50° N, 20-km elevation. MIT two-dimensional model (Cunnold *et al.*, 1974).

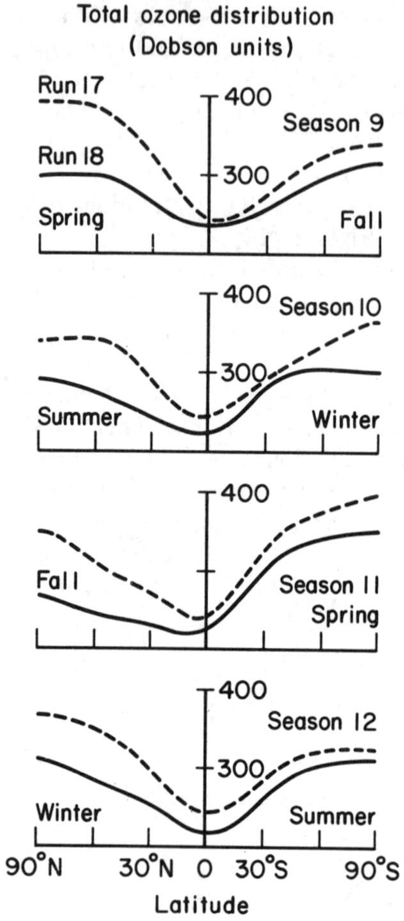

FIGURE A.45 Results of MIT three-dimensional global circulation model using NO_x distributions calculated in two-dimensional model. Ozone vertical profiles were calculated for four seasons on the basis of natural NO_x background (Figure A.43) and SST perturbed NO_x (Figure A.44); the average integrated ozone column is shown for the natural background (dashed line) and the perturbed distribution (solid line). The global average ozone reduction of 12 percent is entered in Figure A.37.

spheric transport in this model is apparently due to the particular selection of horizontal exchange coefficients in the two-dimensional model.

5. *Comparison of One-, Two-, and Three-Dimensional Model Results*

On the basis of simple qualitative considerations, one could argue that the effect of artificial NO_x on ozone might be greater (or less) if one considers three-dimensional air motions compared with the relatively simple one-dimensional models. The limited results now available from the complex two- and three-dimensional models are especially valuable in showing that there is no drastic difference one way or the other in

TABLE A.5 Comparison of Calculated Ozone Reductions

Model Dimension	Percent Ozone Reduction (Global average) As Reported	Recalc. by Chang	Reference
Box	23	—	Johnston (1971)
1	8.0	8	Chang (1973)
1	8.5	9	Crutzen (1973a)
1	11	15	McElroy *et al.* (1974)
1	19	—	Hunten (1974)
1	22	12	Shimazaki and Ogawa (1974)
2	12	—	Hesstvedt (1974a)
	9.4	—	Vupputuri (1974)
2	16	—	Widhopf and Taylor (1974)
3	12	—	Cunnold *et al.* (1974)

the predicted effect of NO_x on ozone between the simple and complex models. For 20-km insertion of 1.8 MT of NO_x yr^{-1}, Table A.5 compares global average results.

For a geophysical problem of this complexity, agreement within a factor of 2 is to be regarded as good agreement, indeed. The two- and three-dimensional models fall within the spread of predictions by the one-dimensional models. Further, most of the spread is directly traceable to the particular choices for eddy-transfer coefficients. The calculations are for the circled point in Figure A.31; it can be seen that after the year 2000 much larger SST fleets might be expected than the one corresponding to these calculations.

6. *Calibration of Models with* NO_x *Insertion Experiments*

The completeness of any model must always be open to doubt. Direct injection experiments in which large amounts of nitrogen oxides are inserted in the stratosphere appear to be hopelessly complicated and expensive. For these reasons, there is interest in the stratospheric effects of nuclear bombs and the interaction of cosmic rays with the atmosphere as a means of studying the effects of external influences on the concentration of ozone.

(a) *Nuclear Bombs* Nuclear bombs form nitric oxide in the atmosphere (Foley and Ruderman, 1973). There is a direct proportionality between bomb-produced NO and bomb-produced carbon-14. Excess carbon-14 was observed in the stratosphere by balloons and aircraft

(Telegadas, 1971), and the distribution of NO in the stratosphere as inferred from observed carbon-14 is given in Figure A.46 for January 1963 (just after the 1961–1962 test series was completed) and for January 1964 and January 1965 (Johnston, 1974). These clouds of nitric oxide lay lower in the stratosphere than model prediction for NO_x from SST exhaust (see, for example, Figure A.42). The total

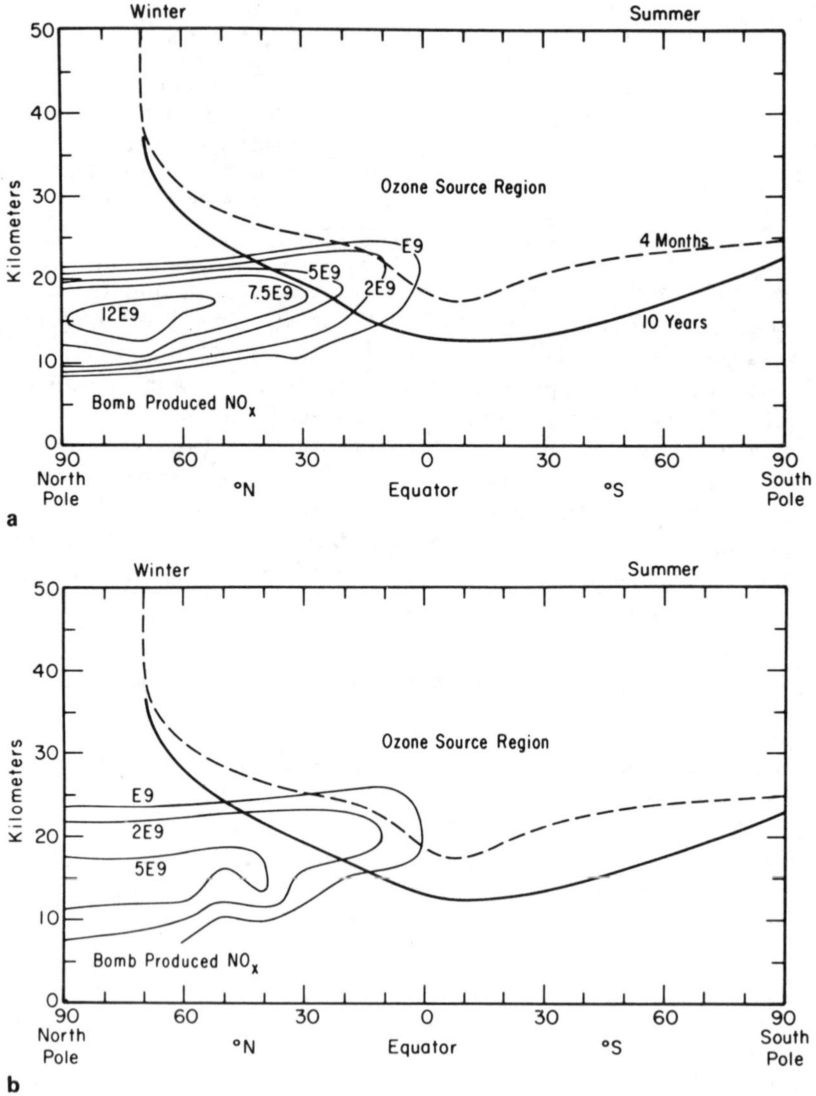

number of nitrogen oxide molecules is comparable with that for one year's operation of several hundred Concordes, but the distribution, Figure A.46, is not conducive to large ozone reductions (Johnston *et al.*, 1973). Chang *et al.* (1973) calculated the expected ozone reduction from the nuclear bomb tests from 1957 to 1971 (Figure A.47). Their maximum predicted ozone reduction is 4 percent.

Available data on ozone neither confirm nor refute the existence of such an effect (Johnston *et al.*, 1973; London and Kelley, 1974). A slow rise during the 1960's is consistent with a slow purging of NO_x from the stratosphere.

(b) *Cosmic Rays* Willett (1962) found that total ozone seemed to be strongly correlated with the 11-year sunspot cycle. Using total-ozone data for another 10 years, Christie (1973) re-examined Willett's correlation. Angell and Korshover (1973), using a different method of treating the data, found a very strong correlation of total ozone with the sunspot cycle. These studies were statistical in nature; there was no apparent physical explanation.

Warneck (1972) pointed out that cosmic rays penetrate the strato-

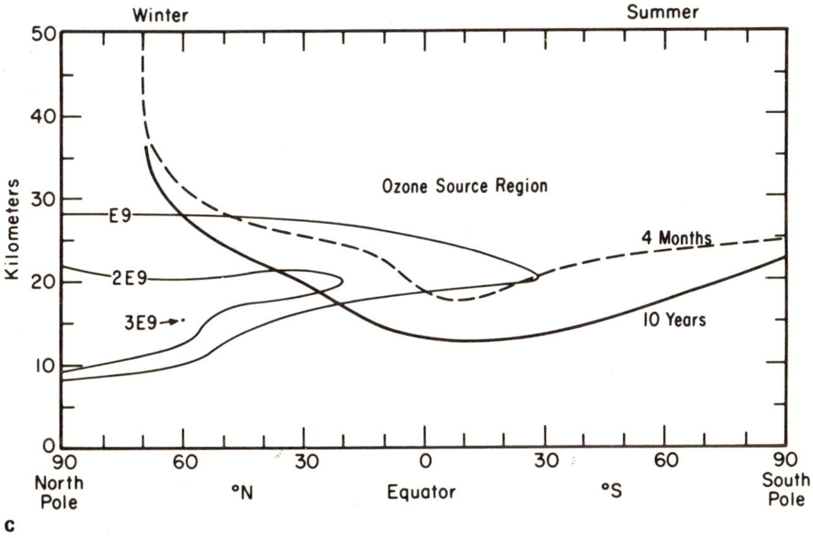

FIGURE A.46 Nuclear-bomb-produced NO_x (molecules cm^{-3}) based on observed carbon-14 (compare Figure A.20). (a) January 1963, (b) January 1964, (c) January 1965. Ozone replacement times are indicated on the figure (compare Figure A.23).

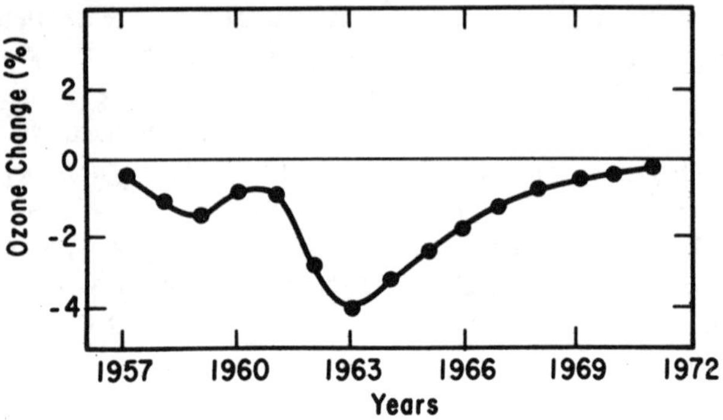

FIGURE A.47 Calculated reduction of ozone in the northern hemisphere as a result of nuclear bomb tests (compare Figure A.20 for times of tests). The maximum calculated ozone reduction was 4 percent in 1963, falling to 2 percent by early 1966 (Chang, 1973).

sphere and form some nitric oxide there. Brasseur and Nicolet (1973) discussed the question and showed that the global rate of production of NO_x from this mechanism is small compared with that from nitrous oxide. Ruderman and Chamberlain (1973) gave the physical interpretation for the correlation of ozone with sunspots in terms of cosmic-ray formation of nitric oxide. The following analysis is based on the work of Ruderman and Chamberlain.

Cosmic rays are steered by the earth's magnetic field, being deflected from the equator and focused into polar regions. The cosmic rays are largely stopped in the stratosphere and uppermost troposphere; ion pairs, electrons, and atoms are produced. The formation of ion pairs from cosmic rays has been measured as a function of latitude and elevation for several decades (Neher, 1971). During the period of maximum sunspots, the stream of charged particles from the sun deflects cosmic rays from the inner solar system. Ruderman and Chamberlain pointed out the cause and effect sequence: the solar cycle affects the flux of cosmic rays into the earth's atmosphere; the cosmic rays produce ions and nitric oxide in the polar stratosphere; the nitric oxide strongly and promptly affects ozone in the upper polar stratosphere, and, upon transport and dilution, it weakly and somewhat later affects ozone in temperate and tropical zones. They were successful in explaining the phase shift between the sunspot cycle and total ozone, and they explained the decreasing amplitude of the effect as one moves from polar to equatorial zones. The location and magnitude of the cosmic-ray-produced nitric oxide are given by Figure A.29.

The correlation of sunspot number with global ozone, as deduced from the data by Christie, for the period 1930–1970 is given by the first two panels of Figure A.48 (compare Figure A.8). The second panel of Figure A.48 shows the annual variation of ozone and an 11-year running average. There was a clear peak in total ozone in 1941 and in 1952. However, the expected peak in 1963 was absent. Chang's calculated ozone reduction from nuclear bombs is inverted and added to Christie's smoothed, global average ozone in the third panel of Figure A.48. With this calculated correction for the perturbation of ozone by nuclear bombs, Christie's analysis of the ozone data shows peak ozone in 1941, 1952, 1963 and an approach to a possible peak in

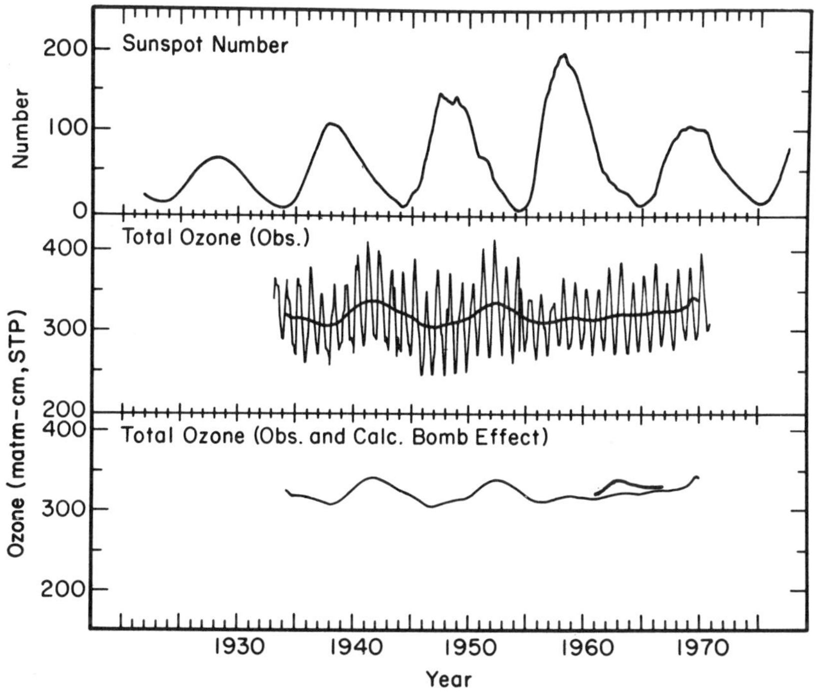

FIGURE A.48 Average ozone and sunspot number (Christie, 1973; compare Figure A.8) with Chang's calculated (Figure A.47) nuclear bomb effect inverted and added. In the third panel, with Chang's correction, there are ozone maxima in 1941, 1952, and 1963.

1974. This correlation of ozone with the 11-year solar cycle and its physical interpretation give a partial confirmation of the model-predicted effect of nitrogen oxides on ozone.

C. PERTURBATION OF THE STRATOSPHERE BY WATER

In the stratosphere above 40 km, the catalytic cycle based on HO_x free radicals

$$
\begin{array}{ll}
(12) & HOO + O \rightarrow HO + O_2 \\
(11) & HO + O \rightarrow H + O_2 \\
(14) & H + O_2 + M \rightarrow HOO + M \\
\hline
\text{NET:} & O + O \rightarrow O_2
\end{array}
$$

occurs at a rapid rate and at a rate comparable with the O_x and NO_x mechanisms (Figure A.26). Between 15 and 20 km the catalytic cycle

$$
\begin{array}{ll}
(9) & H + O_3 \rightarrow HOO + O_2 \\
(10) & HOO + O_3 \rightarrow HO + O_2 + O_2 \\
\hline
\text{NET:} & 2\ O_3 \rightarrow 3\ O_2
\end{array}
$$

occurs at a very slow rate, but at a rate comparable with the NO_x mechanism at those elevations. Between 20 and 40 km the destruction of ozone by H, HO, and HOO occurs at a rate slow compared with the NO_x cycle. In this region, however, hydroxyl radicals produce and destroy nitric acid:

$$
\begin{array}{ll}
(22) & HO + NO_2 + M \rightarrow HNO_3 + M \\
(24) & HO + HNO_3 \rightarrow H_2O + NO_3
\end{array}
$$

The photolysis of nitric acid regenerates hydroxyl radicals:

$$
(23) \quad HNO_3 + h\nu \rightarrow HO + NO_2
$$

The net effect is to bind some of the active forms of NO_x, NO and NO_2, into the inactive nitric acid. Thus HO_x destroys ozone in two regions of the stratosphere by two different cycles, and in another part of the stratosphere it ties up the active ozone-destroying NO_x catalysts into an inactive form. An increase in stratospheric water could cause either an increase or decrease in ozone, depending on the water distribution.

Some model calculations have been made for arbitrarily assumed large increases in stratospheric water content. McElroy *et al.* (1974) found a 0.1 percent increase of ozone when stratospheric water was increased by 24 percent. Crutzen (1974b) found a 1.5 percent increase of ozone when stratospheric water was doubled. Vupputuri (1974), on the other hand, found a 3.5 percent decrease of ozone when stratospheric water was doubled, but apparently he did not include the interaction of the NO_x and HO_x systems through the nitric acid reactions.

The water produced in one year by a large SST fleet, which burns 10^{11} kg of fuel per year, for example, would produce about one tenth as much water as that in the global stratosphere. Considering a 2½-year stratospheric residence time and Eq. (7), the stratospheric water vapor would be increased by 25 percent. According to McElroy, this increase in water would increase ozone by only 0.1 percent. However, the reduction of ozone by nitric oxide from a fleet of the size expected by 1990 (with an NO_2 emission index of 18) is a global average of about 10 to 20 percent, which far outweighs any effect of water on ozone. On the other hand, large increases in stratospheric water, especially that introduced in the middle or upper stratosphere by hypersonic transports (HST) pose problems of radiation balance (Ramanathan, 1974).

D. PERTURBATION OF THE STRATOSPHERE BY CHLORINE COMPOUNDS

The following is a preliminary discussion of two related problems that are outside the original charge of the Committee. In both cases, the results we have been able to obtain suggest further study, which we could not undertake in time to meet the deadline for this report. We recommend that this work be undertaken by a special committee, either a reorganized CIC or a new one.

1. *Chemistry*

As discussed above (Section II.B.3), free chlorine destroys ozone by a cycle involving Cl and ClO. The principal sink of free chlorine is formation of HCl, which is only temporary; eventually the chlorine must be physically removed to the troposphere. Wofsy and McElroy (1974) have summarized the chemistry and given a relation between odd-chlorine mixing ratio and ozone reduction. A recent revision of a rate coefficient has changed the proportionality, raising it by a factor of

1.5 (Wofsy *et al.,* 1974). The percent ozone reduction is therefore given by

$$\delta = 3.6 f \text{ (odd Cl)}, \tag{18}$$

when f is in ppb.

2. *Space Shuttle*

The solid-fuel boosters for the Shuttle use ammonium perchlorate as the oxidizer, and the exhaust therefore contains odd chlorine. In the Environmental Impact Statement, NASA has given the expected source strength for each 3-km layer between 12 and 45 km. Distributed globally, the amounts for 50 launches per year are given in Table A.5 as Q (chlorine atoms cm^{-2} sec^{-1}). In the third column, these values are summed from the top to level i, to give the steady-state downward fluxes ϕ at the bottom of each layer.

A one-dimensional transport calculation then gives mixing ratios as a function of height. The last two columns in Table A.6 show the results for two assumptions: spreading over the whole earth and spreading over the northern hemisphere only. The latter is regarded as much more probable.

McElroy *et al.* (1974) argue that the chemistry at 30 km controls the

TABLE A.6 Distribution with Height of Odd Chlorine from 50 Shuttle Launches per Year

Range (km)	Source (10^4 cm^{-2} sec^{-1})	Flux (10^4 cm^{-2} sec^{-1})	f (ppb)[a]	f (ppb)[b]
12–15	5.7	34.4	0.008^{9}	0.018
15–18	4.9	28.7	0.017^{7}	0.035
18–21	4.4	23.8	0.026	0.052
21–24	4.1	19.4	0.035	0.070
24–27	3.7	15.3	0.042	0.084
27–30	3.4	11.6	0.049	0.098
30–33	3.1	8.2	0.055	0.110
33–36	2.5	5.1	0.059	0.118
36–39	1.8	2.5	0.062	0.124
39–41	0.75	0.75	0.063	0.126

[a] Parts per billion (10^{-9}) by volume for the whole earth.
[b] Parts per billion (10^{-9}) by volume for the northern hemisphere.

ozone density at all lower heights. To a first approximation, this height may therefore be chosen to represent the effect on the whole ozone layer. From the last column, we therefore take an odd-chlorine mixing ratio of 0.1 ppb. Substitution in Wofsy's and McElroy's equation (corrected for the increased rate coefficient) gives an ozone reduction of 0.36 percent.

In view of the rough nature of this estimate, and the uncertainties in the transport and chemistry, the value could be as great as 1 percent. The effect of Shuttle launches is thus potentially serious enough to require a more detailed study.

3. *Chlorofluoromethanes (Freons)*

Molina and Rowland (1974) showed that current use and release of chlorofluoromethanes (e.g., Freons) is at a global rate that is within about a factor of 2 of the natural rate of formation of NO from N_2O; it requires about a hundred years to reach a steady state between chlorofluoromethane release at the ground and stratospheric destruction; and their use has been increasing rapidly. Since ClO is about five times as effective as NO_2, per molecule, in destroying ozone, Molina and Rowland anticipate a probable large (10 percent more or less) future reduction of stratospheric ozone from this source.

This conclusion is generally supported by two other papers (Crutzen, 1974a; Cicerone *et al.,* 1974). A more detailed study, in which the time-dependent problem is followed in detail, has been made by Wofsy *et al.* (1974). Continued release of Freons at the present rate would result in an ozone reduction of 13 percent late in the next century. Use at an accelerated rate (growth of 10 percent per year) would give a 37 percent reduction in 2014. If release is terminated at some time, the biggest effect on ozone occurs about 5 years later, and the recovery has a half-life of some 50 years.

These results represent accurately our present understanding of the subject. Unless an unexpected missing factor is turned up soon, drastic action will have to be taken within a year or two. It should be pointed out that the transport problem for Freons and their products is much less sensitive to the assumed eddy coefficients than is the case for exhaust products. Just as with production of odd nitrogen from natural N_2O, the rate of transport affects only the details by which one form of chlorine gives way to another; the final mixing ratio is essentially unaffected.

References

Ackerman, M., and C. Muller. 1972. Stratospheric nitrogen dioxide from infrared absorption spectra, *Nature* 240:300–301.

Ackerman, M., and C. Muller. 1973. Stratospheric methane and nitrogen dioxide from infrared spectra, *Pure Appl. Geophys.* pp. 1325–1335.

Ackerman, M., J. C. Fontanella, D. Frimout, A. Girard, L. Gramont, N. Louisnard, C. Muller, and D. Nevejans. 1973. Recent spectra of NO and NO_2, *Aeronom. Acta* 120:1–13.

Aldaz, L. 1969. Flux measurements of atmospheric ozone over land and water, *J. Geophys. Res.* 74:6943–6946.

Angell, J. L., and J. Korshover. 1973. Quasi-biennial and long-term fluctuations in total ozone, *Mon. Weather Rev.* 101:426–443.

Bates, D. R., and M. Nicolet. 1950. The photochemistry of atmospheric water vapor, *J. Geophys. Res.* 55:301–327.

Bauer, E. 1973. *Physics of the Stratosphere.* Inst. for Defense Analyses. Arlington, Va.

Brasseur, G., and S. Cieslik. 1973. On the behavior of nitrogen oxides in the stratosphere, *Pure Appl. Geophys.* 106–108:1431–1438.

Brasseur, G., and M. Nicolet. 1973. Chemospheric processes of nitric oxide in the mesophere and stratosphere, *Planet. Space Sci.* 21:939.

Brewer, A. W., and A. W. Wilson, 1968. The regions of formation of atmospheric ozone, *R. Meteorol. Soc. Quart. J.* 94:249–265.

Brewer, A. W., C. T. McElroy, and J. B. Kerr. 1973. Nitrogen dioxide concentrations in the atmosphere, *Nature* 246:129–133.

Brewer, A. W., C. T. McElroy, and J. G. Kerr. 1974. Spectrophotometric nitrogen dioxide measurements, *Proc. of the Third Conf. on the CIAP,* Cambridge, Mass. DOT-TSC-OST-74-15, pp. 257–263.

Brousaides and Morrissey. 1967. Stratospheric humidity sensing with alpha radiation hygrometer, *AFCRL–67–0604.*

Canadian Department of Transport, Meteorological Branch. 1960–1970. *Ozone Data for the World.* Vols. 1–11. Toronto, Canada.

Chameides, W., and J. C. G. Walker. 1973. A photochemical theory of tropospheric ozone, *J. Geophys. Res.* 78:8751–8760.

Chang, J. S. 1973. Global transport and kinetics model, *UCRL-51336.* Lawrence Livermore Lab. Rep.

Chang, J. S. 1974. Simulations, perturbations, and interpretations, *Proc. of the Third Conf. on the CIAP,* Cambridge, Mass. DOT-TSC-OST-74-15, pp. 330–341.

Chang, J. S., and W. Deuwer. 1973. On the possible effect of NO_x injection in the stratosphere due to past atmospheric nuclear weapons test, AIAA/AMS Int. Conf. Environmental Impact Aerospace Ops. in High Atmosphere. June 11–13.

Chang, J. S., and H. S. Johnston. 1974. The effect of NO_x effluents on ozone, *Proc. of the Third Conf. on the CIAP,* Cambridge, Mass. DOT-TSC-OST-74-15, pp. 323–329.

Chang, J. S., A. C. Hindmarsh, and N. K. Madsen. 1973. Simulation of chemical kinetics transport in the stratosphere, *UCRL-74823.* Lawrence Livermore Lab. Rep.

Chapman, S. 1930. A theory of upper-atmosphere ozone, *Mem. R. Meteorol. Soc.* 3:103–125.

Christie, A. D. 1973. Secular or cyclic change in ozone, *Pure Appl. Geophys.* 106–108:1000–1009.

Cicerone, R. J., R. S. Stolarski, and S. Walters. 1974. Stratospheric ozone destruction by manmade chlorofluoromethanes, *Science* 185:1165.

Crutcher, H. L. 1969. Temperature and humidity in the troposphere, *World Survey of Climatology*. D. R. Rex, ed. Elsevier, Amsterdam-London-New York, pp. 45–83.

Crutzen, P. J. 1970. The influence of nitrogen oxides on the atmospheric ozone content, *R. Meteorol. Soc. Quart. J.* 96:320–325.

Crutzen, P. J. 1971. Ozone production rates in an oxygen-hydrogen-nitrogen oxide atmosphere, *J. Geophys. Res.* 76:7311–7327.

Crutzen, P. J. 1972a. SSTs—a threat to the earth's ozone shield, *Ambio* 1:41–51.

Crutzen, P. J. 1972b. The photochemistry of the stratosphere with special attention given to the effects of NO_x emitted by supersonic aircraft, *Proc. of the Surv. Conf. on the CIAP,* Cambridge, Mass., DOT-TSC-OST-72-13, pp. 8–89.

Crutzen, P. J. 1973a. A review of upper atmospheric photochemistry, paper given at IAGA Symposium on Aeronomic Processes in Stratosphere and Mesophere, Kyoto, Japan, Sept.

Crutzen, P. J. 1973b. A discussion of the chemistry of some minor constituents in the stratosphere and troposphere, *Pure Appl. Geophys.* 106–108:1385–1399.

Crutzen, P. J. 1974a. Estimates of possible future ozone reductions from continued use of fluoro-chloro-methanes, *Geophys. Res. Lett.* 1:205.

Crutzen, P. J. 1974b. A review of upper atmospheric photochemistry, *Can. J. Chem.* 52:1569–1581.

Cumming, C., and R. P. Lowe. 1973. Balloon-borne spectroscopic measurement of stratospheric methane, *J. Geophys. Res.* 78:5259–5264.

Cunnold, D. M., F. N. Alyea, N. A. Phillips, and R. G. Prinn. 1974. First results of a general circulation model applied to the SST-NO_x problem, paper given at AIAA/AMS 2nd Internatl. Conf. Environmental Impact of Aerospace Ops. in High Atmosphere, San Diego, Calif., July 8–10.

Dickinson, R. E. 1973. Motions in the stratosphere, *Proc. of the Surv. Conf. on the CIAP,* Cambridge, Mass., DOT-TSC-OST-72-13, pp. 148–161.

Dütsch, H. U. 1969. Atmospheric ozone and ultraviolet radiation, *World Survey of Climatology*. D. F. Rex, ed. Elsevier, Amsterdam-London-New York, pp. 383–432.

Dütsch, H. U. 1971. Photochemistry of atmospheric ozone, *Advances in Geophysics,* Vol. 15. Academic Press, New York, pp. 219–232.

Ehhalt, D. H. 1974. Sampling of trace constituents, *Can. J. Chem.* 52:1510–1518.

Ehhalt, D. H., L. E. Heidt, and E. A. Martell. 1972. The concentration of atmospheric methane between 44 and 62 kilometers altitude, *J. Geophys. Res.* 77: 2193–2196.

Ehhalt, D. H., L. E. Heidt, R. H. Lueb, and N. Roper. 1974. Vertical profiles of CH_4, H_2, CO, N_2O and CO_2 in the stratosphere, *Proc. of the Third Conf. on the CIAP,* Cambridge, Mass., DOT-TSC-OST-74-15, pp. 153–160.

Evans, W. F. J. 1974. Rocket measurements of water vapor in the stratosphere (draft ms.). Atmospheric Environment Service, Downview, Ontario.

Fabian, P., and C. Junge. 1970. Global rate of ozone destruction at the earth's surface, *Arch. Meteorol. Geophys. Bioklimatol. Ser.* 19:161–172.

Farmer, C. B., O. F. Raper, R. A. Toth, and R. A. Schindler. 1974. Recent results of aircraft infrared observations of the stratosphere, *Proc. of the Third Conf. on the CIAP,* Cambridge, Mass., DOT-TSC-OST-74-15, pp. 234–245.

Feely, H. W., and H. Seitz. 1970. Use of lead 210 as a tracer of transport processes in the stratosphere, *J. Geophys. Res.* 75:2885–2894.

Foley, H. M., and M. A. Ruderman. 1973. Stratospheric nitric oxide production from past nuclear explosions and its relevance to projected SST pollution, *J. Geophys. Res.* 78:4441–4450.

Fontanella, J.-C., A. Girard, L. Gramont, and N. Louisnard. 1974. Vertical distribution of NO, NO_2, and HNO_3 as derived from stratospheric absorption infrared spectra, *Proc. of the Third Conf. on the CIAP,* Cambridge, Mass., DOT-TSC-OST-74-15, pp. 217–233.

Goldman, A., D. G. Murcray, F. H. Murcray, W. J. Williams, J. N. Brooks, and C. M. Bradford. 1973. Vertical distribution of CO in the atmosphere, *J. Geophys. Res.* 75:5273–5283.

Grobecker, A. J. 1974. Research program for assessment of stratospheric pollution, *Acta Astron.* 1:179–224.

Gudiksen, P. H., A. W. Fairhall, and R. J. Reed. 1968. Roles of mean meridional circulation and eddy diffusion in the transport of trace substances in the lower stratosphere, *J. Geophys. Res.* 73:4461.

Hammond, A. L., and T. H. Maugh II. 1974. Stratospheric pollution: multiple threats to earth's ozone, *Science* 186:335.

Hard, T. M. 1974. *Stratospheric Trace-Gas Measurements.* Tech. Rep. Transportation Systems Center, Cambridge, Mass. May.

Harries, J. E., J. R. Birch, J. W. Fleming, N. W. B. Stone, D. G. Moss, N. R. W. Swann, and G. F. Neill. 1974. Studies of stratospheric H_2O, O_3, HNO_3, N_2O, and NO_2 from aircraft, *Proc. of the Third Conf. on the CIAP,* Cambridge, Mass., DOT-TSC-OST-74-15, pp. 197–212.

Hering, W. S., and T. R. Borden, Jr. 1964. Ozonesonde observations over North America, *Environ. Res. Papers No. 38.* Vol. 2. Air Force Cambridge Research Labs., Bedford, Mass. July.

Hesstvedt, E. 1972. The effect of water vapor and nitrogen oxides upon the ozone layer, studied in a photochemical model with meridional transport, paper given at Symposium on Atmospheric Ozone. Arosa, Switzerland. Aug.

Hesstvedt, E. 1973. Effect of supersonic transport upon the ozone layer studied in a two-dimensional photochemical model with transport, *AGARD Conf. Proc. No. 125.* Item 6. London, England.

Hesstvedt, E. 1974a. Reduction of stratosphere ozone from high flying aircraft studied in a two-dimensional photochemical model with transport, *Can. J. Chem.* 52:1592–1598.

Hesstvedt, E. 1974b. Paper given at IAMAP/IAPSO First Special Assembly, Melbourne, Australia. Jan.

Hoshizaki, H., J. W. Myer, and K. O. Redler. 1973. Potential destruction of ozone by HCl in rocket exhausts, *Advances in Aircrafts and Fuels.* Chap. 4. Also Rep. LMC-D-354202. Lockheed, Palo Alto, Calif.

Hunt, B. G. 1966a. The need for a modified photochemical theory of the ozonosphere, *J. Atmos. Sci.* 23:88–95.

Hunt, B. G. 1966b. Photochemistry of ozone in a moist atmosphere. *J. Geophys. Res.* 71:1385–1398.

Hunten, D. M. 1974. Stratospheric pollution and how it dissipates.

Hunten, D. M. 1975. Residence times of aerosols and gases in the stratosphere, *Geophys. Res. Lett.* 2(1):26–28.

Isaken, I. A. 1973. The production and distribution of nitrogen oxides in the lower stratosphere, *Pure Appl. Geophys.* 106–108:1438–1445.

Johnston, H. S. 1968. *Kinetics of Neutral Oxygen Species.* Vol. 20. U.S. Department of Commerce, National Standard Reference Data System, National Bureau of Standards.

Johnston, H. S. 1971. Reduction of stratospheric ozone by nitrogen oxide catalysts from supersonic transport exhaust, *Science* 173:517–522; Catalytic reduction of stratospheric ozone by nitrogen oxides. UCRL Rep. No. 20568, pp. 1–106.

Johnston, H. S. 1974. Supersonic aircraft and the ozone layer, *Environ. Change* 2:339–350.

Johnston, H. S., and E. Quitevis. 1974. The oxides of nitrogen with respect to smog, supersonic transports, and global methane, paper given at 5th Internatl. Conf. of Radiation Res. Seattle, Wash. July 14–20.

Johnston, H. S., and G. Whitten. 1973. Instantaneous photochemical rates in the global stratosphere, *Pure Appl. Geophys.* 106–108:1468–1489.

Johnston, H. S., G. Whitten, and J. Birks. 1973. The effect of nuclear explosions on stratospheric nitric oxide and ozone, *J. Geophys. Res.* 78:6107–6135.

Johnston, H. S., D. Kattenhorn, and G. Whitten. 1974. Implications of carbon-14 data for the period 1963–1965 for models of stratospheric motion and the reduction of ozone by SSTs, Report to CIAP and CIC. October. 33 pp.

Komhyr, W. D., E. W. Barrett, G. Slocum, and H. K. Weickman. 1971. Atmospheric ozone increase during the 1960's, *Nature* 232:390–391.

Krueger, A. J. 1973. *The Mean Ozone Distribution from Several Series of Rocket Soundings to 52 km at Latitudes from 58°S to 64°N.* Preprint X-651-73-67. Goddard Space Flight Center, Greenbelt, Md. February.

Krueger, A. J., and A. Minzner. 1973. A proposed mid-latitude ozone model for the U.S. Standard Atmosphere (Summary).

Lazrus, A. L., and B. W. Gandrud. 1974. Progress report on distribution of stratosphere, *Proc. of the Third Conf. on the CIAP,* Cambridge, Mass., DOT-TSC-OST-74-15, pp. 161–167.

Loewenstein, M., J. P. Paddock, I. G. Poppoff, and H. F. Savage. 1974. In situ NO and O_3 measurements in the lower stratosphere from a U-2 aircraft, *Proc. of the Third Conf. on the CIAP,* Cambridge, Mass., DOT-TSC-OST-74-15, pp. 213–215.

London, J., and J. Kelley, 1974. Global trends in atmospheric ozone, *Science* 184:987–989.

London, J., and J. H. Park. 1973. The interaction of ozone photochemistry and dynamics in the stratosphere—a three-dimensional atmospheric model, paper given at IAGA Symposium on Aeronomic Processes in Stratosphere and Mesosphere. Kyoto, Japan. September.

Lovill, J. E. 1972. Characteristics of the general circulation of the atmosphere and the global distribution of total ozone as determined by the Nimbus III satellite infrared interferometer spectrometer, *Atmospheric Science Paper No. 180.* Dept. Atmospheric Science, Colorado State U., Fort Collins, Colo. December.

McConnell, J. C., and M. B. McElroy. 1973. *J. Atmos. Sci.* 30:1465.

McElroy, M. B., and J. C. McConnell. 1971. Nitrous oxide: A natural source of stratospheric NO, *J. Atmos. Sci.* 28:1095–1098.

McElroy, M. B., S. Wofsy, J. Penner, and J. McConnell. 1974. Atmospheric ozone: Possible impact of stratospheric aviation, *J. Atmos. Sci.* 31:287–300.

Mahlman, J. D. 1973a. A three-dimensional stratospheric point-source experiment and its implications for dispersion of affluent from a fleet of supersonic aircraft, paper given at AIAA/AMS Int. Conf. Environmental Impact Aerospace Ops. in High Atmosphere. AIAA Paper 72-528. Denver, Colo. June.

Mahlman, J. D. 1973b. Preliminary results from a three-dimensional, general-circulation/tracer model, *Proc. of the Second Conf. on the CIAP,* Cambridge, Mass., DOT-TSC-OST-73-4, pp. 321–337.

Mastenbrook, H. J. 1971. The variability of water vapor in the stratosphere, *J. Atmos. Sci.* 28:1495–1501.

Molina, M. J., and F. S. Rowland. 1974. Stratospheric sink for chlorofluoromethanes—chlorine atom-catalysed destruction of ozone, *Nature* 249:810–812.

Murcray, D. G., A. Goldman, A. Csoeke-Poeckh, F. H. Murcray, W. J. Williams, and R. N. Stocker. 1973a. Nitric acid distribution in the stratosphere, *J. Geophys. Res.* 78:7033–7038.

Murcray, D. G., A. Goldman, F. H. Murcray, W. J. Williams, J. N. Brooks, and D. B. Barker. 1973b. Vertical distribution of minor atmospheric constituents as derived from airborne measurements of atmospheric emission and absorption spectra, *Proc. of the Second Conf. on the CIAP,* DOT-TSC-OST-73-4. pp. 86–98.

Murcray, D. G., A. Goldman, J. N. Brooks, J. Van Allen, R. N. Stocker, J. J. Kosters, and D. B. Barker. 1974. Recent results of stratospheric trace-gas measurements from balloon-borne spectrometers, *Proc. of the Third Conf. on the CIAP,* Cambridge, Mass., DOT-TSC-OST-74-15, pp. 184–192.

Neher, H. V. 1971. Cosmic rays at high latitudes and altitudes covering four solar maxima, *J. Geophys. Res.* 76:1637–1651.

Newell, R. E. 1969. Radioactive contamination of the upper atmosphere, *Progr. Nucl. Energy.* Series XII, Health, Physics, Vol. 2. Pergamon, Oxford. p. 538.

Nicolet, M. 1970. Ozone and hydrogen reactions, *Ann. Geophys.* 26:531–546.

Nicolet, M. 1971. Aeronomic reactions of hydrogen and ozone, *Mesospheric Models and Related Experiments.* G. Flocco, ed. D. Reidel, Dordrecht, Holland, pp. 1–51.

Nicolet, M., and E. Vergison. 1971. L'Oxyde azoteux dans la stratosphere, *Aeronom. Acta* 90:1–16.

Patel, C. K. N. 1974. Spectroscopic measurements of stratospheric nitric oxide and water vapor, *Science* 184:1173–1174.

Ramanathan, V. 1974. A simplified stratospheric radiative transfer model: Theoretical estimates of the thermal structure of the basic and perturbed stratosphere, paper given at AIAA/AMS 2nd Int. Conf. Environmental Impact Aerospace Ops. in High Atmosphere. San Diego, Calif. July 8–10.

Reed, R. J., and K. E. German. 1965. A contribution to the problem of stratospheric diffusion by large scale mixing, *Mon. Weather Rev.* 93:313.

Ridley, B. A., H. I. Schiff, A. Shaw, and L. R. Megill. 1974. In-situ measurements of NO in the stratosphere using chemiluminescence, *Proc. of the Third Conf. of the CIAP,* Cambridge, Mass., DOT-TSC-OST-74-15, pp. 193–196.

Ruderman, M. A., and J. W. Chamberlain. 1973. *Origin of the Sunspot Modulation of Ozone: Its Implications for Stratospheric NO Injection.* JSS-73-18-3. Inst. for Defense Analyses.

SCEP. 1970. *Man's Impact on the Global Environment, Study of Critical Environmental Problems.* MIT Press, Cambridge, Mass.

Scholtz, T. G., D. H. Ehhalt, L. E. Heidt, and E. A. Martell. 1970. Water vapor, molecular hydrogen, methane, and tritium concentrations near the stratopause, *J. Geophys. Res.* 75:3049–3054.

Schutz, K., C. Junge, R. Beck, and B. Albrecht. 1970. Studies of atmospheric N_2O, *J. Geophys. Res.* 75:2230–2246.

Seiler, W., and P. Warneck. 1972. Decrease of the carbon monoxide mixing ratio at the tropopause, *J. Geophys. Res.* 77:3204–3214.

Shimazaki, T., and T. Ogawa. 1974. Theoretical models of minor constituents' distributions in the stratosphere and the impacts of the SST exhaust gases, paper given at IAMAP/IAPSO 1st Special Assemblies. Melbourne, Australia.

Sissenwine, M. 1969. Standard and supplemental atmospheres, *World Survey of Climatology*. Vol. 4. D. F. Tex, ed. Elsevier, Amsterdam-London-New York, pp. 5–44.

Stewart, R. W. 1973. Response of stratospheric ozone to the simulated injection of nitric oxide, paper given at Fall AGU Mtg. San Francisco, Calif.

Stewart, R. W., and M. I. Hoffert. 1973. Stratospheric contamination experiments with a one-dimensional atmospheric model, paper given at AIAA/AMS Int. Conf. Environmental Impact Aerospace Ops. in High Atmosphere. Denver, Colo.

Stolarski, R. S., and R. J. Cicerone. 1974. Stratospheric chlorine: Possible sink for ozone, *Can. J. Chem.* 52:1610–1615.

Telegadas, K. 1967. The seasonal stratospheric distribution of cadmium-109, plutonium-238 and strontium-90, *Health and Safety Lab. Rep. 184*. pp. 53–118.

Telegadas, K. 1971. The seasonal stratospheric distribution and inventories of excess carbon-14 from March 1955 to July 1969, *Health and Safety Lab. Rep. 243*. U.S. AEC. pp. 3–86.

Telegadas, K., and R. J. List. 1969. Are particulate radioactive tracers indicative of stratospheric motions, *J. Geophys. Res.* 74:1339–1350.

Tiefenau, H., and P. Fabian. 1972. The specific ozone destruction at the ocean surface and its dependence on horizontal wind velocity from profile measurements, *Arch. Meteorol. Geophys. Bioklimatol. Ser.* A21:399–412.

Toth, R. A., C. B. Farmer, R. A. Schindler, and O. F. Raper. 1974. Aspects of infrared detection of nitric oxide in the stratosphere. Paper SA 103. AGU Meeting. Washington, D.C. Apr. 12.

Vupputuri, R. K. 1974. The role of stratospheric pollutant gases (H_2O, NO_x) in the ozone balance and its implications for the seasonal climate of the stratosphere, paper given at IAMAP/IAPSO 1st Special Assemblies. Melbourne, Australia.

Warneck, P. 1972. Cosmic rays as a source of odd nitrogen in the stratosphere, *J. Geophys. Res.* 77:6589–6591.

Whitten, R. C., and R. P. Turco. 1973. A model of studying the effects of injecting contaminants into the stratosphere and mesophere, paper given at AIAA/AMS Internatl. Conf. Environmental Impact Aerospace Ops. in High Atmosphere. AIAA Paper 73-539. Denver, Colo.

Whitten, R. C., and R. P. Turco. 1974. The effect of SST emission on the earth's ozone layer, paper given at IAMAP/IAPSO 1st Special Assemblies. Melbourne, Australia.

Widhopf, G. F. 1974. Distribution of trace species in the stratosphere and the effect of SST pollutants, paper given at AGU Mtg., San Francisco, Calif. September 12.

Widhopf, G. F., and T. D. Taylor. 1974. Numerical experiments on stratospheric meridional ozone distributions using parameterized two-dimensional model, *Proc. of the Third Conf. on the CIAP,* Cambridge, Mass., DOT-TSC-OST-74-15, pp. 376–389.

Willett, H. C. 1962. The relationship of total atmospheric ozone to the sunspot cycle, *J. Geophys. Res.* 67:661–670.

Wofsy, S. C., and M. B. McElroy. 1973. On vertical mixing in the upper stratosphere, *J. Geophys. Res.* 78:2619–2624.

Wofsy, S. C., and M. B. McElroy. 1974. HO_x, NO_x, and ClO_x: Their role in atmospheric photochemistry, *Can. J. Chem.* 52:1582–1591.

Wofsy, S. C., M. B. McElroy, and N. D. Sze. 1975. Freon consumption: Implications for atmospheric ozone, *Science* 187:535–537.

Appendix B

Solar Ultraviolet Flux at the Earth's Surface

The solar uv flux that reaches the surface of the earth is a function of the solar spectral irradiance at the top of the atmosphere and the absorption and scattering by the atmosphere. The spectral irradiance of the sun above the atmosphere has been measured by several groups, and a summary of the data is given in Figure B.1. The solid line on this figure is the estimated best fit to the data obtained from CIAP Monograph 5.

Molecular ozone has a strong absorption band in the uv centered at 250 nm and extending beyond 350 nm. The absorption coefficient falls off rapidly with wavelength, as shown in Figure B.2, and the attenuation of the incident solar flux is therefore a strong function of wavelength. Figure B.3 plots some calculated surface uv fluxes versus wavelength in the direct solar beam for several total ozone column densities. The fractional change in uv radiation is seen to be larger for the shorter wavelengths.

In addition to the absorption of the incident solar flux by ozone, one must also take into account multiple Rayleigh scattering by atmospheric molecules; scattering by particles and aerosols; reflection, scattering, and attenuation by clouds; haze and smog near the ground; and reflection from the ground. The computation of the direct and

---

This Appendix presents the findings of the Panel on Atmospheric Physics and Chemistry: F. Johnson, *Chairman;* J. Friend, *Vice Chairman;* J. Chamberlain, D. Hunten, H. Johnston, F. Kaufman, H. Landsberg, and H. Panofsky. Consultants to the Panel: R. Hudson, C. Leith, E. Reiter, and G. Robinson.

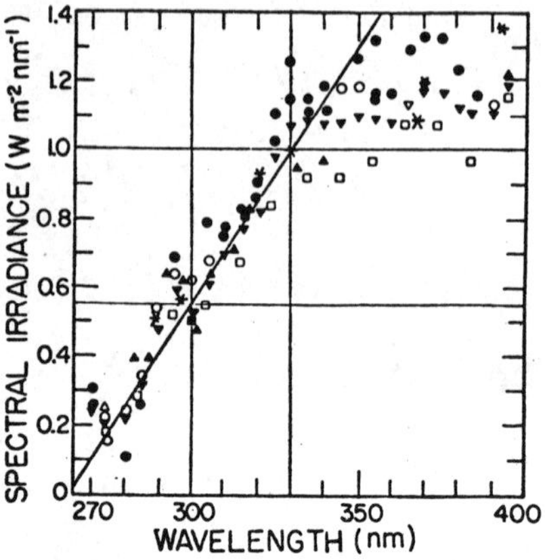

FIGURE B.1 Plot of reduced measurement points from pertinent references. Key: Eppley/JPL, *; NRL, ○; NCAR, ●; GSFC/MUSE: BUV, ▲; NASA 1971, ▼; Labs and Neckel, □. (From CIAP Monograph 5, Draft, p. 228.)

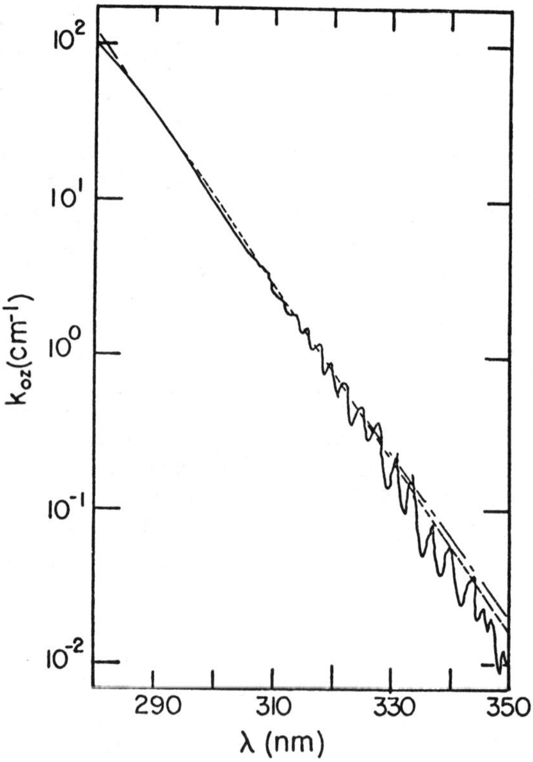

FIGURE B.2 Ozone absorption coefficient. The solid curve is the data of E. C. Y. Inn and Y. Tanaka, *J. Opt. Soc. Am. 43,* 870 (1953); the dashed line is the representation by C. Fabry and H. Buisson, *Compt. Rend. 192,* 457 (1931); and the dash–dot line is the representation $k_o \exp[-(\lambda-\lambda_o)/d_o]$.

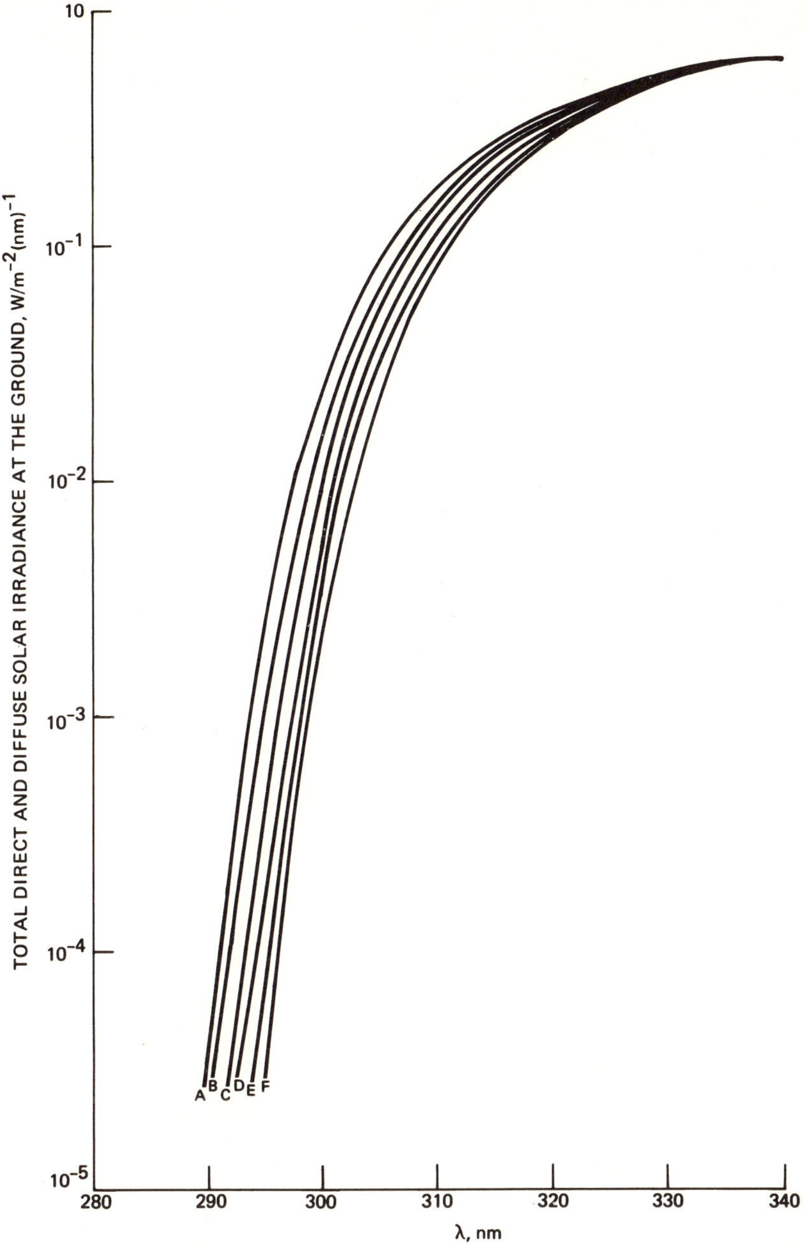

FIGURE B.3 Absorption of solar uv radiation by ozone for solar elevation 60° and a specified model atmosphere. A, 0.20 cm O_3; B, 0.24 cm O_3; C, 0.28 cm O_3; D, 0.32 cm O_3; E, 0.36 cm O_3; F, 0.40 cm O_3.

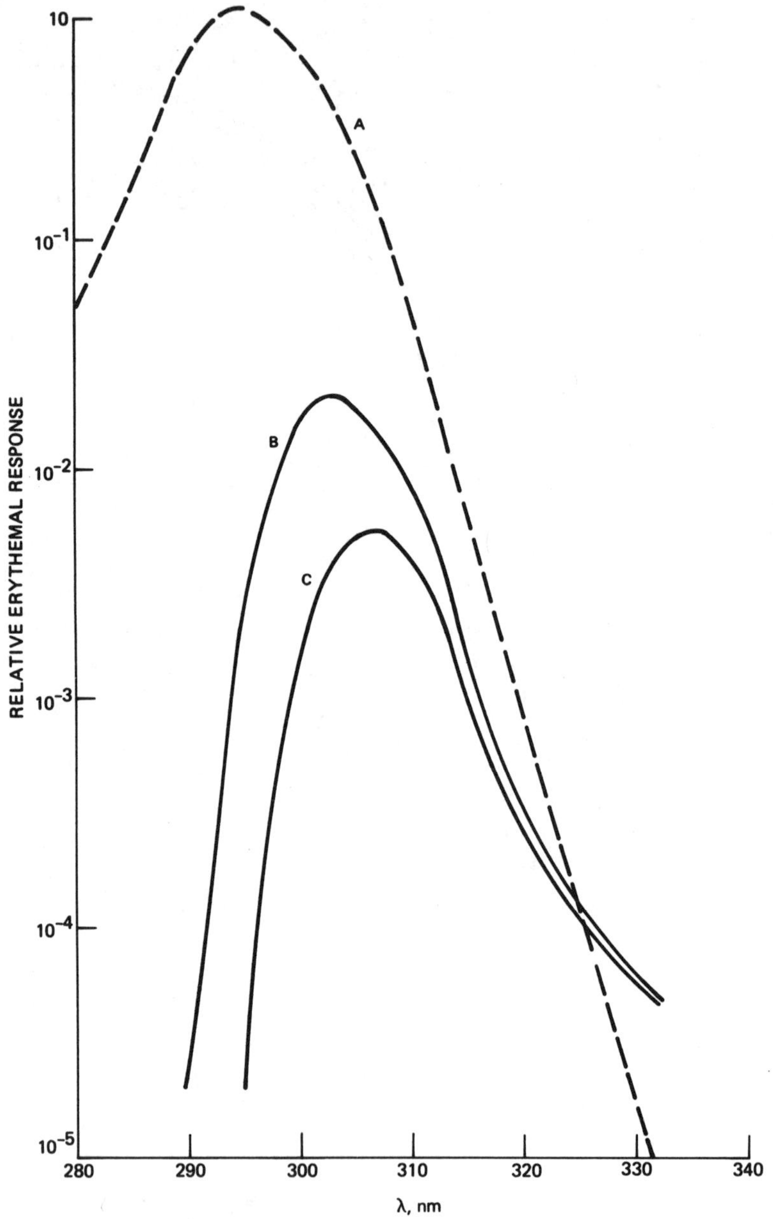

FIGURE B.4 Relative erythemal efficiency of solar radiation at the ground for different atmospheric ozone content. A, relative erythemal response; B, erythemal efficiency of radiation, O_3, 0.20 cm; C, erythemal efficiency of radiation, O_3, 0.40 cm. (Standard atmosphere, solar elevation 60°.)

diffuse radiation that impinges on a surface near the ground is well understood in principle, although difficult to perform in practice. There are few experimental data with which to verify these computed fluxes. Observations have shown that large local variations exist in uv irradiance at the surface. However, it is felt that relative changes in the uv flux at the surface, due to relative changes in the stratospheric ozone column, can be calculated with some precision, as any concomitant change due to scattering or reflection or the like will closely follow the primary change.

In order to determine the medical impact of a change in ozone amount, these solar fluxes must be weighted with a suitable response function for human skin. (A skin-cancer response curve would be preferable, but, lacking this, the erythemal response curve or a DNA response curve may be used.) For example, the erythemal data that have been used by many workers are those of Coblentz and Stair [*J. Res. NBS* 12:13 (1934)], and their results are shown as the dotted curve in Figure B.4. Figure B.4 also shows the results of folding the erythemal response of Coblentz and Stair with the solar flux spectrum at the earth for total columns of ozone of 0.2 and 0.4 cm, a range of ± 30 percent around the mean value of 0.3 cm. The curves in Figure B.4 peak at about 305 nm, and thus the changes in solar flux at this wavelength tend to dominate the overall change in the integral under the total curve. Table B.1 presents results, taken from the absolute calculations given in the CIAP monographs, of the relative change in this integral versus a 10 percent change in the total ozone column. To a close degree of approximation, the percentage change of erythemal ultraviolet is 2.5 times the percentage decrease of ozone during those months when erythemal radiation is appreciable.

There are two approaches in use for determining the flux of uv-B and its changes at different places and times. The direct method uses an instrument—the Robertson meter—filtered to have a wavelength response similar to the erythemal curve. The transmission of the Robertson filter does not decrease with increasing wavelength as rapidly as

TABLE B.1 Perentage Increase in Erythemal Dose for a 10 Percent Decrease in Ozone Column

	Month			
Latitude	Jan.	Apr.	July	Oct.
30° N	27	25	24	24
50° N	46	25	24	26

the erythemal sensitivity function. A recommended alternative is to compute the uv-B flux from the known properties of ozone and the amounts measured or interpolated from the existing observational network. The effect of clouds and scattering in the lower atmosphere must then be included. Individual values obtained in this way are not particularly accurate; but long-term means and trends, which are the important things for medical and biological problems, will be more precise than those obtained from flux measurements. In particular, the *changes* in uv-B from a given *change* of mean ozone, all other factors being equal, can be computed accurately. The computational approach to the flux of uv-B has been used in the estimates of the effects of ozone reduction on skin cancer among light-skinned people. These estimates used a response curve intermediate between that for erythema and for DNA action (see Appendix C).

Using a similar method, Setlow [see R. B. Setlow, *Proc. Nat. Acad. Sci. U.S.* 71:3363–3366 (1974)] has obtained values ranging from 2.0 to 2.3 using the DNA absorption spectrum as the action spectrum and values from 1.7 to 1.9 using the erythemal action spectrum. A value of 2 is used elsewhere in this report to estimate percentage increases in biologically effective radiation from percentage ozone decreases.

Check this by observation
i.e. δ(EER) vs δ(~~total~~ stratospheric O_3 column)

Appendix C

Estimates of Increases in Skin Cancer Due to Increases in Ultraviolet Radiation Caused by Reducing Stratospheric Ozone

I. Summary

Following a brief introduction, the evidence for the role of ultraviolet (uv) radiation as a causal factor for skin cancer in man—for malignant melanoma and for nonmelanoma skin cancer—is reviewed in Section III. We conclude that the evidence for uv radiation as a causal factor exists for malignant melanoma and is strong for nonmelanoma skin cancer. For both kinds of skin cancer there is a clear association of increasing incidence with the increase in intensity of uv radiation that occurs with decreasing latitude.

The increase in uv radiation that will result from decreasing stratospheric ozone is known from other studies. We use the empirical association between skin-cancer incidence (and mortality) and intensity of uv radiation to estimate the percentage increase in skin cancer that

---

This Appendix is an Interim Report of the Panel to Review Statistics on Skin Cancer of the NRC Committee on National Statistics. The Final Report of the Panel will be issued when further data arrive. The members of the Panel are *Chairman:* Elizabeth L. Scott, Department of Statistics, University of California, Berkeley; Peter Bloomfield, Department of Statistics, Princeton University; Philip Cole, Department of Epidemiology, Harvard School of Public Health; Richard A. Craig, Department of Meteorology, Florida State University; Harley A. Haynes, Department of Dermatology, Harvard Medical School; Richard B. Setlow, Department of Biology, Brookhaven National Laboratory. *Staff:* Miron L. Straf, Research Associate; Theodore D. Woolsey, Consultant.

would result from the increase in uv radiation due to a specified reduction of ozone. The simple model estimates the logarithm of incidence rate (or of mortality) as a linear function of the intensity of uv radiation. We consider the difficulties inherent in the simple model to be chiefly that skin cancer is underreported; that the incidence of skin cancer, and especially of melanoma, is increasing over the years and there is good evidence that these increases are due to changing life-styles, such as more exposure of women's legs; that incidence data are available for only a few localities in the United States; that the precise wavelengths of uv radiation that cause skin cancer and their relative importance are not known, and although there is reason to believe that the action spectrum of erythema (sunburn) or of DNA damage may be appropriate (most of the results reported here refer to an action spectrum intermediate between these two), there is also evidence that the most important wavelengths are even shorter, which would lead to even higher predicted increases in skin cancer due to ozone reduction due to the sharp wavelength dependence of the ozone shield; and that the consequences of increased intensities of uv radiation cannot be measured just by comparing the rates of skin cancer in different localities, since not only will the intensity of uv radiation depend on locale but so will other factors such as clothing. These difficulties affect the estimate of the increase in skin cancer resulting from a reduction in ozone, and we have tried to take them into account in more realistic, and thus more complicated, models. Rates from more localities will soon be released from a just completed survey; these are needed to verify the more complicated models.

Nevertheless, we can already obtain estimates of the percentage increase in skin cancer that would result from a reduction in the ozone shield. The estimates presented here are conservative in the sense that, under other assumptions than we have used, the predicted increases would be larger. Averaged over the United States, the effect of a 10 percent decrease in ozone will be about a 30 percent increase in the incidence of skin cancer, both melanoma and nonmelanoma, and about a 20 percent increase in malignant melanoma mortality. If the ozone reduction is more pronounced, the resulting percentage increases in skin cancer and in melanoma are much more striking, increasing exponentially.

II. Introduction

The Committee on National Statistics appointed the Panel to Review Statistics on Skin Cancer at the suggestion of the National Cancer Institute and of the Climatic Impact Committee of the National Research Council. Two charges were made to our panel:

1. to study the two surveys of skin cancer made by the Third National Cancer Survey of the National Cancer Institute and by the Department of Epidemiology of the M. D. Anderson Hospital and Tumor Institute, particularly to investigate apparent discrepancies between their reported incidences of skin cancer, and to recommend methods to improve the collection of data on skin cancer; and
2. to study the effect on the incidence on skin cancer of increased intensities of ultraviolet (uv) radiation that may result from ozone reduction caused by emissions from high-altitude aircraft. This Appendix covers the latter study.

The contemplated mechanism is that aircraft emissions in the stratosphere may decrease there the ambient ozone that protects all living things from uv radiation. In order to predict what increases in skin cancer would result from assumed relative reductions in stratospheric ozone, we need a quantitative description of the relationship between intensities of uv radiation and reported incidence, prevalence, and mortality of skin cancer, both malignant melanoma and other skin cancer. This relation could be used to predict effects on skin cancer of any potential changes in the natural protection from uv radiation provided by stratospheric ozone. In addition to emissions by high-altitude aircraft, man-made chlorofluoromethanes used as aerosol propellants and refrigerants, multiple nuclear detonations, and propellants proposed for the Space Shuttle are examples of sources of substances that can reduce stratospheric ozone.

Our predictions are primarily of the effects of reductions of stratospheric ozone on increases in skin cancer among whites, as skin cancer in which uv radiation appears to be a causal factor has a very low incidence among dark-skinned races. Moreover, we have not studied the many other skin problems that appear to increase with increases in intensities of and exposure to uv radiation. These other problems include skin aging, acute sunburn, actinic keratoses, keratoacanthomas, and other skin manifestations. In addition, there are effects of increased intensities of uv radiation on climate, ecology, and agriculture.

III. Skin Cancer in Man: The Evidence for the Role of Ultraviolet Radiation as a Causal Factor

A. FACTORS IN THE RELATIONSHIP OF SUNLIGHT TO SKIN CANCER IN MAN

Classical evidence in support of the role of sunlight, particularly of ultraviolet (uv) radiation, as a causal factor in human skin cancer comprises a number of factors (Emmett, 1973; Urbach, Rose, and Bonnem, 1972; Urbach, Berger, Robertson, and Davies, 1974). Most of these factors may be summarized by six associations of skin cancer:

1. *Association with exposed areas of the skin.* Among Caucasians, skin cancers occur most frequently on parts of the body most exposed to sunlight—the head, neck, arms, and hands and the legs of women.

2. *Association with protection against ultraviolet radiation.* Among races with dark skin, in which pigment filters uv radiation, there is very little skin cancer and the disease does not predominantly occur in areas of the skin exposed to the sun. Sunburn and skin cancer arise in the same tissue, and uv radiation is known to cause sunburn. It appears that those who are more susceptible to skin cancer sunburn more easily. Caucasians of Celtic origin are more susceptible to both skin cancer and sunburn; those of Spanish origin, less. Lotions that provide protection from uv radiation appear to inhibit development of solar keratoses, which are often precursors of skin cancer.

3. *Association with the amount of exposure to the sun.* Among Caucasians, there appears to be a greater prevalence of skin cancer among those who spend more time outdoors.

4. *Association with intensity of solar insolation.* As one approaches the equator, measures of incidence of skin cancer among white-skinned people increase, as do the amount of solar radiation and intensity of uv radiation.

5. *Association of ultraviolet radiation in laboratory studies.* Skin cancer can be produced in mice with repeated doses of uv radiation in the same spectral range that produces sunburn in human skin.

6. *Association with insufficient ability to repair DNA damaged by ultraviolet radiation.* Those with the recessively inherited disease xeroderma pigmentosum, who consequently have a defect in the repair of DNA damaged by uv radiation, have a predisposition to the development of skin cancer. Such victims are often photosensitive, and their tumors appear to be induced by sun exposure. They usually die as a result of skin malignancies before reaching adult life.

B. VARIETIES OF SKIN CANCER

We briefly discuss the two varieties of skin cancer, malignant melanoma and nonmelanoma skin cancer, and some factors of their relationship to uv radiation.

Malignant melanoma is the more virulent but relatively rare form of skin cancer. The disease is fatal, with a median survival time of about 7 years (Axtell, Cutler, and Myers, 1972). Over half of those afflicted are under 55 years of age (Axtell, Cutler, and Myers, 1972; National Cancer Institute, Third National Cancer Survey Advanced Three Year Report, 1974). The average annual age-adjusted number of reported new cases per 100,000 white population among metropolitan areas of the United States reported by the Third National Cancer Survey (National Cancer Institute, Third National Cancer Survey Advanced Three Year Report, 1974) varies from three in the north to eight in the southwest.

There are a number of reasons to infer an association with sunlight from the pattern of occurrence of malignant melanoma. First, a number of published site distributions show a proportionately large occurrence of melanomas on the upper back in males and the lower legs in females, which are sites that may be uncovered or covered only by light translucent clothing during periods of greater solar insolation (Beardmore, 1972; Davis, Herron, and McLeod, 1966; Lee and Merrill, 1970; Lee and Yongchaiyudha, 1971; Magnus, 1973; Melanoma Clinical Cooperative Group, 1974). The difference between patterns of occurrence for each sex is most striking for these sites, as is shown in Figure C.1. Second, the more exposed sites (head, neck, arms, and hands), which do not constitute much of the total area of the skin, have the greatest relative density of occurrence per unit area of skin. Third, melanomas rarely arise on the buttocks or on genitals—skin that is unexposed to sunlight. Part of the occurrence of melanoma on the trunk and upper extremities may be due to the fact that many melanomas arise in a pre-existing nevus ("black mole"), and these skin nevi are found most frequently there (Macdonald, McGuffee, and White, 1973; McGovern, 1972). Fourth, there is a type of melanoma that arises in lentigo maligna, an apparently precancerous lesion analogous to a solar keratosis; and this type occurs predominantly on exposed areas of the skin (Clark, From, Bernadino, and Mihm, 1969; Clark and Mihm, 1969; Ward, 1967). From 8 to 15 percent of cases of malignant melanoma of the skin reported in hospital studies (Clark, From, Bernadino, and Mihm, 1969; McGovern, 1972; Melanoma Clinical Cooperative Group, 1974) are classified as lentigo maligna melanoma. Fifth, from an analysis of

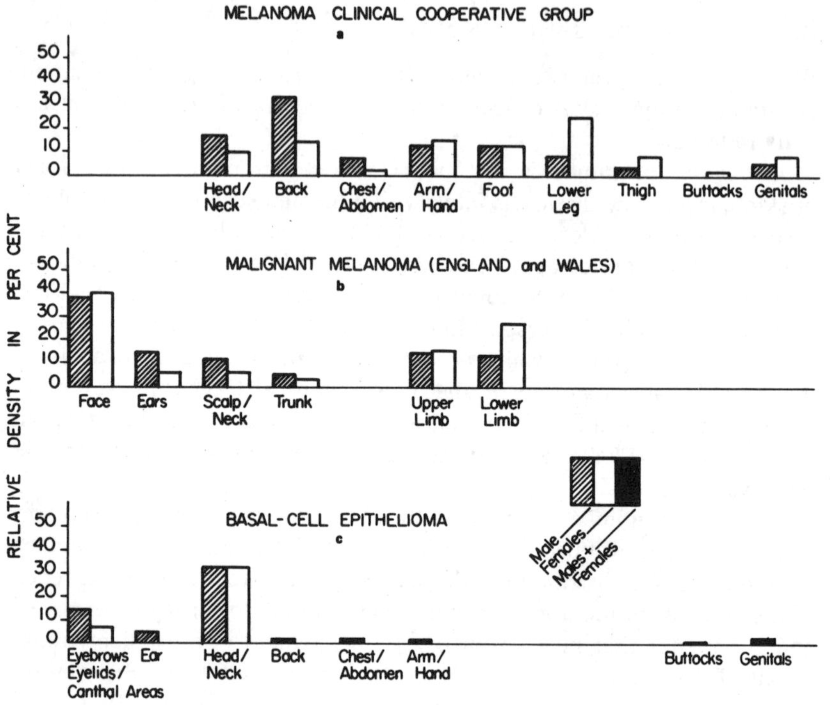

FIGURE C.1 Relative density of skin cancer per unit area of skin. (a) Relative density of malignant melanoma for 182 males and 187 females diagnosed between September 1, 1972, and August 1, 1974 (Melanoma Clinical Cooperative Group, 1974). (b) Relative density of malignant melanoma of 315 males and 549 females in England and Wales (Lee and Yongchaiyudha, 1971). (c) Relative density of 840 histologically diagnosed basal-cell skin cancers (Brodkin, Kopf, and Andrade, 1969).

melanoma incidence in Norway, Magnus (1973) shows that, for each age group, cohorts born more recently have a markedly greater incidence of melanoma of the neck, trunk, and lower limb than cohorts of earlier generations. Such an increased incidence of melanoma on the face and foot is only slight. This result may be attributed to the increased exposure to sunlight of the former sites compared with the latter ones among more recent generations. What is more, this is evidence that the observed increase is not due to a systematic change in diagnosis. Magnus further shows that the difference in the frequencies of occurrence on the lower limb between males and females increases with decreasing latitude.

In an analysis of melanoma in Israel, which has intense periods of sunshine and is located on the same latitude as central Texas, Movshovitz and Modan (1973) show that, among those of European extraction, there is a high incidence of melanoma among the native-born, an intermediate rate among the veteran foreign-born, and a lower rate among more recent European-born immigrants.

Melanoma is very rare in heavily pigmented populations. Further, melanomas occur early in patients with xeroderma pigmentosum (Moore and Iverson, 1954; Lynch, Anderson, Smith, Howell, and Krush, 1967; Robbins, Kraemer, Lutzner, Festoff, and Coon, 1974), indicating an association with lack of defense from uv radiation.

In laboratory experiments with mice, malignant melanoma has been produced only by applying uv radiation in conjunction with chemical skin applications that are benign in themselves (Epstein, Epstein, and Nakai, 1967).

Strong evidence of the role of sunlight in the etiology of malignant melanoma is the clear relationship of changes in melanoma incidence and also mortality with changes in latitude of location. The lower half of Figure C.2 shows plots and regression lines for melanoma incidence and mortality with common slopes for each sex. These rates are reported average annual age-adjusted incidence of melanoma from the Third National Cancer Survey (National Cancer Institute, Third National Cancer Survey Advanced Three Year Report, 1974) and average (1950–1969) annual age-adjusted rates of reported mortality of melanoma (Mason and McKay, 1974). The factors associated with latitude may be many, but, in light of the evidence discussed above, it is reasonable to assume that the increase in intensities of uv radiation with decreasing latitude is a causal factor. In another section, we present our analysis of the relationship of incidence and mortality rates of malignant melanoma of the skin to uv radiation.

Nonmelanoma skin cancer is the most common of all cancers in humans. It principally occurs as either basal-cell skin cancer, the most common type, or squamous-cell skin cancer. Most cases that are seen and treated are done so in physicians' offices and do not require hospitalization. If treated early, the disease is rarely fatal, but squamous-cell skin cancer may infrequently metastasize to other parts of the body and cause death.

Most published reports show that nonmelanoma skin cancer occurs principally on the exposed areas of the skin; for the most part, on the head and neck, as shown in Figure C.1 (Brodkin, Kopf, and Andrade, 1969; Gordon, Silverstone, and Smithurst, 1972; Silverstone and Gordon, 1966; Swanbeck and Hillström, 1971; Urbach, 1969, 1971). Among skin cancers that develop on the head and neck, squamous-cell

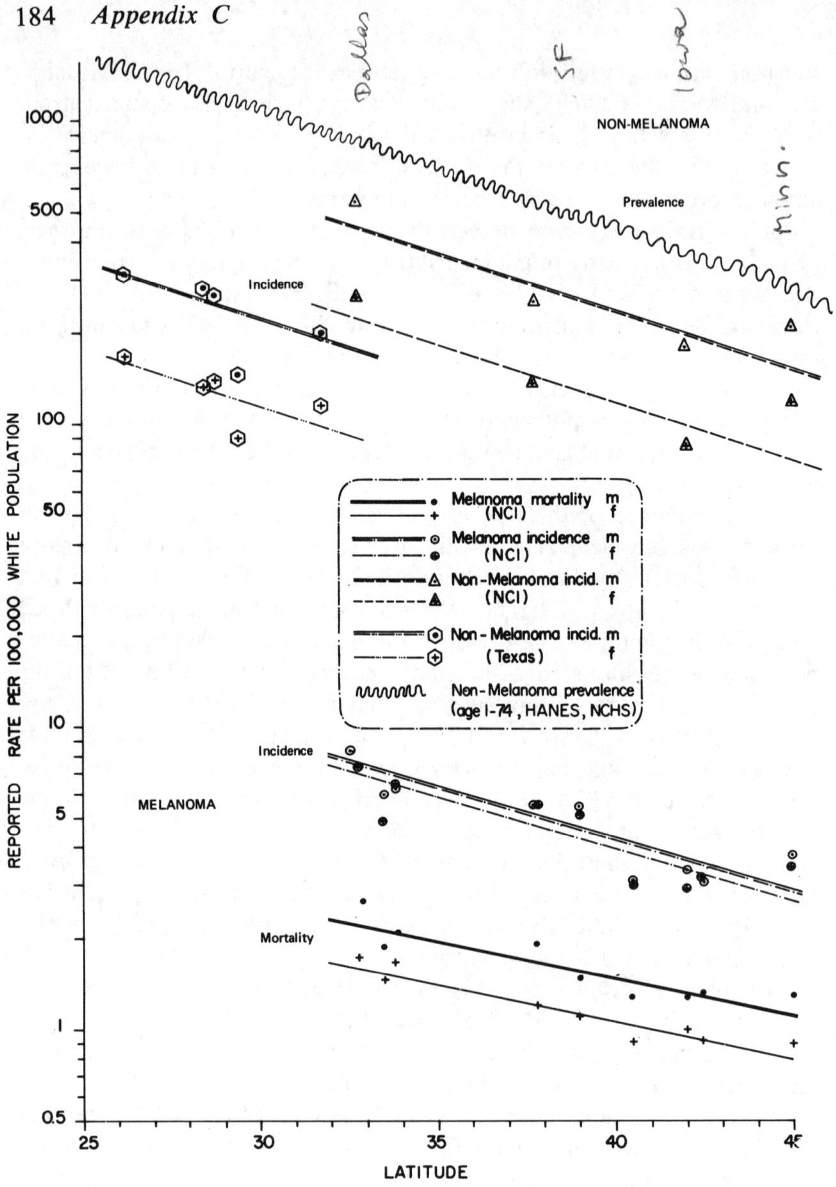

FIGURE C.2 Reported skin-cancer rates among whites as a function of latitude. [Sources: melanoma mortality from Mason and McKay (1974), melanoma incidence from National Cancer Institute (1974), nonmelanoma skin-cancer incidence (NCI) from Scotto, Kopf, and Urbach (1974), nonmelanoma skin-cancer incidence (Texas) from Macdonald (1974a), and prevalence of nonmelanoma skin cancer based on preliminary data from the Health and Nutrition Examination Survey of the National Center for Health Statistics (McDowell, 1974).]

skin cancers occur primarily on sites that are most exposed to radiation, whereas about one third of those basal-cell skin cancers occur on areas normally slightly exposed to radiation (Urbach, 1969; Urbach, 1971). Figures C.3(a) and C.3(b) show the site distributions of 840 basal-cell skin cancers diagnosed histologically in the study of Brodkin, Kopf, and

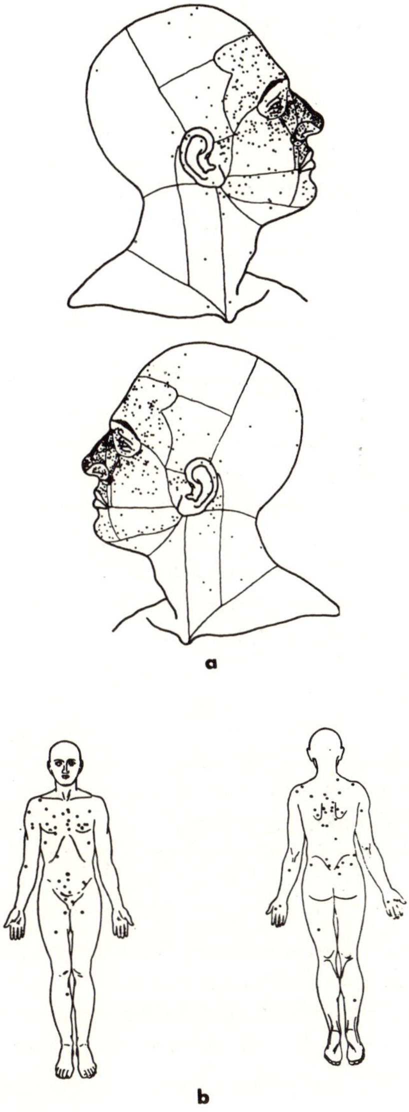

FIGURE C.3 (a) Location of basal-cell skin cancers of the head and neck from Brodkin, Kopf, and Andrade (1969). Of 840 basal-cell skin cancers in the study, all but 74, or 8.8 percent, occurred on the head and neck. (b) Location of the 74 basal-cell skin cancers of the torso and extremities from Brodkin, Kopf, and Andrade (1969).

Andrade (1969). All but 74, or 8.8 percent, of these cancers occurred on the head and neck.

Nonmelanoma skin cancer occurs more frequently among those who spend more time out of doors (Atkin, Fenning Heady, Kennaway, and Kennaway, 1949; Gellin, Kopf, and Garfinkel, 1969; Urbach, Rose, and Bonnem, 1972), which may account for the fact that observed incidence is higher in males than in females (Scotto, Kopf, and Urbach, 1974).

The disease is rare in heavily pigmented races; when it does occur, the distribution of squamous-cell skin cancers is not on exposed areas, but that of basal-cell skin cancers is (Emmett, 1973). Moreover, squamous-cell skin cancer is very much more common among albino members of dark-skinned races than among their counterparts (Higginsen and Oettlé, 1960; McGovern and Lane Brown, 1969; Oettlé, 1963, 1964). In addition, nonmelanoma skin cancer occurs early in patients with xeroderma pigmentosum, who have an insufficient ability to repair DNA damaged by uv radiation (Cleaver, 1970; Moore and Iverson, 1954; Reed, Landing, Sugarman, Cleaver, and Melnyk, 1969; Robbins, Kraemer, Lutzner, Festoff, and Coon, 1974).

In laboratory experiments with mice, squamous-cell skin cancer but not basal-cell has been induced with uv radiation (Blum, 1959, 1974; Emmett, 1973). Blum has conducted myriad laboratory experiments providing many details of the experimental induction of cancer by uv radiation. Further, there is evidence that a chemical that is formed *in vitro* in human skin on exposure to uv radiation is a carcinogen in experimental animals (Black and Lo, 1971) and that the formation of this chemical *in vivo* in the skin of mice exposed to uv radiation increases with the dose of radiation prior to the appearance of squamous-cell skin cancer (Black and Douglas, 1973).

An increasing incidence of nonmelanoma skin cancer is associated with a decreasing latitude of location and, thus, with increasing intensities of uv radiation. The upper half of Figure C.2 shows reported average annual age-adjusted incidence rates of nonmelanoma skin cancer from four areas of the Third National Cancer Survey (Scotto, Kopf, and Urbach, 1974) and from a survey of five regions in Texas by the Department of Epidemiology of the M. D. Anderson Hospital and Tumor Institute (Macdonald, 1974a) with respect to average latitude. Regression lines, all with a common slope, are shown for the reported incidence and for estimated prevalence of nonmelanoma skin cancer based on preliminary data from the Health and Nutrition Examination Survey (HANES) of the National Center for Health Statistics (McDowell, 1974). Because skin cancer is grossly underreported, we consider regression lines for each survey separately but fitted with a common slope. Mortality rates are not shown because, not only is mortality from nonmelanoma skin cancer unusual, but, as we shall discuss later, it reflects the standard of medical care and is confounded by errors of classification on death certificates. The relationship of incidence with latitude is corroborated by other studies (Auerbach, 1961; Ringertz, 1970; Gordon, Silverstone, and Smithurst, 1972).

We conclude that the evidence for uv radiation as a causal factor in nonmelanoma skin cancer is strong. In another section, we shall present our analysis of the relationship of reported incidence and prevalence rates of nonmelanoma skin cancer to uv radiation.

IV. Estimation of Effect of Ozone Reduction on Increase in Skin Cancer among Whites

A. THE USE OF AN EMPIRICAL RELATION FOR PREDICTIONS

As described earlier, the observed incidence of skin cancer among whites increases with an approach toward the equator as does the intensity of ultraviolet (uv) radiation. The latter increases partly because of the shorter path of sunlight through the ozone and partly because of the smaller amount of ozone at latitudes closer to the equator. These facts suggest an empirical study of the relations between the observed skin-cancer incidence (or mortality) among whites and the intensity of uv radiation for a number of locations. We know how ozone reduction increases the intensity of uv radiation; if we also knew the empirical association between skin-cancer incidence (or mortality) and uv radiation, we could then estimate the percentage increase in skin cancer that would result from a reduction of ozone. Justification for using an empirical association for predictions is the laboratory and clinical evidence of a causal relation between uv radiation and skin cancer.

B. SOME PROBLEMS

There are, however, many problems in establishing a quantitative relationship that can be used for such predictions. One important problem is that the action spectrum that gives the relative effectiveness of various wavelengths of uv radiation in producing skin cancer in man is unknown, both for malignant melanoma and other skin cancers. Sunburn, which is a readily observable damage to the skin of man, appears to come in the natural environment from solar uv radiation at wavelengths in the narrow band from 290 to 320 nanometers (nm). Ultraviolet radiation

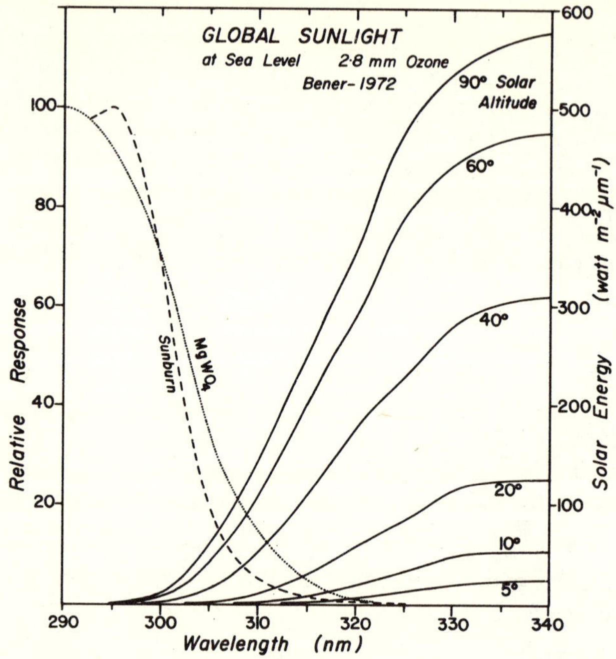

Calculated FIGURE C.4 Intensity of global (direct plus diffuse) solar ultraviolet radiation on a horizontal surface at sea level for different wavelengths and solar angles determined by Bener (1972) from Robertson (1974). Also shown are an erythema (sunburn) action spectrum of Berger, Urbach, and Davies and a magnesium tungstate phosphor action spectrum (Urbach, Berger, Robertson, and Davies, 1974).

at wavelengths below 290 nm is effectively absorbed by the atmosphere with present levels of ozone (Gates, 1966; Koller, 1965). Figure C.4 from Robertson (1974), which is based on calculations of Bener (1972) shows the intensity of direct solar uv radiation at sea level as a function of wavelength and solar angle. Absorption of uv radiation by the atmosphere is greater at shorter wavelengths; as the sun approaches the horizon, wavelengths shorter than 295 nm are reduced to negligible intensity. At wavelengths above 320 nm, uv radiation does not appear to cause sunburn in man (Robertson, 1968). Moreover, evidence from experimental studies of mice shows that uv radiation at wavelengths longer than 320 nm are ineffective in inducing skin cancer (Blum, 1959; Urbach, Rose, and Bonnem, 1972; Freeman, Hudson, and Carnes,

1970; Freeman, 1974), although exposure to longer wavelengths may accentuate the effects of shorter ones (Forbes, 1973).

An erythema action spectrum is commonly used as the putative skin-cancer action spectrum. Evidence from experimental studies of mice is incomplete but not in disagreement with this use (Freeman, Hudson, and Carnes, 1970; Freeman, 1974). Skin cancer and erythema arise in the same tissue, and those who develop skin cancer are noticed to have a higher than average susceptibility to sunburn (Urbach, Rose, and Bonnem, 1972; Gellin, Kopf, and Garfinkel, 1969). Within the band 290 to 320 nm, the erythema action spectrum is similar to that of damage to DNA by absorption of UV radiation, and there are arguments for using the latter as a skin-cancer action spectrum (Setlow, 1974). There is much uncertainty, however, in the determination of an erythema action spectrum: considerable variation exists among individuals tested and between average results of different experimenters. Figure C.5 from Robertson (1968) shows individual variations in erythema action spectra determined by Magnus (1964), and Figure C.6 from Robertson (1968) shows erythema action spectra of three other investigations (Coblentz and Stair, 1934; Everett, Olson, and Sayre, 1965; Freeman, Owens, Knox, and Hudson, 1966) with an average of those obtained by Magnus. In spite of this uncertainty, there appears to be agreement that a threshold to the response of erythema exists at about 320 nm

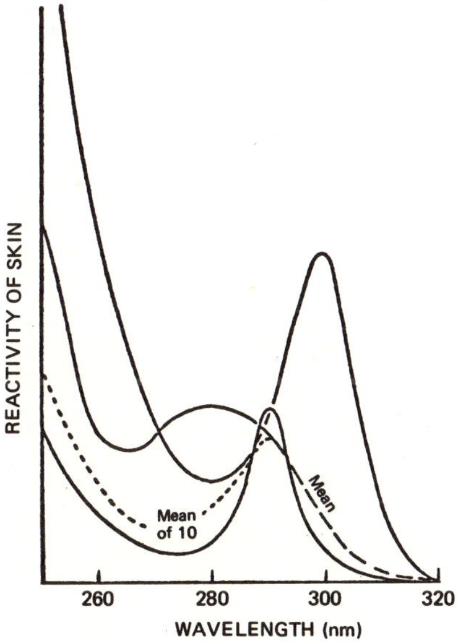

FIGURE C.5 Examples of individual variation in erythema action spectrum determined by Magnus (1964) from Robertson (1968).

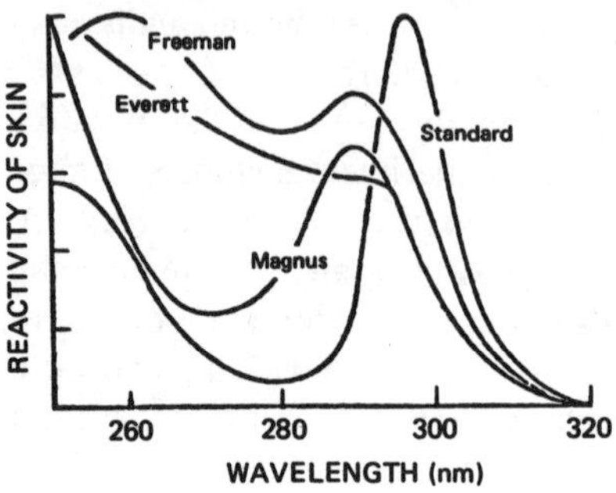

FIGURE C.6 Various determined erythema action spectra: Everett, Olson, and Sayre (1965); Freeman, Owens, Knox, and Hudson (1966); Magnus (1964); and an early accepted standard (Coblentz and Stair, 1934) from Robertson (1968).

and that uv radiation at wavelengths within the range 290 to 300 nm is relatively more effective than that at wavelengths longer than 300 nm (Emmett, 1973; Robertson, 1968; Urbach, Berger, Robertson, and Davies, 1974). There is also agreement that uv radiation at wavelengths within 250 to 260 nm has a large relative effectiveness in inducing erythema. Tumors have occurred in mice exposed to uv radiation at wavelengths in this range (Blum and Lippincott, 1942). These shorter wavelengths, however, are more effectively attenuated by white skin (Kirby-Smith, Blum, and Grady, 1942).

In our analysis, we have used a number of weighted averages of intensities of uv radiation at four wavelengths within 296 to 312 nm. In one of these weighted averages, the weights correspond to an action spectrum intermediate to one of erythema and one for affecting DNA. Ultraviolet radiation at wavelengths shorter than those we have used, however, may be important in predicting increases in skin cancer associated with ozone reduction, even though such radiation reaches the surface of the earth at lower intensities. The effect of ozone reduction is to increase the intensity of radiation at a shorter wavelength relative to the variation in intensity over the United States much more than that of a longer wavelength. (Loosely speaking, a given percentage of ozone reduction in an area has the effect of increasing the intensity of uv radiation to that present in another area toward the south. For shorter wavelengths, such a comparable area would be further south than one for longer wavelengths.) Thus, if shorter wavelengths than those we have used in establishing a relationship between uv radiation and skin-cancer incidence or mortality are effective in inducing skin cancer in a natural environment or if the relative carcinogenic effectiveness of the shorter wavelengths that we have used is greater than that attributed to them, then estimates of relative increases in skin cancer associated with ozone reduction would be higher than those we obtain from the relationship. What is more, uv radiation at shorter wavelengths, which, in the natural environment, are reduced to negligible intensities by ozone absorption and transmission through skin, may induce skin cancer if present in nonnegligible amounts as a result of ozone reduction. An action spectrum for affecting DNA (Setlow, 1974) shows an increasing relative effectiveness as wavelengths decrease below 290 nm; whereas, an erythema action spectrum (Emmett, 1973; Robertson, 1968) shows a decreasing relative effectiveness. There are also problems with wavelengths longer than 320 nm. Although such wavelengths are ineffective in inducing skin cancer in mice, they may accentuate the effects of shorter wavelengths (Forbes, 1973).

A second problem is that the form of the relationship between uv radiation and skin-cancer incidence (or mortality) is unknown for malignant melanoma and also other skin cancer. What may be relevant is not the simple total intensity of uv radiation but a measure that gives more weight to strong intensities. Intervals between such exposures may also be important. Moreover, the effect of an increased intensity of uv radiation may be not only to increase the number of those who have incurred skin cancer and the frequency of reported cases of skin cancer but also to make those who are susceptible incur the disease at an earlier age. The latter effect, which would be expected if effects of doses of uv radiation on man were cumulative, also results in increased age-adjusted rates. From evidence of experimental studies on mice, Blum (1959) concluded that the effect of uv radiation is without threshold and is cumulative.

The relationship between uv radiation and skin-cancer incidence may involve other physical factors. There are several that vary with location and that may interact with uv radiation in inducing skin cancer. In addition to radiation at wavelengths that in themselves may not be carcinogenic, these include temperature, wind, humidity, and pollutants. For some of these, there is evidence of an interaction in inducing skin damage (Bain, Rusch, and Kline, 1943; Freeman and Knox, 1964; Owens, Knox, Hudson, and Troll, 1974). In our linear models for analysis of transformations of age-adjusted and age-specific rates of skin cancer, we have included variables for average intensity of uv radiation, long-wavelength solar radiation, temperature–humidity indices, and some interaction effects.

A third problem is that the available data needed to make predictions are inadequate. To begin with, nonmelanoma skin cancer is grossly underreported. In a probability sample of the Health and Nutrition Examination Survey (HANES) of the National Center for Health Statistics, respondents were asked to report whether a case of skin cancer had been diagnosed previously but was suppressed or in remission. A population prevalence estimate based on those in the sample who were examined and found to have microscopically confirmed skin cancer, however, is about 8 times as much as an estimate based on those so reporting a previous diagnosis (McDowell, 1974).

Local attitudes toward treatment affect the number of reported incidences. In a survey of skin-cancer prevalence based on a probability sample of households of a predominantly rural county in Tennessee, a followup after a year revealed that only 17 of the 43 persons with clinical diagnoses of skin cancer by a dermatologist were known to have visited their doctors' offices specifically to inquire about their skin lesions after being advised to do so (Zagula-Mally, Rosenberg, and Kashgarian, 1974).

Even if a nonmelanoma skin cancer is clinically diagnosed, a not unusual approach is to treat the lesion without histologic examination. Such lesions may not be reported. Even if special efforts are made to find these cases from doctors' records, there are difficulties in making the survey complete. Further, because of the lack of histologic proof of cancer, the cases obtained will be a confusion of those of skin cancer and those of benign growths that are not skin cancer. Data on incidence of malignant melanoma of the skin are of much better quality than those of nonmelanoma skin cancer because of more complete reporting and a higher percentage of histologic proof.

In addition to the incomplete reporting of the disease, there may be a seasonal variation in the number of those who seek treatment (Owen, Fliegelman, Jetton, and Musgrave, 1974). Patients seeking treatment more often during the summer months would also contribute to underestimation of annual incidence from the six-month skin-cancer supplement of the Third National Cancer Survey, which is one of our sources of data.

Another inadequacy of available data is that there are too few observations, particularly of nonmelanoma skin cancer, that are obtained with a common methodology for which an association with uv radiation is not obscured by variability due to other factors. There has been relatively little attention given to collecting statistics on nonmelanoma skin cancer; consequently, morbidity data are available for very few areas. Most cancer registries exclude nonmelanoma skin cancer; and, in those that include this disease, reporting is often acknowledged to be poor. It is fruitless to pool morbidity data of many countries into a common analysis in order to make predictions. Scatterplots of reported incidence rates both of malignant melanoma and nonmelanoma skin cancer from selected registries in the compendium *Cancer Incidence in Five Continents* (Doll, Muir, and Waterhouse, 1970) versus average latitude revealed much variability. Countries differ in genetic background and ethnic composition of their populations; the quality of medical care and type of health care delivery systems; and culture, lifestyle, principal occupations, and other aspects related to the amount of exposure to sunlight. Countries differ also in what they choose to record as skin cancer. All of these differences contribute to the observed variability. We also examined some reports of surveys in Australia (Beardmore, 1972; Carmichael, 1962; Carmichael and Silverstone, 1961; Davis, Herron, and McLeod, 1966; Gordon and Silverstone, 1973; Gordon, Silverstone, and Smithurst, 1972; Herron, 1969; Silverstone, Campbell, Hosking, Lang, and Richardson, 1963; Silverstone and Gordon, 1966; Silverstone and Searle, 1970). Although the pattern corroborates that of increasing incidence toward the equator, it is difficult to incorporate such reported rates with those of other countries or to compare relationships to those obtained with data from other countries because of different definitions and imprecise methodology of some studies in Australia. Reported rates from one study (Carmichael and Silverstone, 1961; Gordon and Silverstone, 1973) appear to be the result of curve-fitting.

For incidence of nonmelanoma skin cancer among areas of the United States, data were available from two sources: the National Cancer Institute's six-month (September 1, 1971, to February 29, 1972) skin-cancer survey of four areas (Scotto, Kopf, and Urbach, 1974), which was a supplement to the Third National Cancer Survey, and a retrospective survey covering five years (1962–1966) of five regions in Texas including a continuing registry in El Paso under the direction of the Department of Epidemiology of the M. D. Anderson Hospital and Tumor Institute (Macdonald, 1974a). Even the reported incidence rates from these sources, however, cannot be combined into a common analysis. The different definitions used, the different methods employed to acquire sources and collect data, the different lengths of time frames and different periods of time, and other different aspects indicate that each survey must be studied separately.* Since the incidence of melanoma is increasing (Burbank, 1971; Axtell, Cutler, and Myers, 1972; Magnus, 1973), and, reportedly, that of nonmelanoma is also increasing, what is required for analysis are observations taken during the same time period as well as with a common methodology.

* A comparison of these two surveys with an assessment of possible effects of these differences in methodology is a separate study of the Panel to Review Statistics on Skin Cancer of the NRC Committee on National Statistics

Of these two sources of data on incidence of nonmelanoma skin cancer, only the skin-cancer supplement to the Third National Cancer Survey covers areas with sufficient variability in uv radiation to make predictions. This supplementary survey, however, covers only four areas: the State of Iowa and the Dallas–Fort Worth, Minneapolis–St. Paul, and San Francisco–Oakland standard metropolitan statistical areas. With so few observations, in which there may be errors of underreporting and seasonal variation, the form of the relationship of nonmelanoma skin-cancer incidence to uv radiation and the effects of many confounding factors remain unknown. All we could do was to use available data of reported skin-cancer incidence obtained in other manners from the Second National Cancer Survey (Auerbach, 1961; Dorn and Cutler, 1958), incidence of malignant melanoma of the skin from the Third National Cancer Survey (National Cancer Institute, Third National Cancer Survey Advanced Three Year Report, 1974), skin-cancer mortality (Mason and McKay, 1974), and prevalence of nonmelanoma skin cancer and moderate or severe solar keratosis from the Health and Nutrition Examination Survey (McDowell, 1974) in order to corroborate general models describing relationships of skin-cancer incidence, mortality, or prevalence rates to uv radiation. (For details of these sources, see the Addendum on The Variables.)

Although skin-cancer mortality data are available for almost every county in the United States (Mason and McKay, 1974), nonmelanoma skin-cancer mortality data are unreliable. Before the 1968 U.S. revision of the International Classification of Diseases, the word "skin" was not required on death certificates; thus, expressions such as "carcinoma of face" and "carcinoma, squamous cell" would result in the classification of nonmelanoma skin cancer. Such sparse and ambiguous information on the death certificates has been found to lead to the inflation of the number of nonmelanoma classifications (Dunn, Levin, Linden, and Harzfeld, 1965). This effect can be expected to be larger when the quality of medical care is low. In fact, when we used the number of doctors per capita as a rough index of medical care, we found an association with this index only in the case of nonmelanoma skin-cancer mortality. The tendency for an increase in nonmelanoma mortality as the number of doctors decreases may also reflect the fact that, if treated reasonably promptly, nonmelanoma skin cancer need not be, in most cases, a fatal disease. Moreover, an actinic etiology of squamous-cell carcinoma carries a favorable prognosis: when developed on sun-damaged skin, it rarely metastasizes (Emmett, 1973; Lund, 1965; Swanbeck and Hillström, 1971; Szymanski, 1963). Analysis of the mortality for malignant melanoma of the skin, however, is useful because the disease is often lethal.

A fourth problem in establishing a quantitative relation between skin-cancer incidence and uv radiation is the concomitant association with amount of exposure to sunlight. The effect of an increased intensity of uv radiation cannot simply be measured by comparing a population with another that is closer to the equator and thus receives a greater intensity of uv radiation. The amount of time exposed to sunlight is one respect in which the populations may differ and which can influence the propensity toward skin cancer. It is plausible that exposure to sunlight also increases with an approach toward the equator. Exposure to the sun varies from person to person even in the same location. It depends on occupation, style of life, and clothing; furthermore, it may depend on behavior over several decades as the effects of uv radiation on man may be cumulative. In any case, such data are not available on the population exposed to risk and only rarely on patients. (Since skin cancer is predominantly a disease of older age groups, surveys that inquire of a patient's occupation show many as retired.)

To approximate the exposure-factor on the average for a location, we have used the total solar radiation at the location. The total solar radiation is of importance also, if there is an interaction between uv and longer wavelengths in inducing skin cancer. Other component measures of exposure-factor, such as humidity and temperature–humidity combinations, were also studied.

The fifth problem is that there is a genetic susceptibility to skin cancer. One agreed-upon example is a strong tendency for Celtic peoples to be susceptible (Lane Brown, Sharpe, Macmillan, and McGovern, 1971; Lane Brown and Melia, 1973). In addition, the observed incidence of skin cancer, both malignant melanoma and nonmelanoma, is lower among Latins than among other Caucasians (Macdonald, Johnson, and Murphy, 1971; Macdonald, 1974b). Unknown differences in genetic susceptibilities to skin cancer may mask or accentuate an effect that would be observed among homogeneous populations. Populations in our study differ in their ethnic compositions. For example, in the Minneapolis–St. Paul SMSA, over 30 percent of the white population aged 65 years and over are of foreign stock from northwest Europe, compared with 12 percent for the Detroit SMSA and 18 percent for the San Francisco–Oakland SMSA. In the Dallas–Fort Worth SMSA, about 7 percent of white residents are Latins; in the San Francisco–Oakland SMSA, about 14 percent.

We have not included a quantitative analysis of the effect of different ethnic compositions on reported incidence rates: measures of incidence rates used among different ethnic groups are unknown, age-specific data of ethnic composition are available for relatively few areas, and there are not enough observations to analyze such an effect in our models meaningfully. We have, however, tabulated some data on the proportion of whites of foreign stock from northwest Europe for a number of areas. (See the Appendix on Proportion of Whites of Foreign Stock from Northwest Europe.*) A large proportion in an area such as in the Minneapolis–St. Paul SMSA explains in part the greater reported incidence rates than those predicted from our models.

Finally, a sixth problem is the determination of an average intensity of uv radiation at each location for the corresponding observed or interpolated mean amount of ozone for each month. The theoretical calculations of intensities of direct and diffuse uv radiation that were made under a number of assumptions by S. V. Venkateswaran were crudely adjusted for tabulated values of reported cloud cover. (See the Addendum on The Variables.) At present very little is known about the absorptive properties of clouds and other aerosols at wavelengths in the uv range (Venkateswaran, 1974c). In addition to the observed proportion of sky covered by clouds, the type of clouds is important, as clouds differ in their absorptive properties. Moreover, the percentage of uv radiation reflected by the surface was assumed to be the same in all locations and seasons, which is a simplification. Data were available from a few months of operation of Robertson-Berger meters (Urbach, Berger, Robertson, and Davies, 1974), which measure a weighted average of intensities of uv radiation with respect to an erythema action spectrum. If the action spectrum used in these meters is the carcinogenic one, then such measurements would be superior in establishing a relationship to skin-cancer incidence. Since, however, the few observations that are available now cover less than one year, and are subject to large seasonal effects and considerable variation from year to year, they were not used.

C. SOURCES OF DATA

Skin-cancer data from three independent sources were used to describe quantitative relations to ultraviolet radiation: (1) reported incidence of malignant melanoma of the skin among nine areas from the three-year (1969–1971) Third National Cancer Survey (TNCS) (National Cancer Institute, Third National Cancer Survey Advanced Three Year Report, 1974) and reported incidence of nonmelanoma skin cancer from the six-month (September 1, 1971, to February 29, 1972) skin-cancer supplement to the TNCS (Scotto, Kopf, and Urbach, 1974); (2) National Center for Health Statistics (NCHS) compilations of aver-

* See Note preceding the References.

age (1950–1969) annual age-adjusted rates of reported mortality of malignant melanoma and nonmelanoma skin cancer published by the National Cancer Institute (Mason and McKay, 1974); and (3) estimates of prevalence of nonmelanoma skin cancer and of moderate or severe actinic keratosis, which is often a precursor of skin cancer, based on preliminary data from the Health and Nutrition Examination Survey (HANES) of the NCHS (McDowell, 1974). The preliminary HANES data were early compilations of observations from a portion of the areas selected in a probability sample. Areas for which data were obtained were visited between April 1971 and October 1972.

Two other sources of data were useful for corroboration: The Second National Cancer Survey of 10 cities in 1947 and 1948 (Auerbach, 1961; Dorn and Cutler, 1958), in which cases of skin cancer were collected under relatively incomplete methods compared with those used in the skin-cancer supplement of the TNCS, and a five-year (1962–1966) retrospective survey of five regions in Texas by the Department of Epidemiology of the M. D. Anderson Hospital and Tumor Institute (Macdonald, 1974a).

The intensity of uv radiation (the number of photons per square meter per nanometer per second averaged over the year), which is called *uv flux,* was obtained at particular wavelengths for each of the locations of the HANES and the TNCS from theoretical computations by Venkateswaran (1974a, 1974b, 1974c). Based on all presently available ozone data, the computations were made for each month and for selected percentages of ozone reduction: 5, 10, 20, and 50 percent. Analyses were done for individual wavelengths available from Venkateswaran in the range presumed to do the most damage and also for weighted-average wavelengths. For one measure of uv flux, we used a weighted average of the respective intensities of effective (corrected for cloud cover) uv radiation at four wavelengths, 296.7, 300.4, 305.5, and 311.4 nm, with respective weights 0.63, 0.28, 0.07, and 0.02, which correspond to an action spectrum intermediate to that of DNA damage and that of erythema. Our measures of uv flux should be verified by other estimates and especially by direct measurements when these become available. Further details about the sources of data and the variables used, including definitions of the different measures of skin-cancer incidence are given in the Addendum on The Variables.

D. THE GENERAL PROBLEM

We wish to predict increases in skin-cancer incidence (or mortality) that would result from various assumed percentages of ozone reduction.

An analysis of several models led us to phrase predictions in terms of relative increases in rates of skin cancer. An advantage of such predictions is that our computations show them to be independent of age and sex and similar for different measures of incidence and prevalence.

Because it is uv flux that is the putative carcinogen and because there is a variation in uv flux among locations with the same amount of ozone, primarily because of the variation among solar angles of locations, we must first estimate the relative increase in skin-cancer incidence associated with an absolute increase in uv flux. Then, we can determine the increase in uv flux that would result from ozone reduction. One cannot simply predict from this method, however, that a given percentage of ozone reduction would lead to a determined percentage increase in skin-cancer incidence independent of the uv flux at a location. Predictions of relative increases in skin-cancer incidence depend not only on the relative ozone reduction but also on the uv flux (and, in one of our models, on the level of solar radiation) at a location. Nor can one conclude that a given percentage increase in uv flux would lead to a determined percentage increase in skin-cancer incidence independent of the uv flux at a location. It is the absolute increase in uv flux and not a relative increase from which one obtains predictions of relative increases in skin-cancer incidence from an analysis of the correlation of incidence, or a transformation of incidence, with uv flux.

We have studied the relation between skin-cancer incidence (or mortality) and uv radiation separately for each survey. Because of the relatively few observations of incidence of nonmelanoma skin cancer, the underreporting of the disease, and unreliability of nonmelanoma skin-cancer mortality data, we give more credence to our analysis of data of malignant melanoma of the skin. We compare the results of the different surveys in an effort to verify our conclusions.

E. THE RELATION BETWEEN SKIN CANCER AND ULTRAVIOLET RADIATION

The form of the relation describing skin-cancer incidence (or mortality) as a function of uv flux is not known. We have examined plots of the observations available and have fitted several simple functions. We find that, over all, the logarithm of incidence (or of mortality) versus uv flux provides the best fit of a simple function both for melanoma and nonmelanoma (where the scatter is large). For a comparison of some simple functions fitted to incidence data, see the Appendix on Models.*

* See Note preceding the References.

The log incidence (or log mortality) is also appropriate against latitude, as was shown in Figure C.2. Moreover, there is an important advantage to this use of the logarithm transformation: the increase in log incidence associated with an increase in uv flux does not appear to depend on sex or on age. Even though the values of the incidence and mortality of both melanoma and nonmelanoma depend on sex and age, still the slope of log incidence (and of log mortality) with respect to uv flux is approximately independent of sex and age, depending only on disease. What is more, the slopes with respect to uv flux of incidence of melanoma and of nonmelanoma skin cancer are nearly the same. Figure C.7 shows plots of incidence of melanoma and nonmelanoma skin cancer reported from the Third National Cancer Survey and the TNCS skin-cancer supplement among three age groups with respect to uv flux in photons per square meter per second per nanometer at 305.5 nm. Regression lines in this figure have been drawn with a common slope. The larger incidence of nonmelanoma than of melanoma skin cancer, the increasing incidence with age for both diseases, and the greater incidence of each disease among males, with the exception of melanoma incidence among those 35 to 44 years of age, are all reflected by the position of the ordinates on these regression lines for any given value of uv flux. Each of these parallel lines describes well the pattern of incidence in the respective one of the twelve age by sex by disease groups. However, an examination of the direction of the deviations at each location shows that there is some systematic effect associated with location that has not been allowed for. This fact means that uv flux alone is not an adequate predictor and that a more complex model is called for.

A common slope with respect to uv flux has the important advantage that only one slope need be estimated. As a result, the accuracy of estimation is improved and a simpler prediction provided, the same for both sexes and all ages. What is more, the slope with respect to uv flux of logarithms of different measures of incidence and prevalence are similar. Thus, crude statements of relative increases may apply to all of them.

A further advantage in using the logarithm transformation is that the deviations from the estimated relation in one of our models appear to satisfy the conditions for standard statistical analyses. In addition, there are theoretical reasons for using the log transformation: the complementary log-log transformation developed by Yates (1955) for use in a dose–response relation with irradiation is asymptotically equivalent, as incidence becomes small, to the log transformation. A statistical bias in estimation, however, arises when unbiased estimates

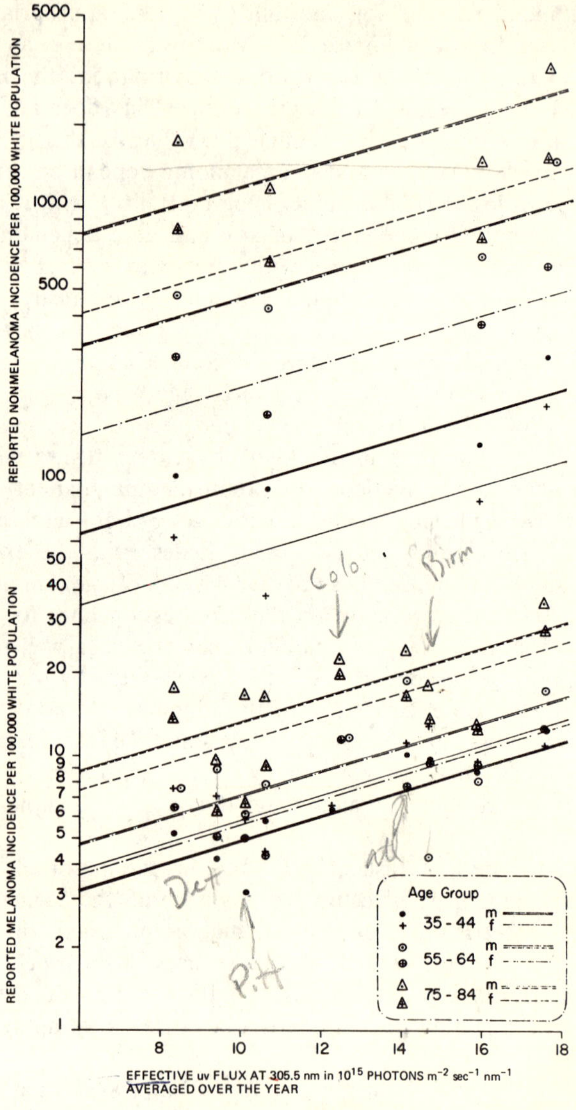

supplement 6 months data

TNCS National Cancer Survey

Corrected for cloud cover

FIGURE C.7 Reported skin-cancer incidence rates per 100,000 white population among three age groups as a function of ultraviolet flux. [Sources: melanoma incidence from National Cancer Institute (1974), nonmelanoma skin-cancer incidence from Scotto (1974) and Scotto, Kopf, and Urbach (1974).] The computations of uv flux used to prepare this figure are not adjusted for altitude.

(or exposure, or temp, humidity, or ethnic comp.)

of changes in log incidence are obtained from an analysis with the log transformation and then transformed into estimates of relative increases in incidence. We have corrected for this bias. (See the Appendix on Methodology.*)

F. MODELS ADOPTED

Two general models are used extensively to describe the relationship between log incidence (or log mortality) and uv flux. Although we use the term models, these are no more than interpolation formulas. (See the Appendix on Models.*) For a measure of uv flux, we have used a weighted average of the respective intensities of average annual altitude-adjusted effective (corrected for cloud cover) uv radiation in photons per square meter per second per nanometer at four wavelengths, 296.7, 300.4, 305.5, and 311.4 nm, with respective weights 0.63, 0.28, 0.07, and 0.02, which correspond to an action spectrum intermediate to that of DNA damage and that of erythema. Analyses were also done for individual wavelengths and other weighted averages.

The simple model predicts log incidence (or log mortality) as a linear function of uv flux with the same increase in log skin-cancer incidence (or mortality) for every age and sex (see Figure C.7). If we let F be our measure of uv flux and μ_{ij} be the log incidence among those of sex i in age group j, then this model implies

$$\mu_{ij} = \alpha_{ij} + \beta F,$$

for some numbers α_{ij} and β. The parameter β is of special interest because it provides the increase in log incidence for a unit increase in uv flux.

The interaction model allows for the effect of different solar radiation (insolation) and for a possible interaction between the effects of solar radiation and of uv flux, as well as the effect of uv flux considered alone. The presence of this interaction means that, in two areas with different values of solar radiation, the changes in log incidence for a common absolute increase in uv flux are not the same.

The interaction model has three variable terms, one for uv flux (F), one for solar radiation (S), and one for the interaction between the two (the product of F times S). In terms of this model,

$$\mu_{ij} = \alpha_{ij} + \beta F + \gamma S + \delta F \cdot S,$$

* See Note preceding the References.

for some numbers α_{ij}, β, γ, and δ. The systematic deviations for some localities, so evident in Figure C.7, which corresponds to the simple model, disappear when the interaction model is used.

We have also used other models that incorporate more predictor variables: comfort (temperature–humidity) indices, measures of urbanicity and the quality of medical care, and effective buying power. In addition, we have compared the results from the simple model with the various combinations of using and not using the logarithm transformation on incidence (or mortality) and on uv flux. (See the Appendix on Models.*)

G. INTERPRETATION OF INTERACTION

The interaction model allows the effect of uv radiation to be different in locations that have different solar radiation. Such a difference may be due to a physical interaction between the climatic and meteorological variables in their effects on skin, or it may be due to an association between solar radiation and the constituents of life style and so to a change in exposure-factor. We give several interpretations in terms of the latter reason in the Appendix on Models.*

H. OTHER FACTORS

The quality of medical care at a location will influence the number of diagnoses of skin cancer at the location and thus the number counted. Moreover, in areas with a lower quality of medical care, many cases of nonmelanoma skin cancer may not be treated reasonably promptly and lead to deaths. Also, in such areas, sparse and ambiguous information may be provided on death certificates that later may be incorrectly coded for nonmelanoma skin cancer. A discussion of problems of nonmelanoma skin-cancer mortality data is given in a study of Dunn, Levin, Linden, and Harzfeld (1965).

It is conceivable that part of the increase in skin-cancer incidence with decreasing latitude is confounded with differences in the quality of medical care. We put in an additional term to check this possibility. As an indicator of this quality, we used the number of active physicians and doctors of osteopathy per 1000 residents in the location. The effect of medical care was significant only in the case of nonmelanoma mortality, as discussed earlier. We conclude that nonmelanoma mortality does not provide a usable prediction of the effect of uv radiation due to the apparent confounding.

* See Note preceding the References.

Skin cancer may be the outcome of exposure over several decades. Then we would expect that so-called retirement communities in the southern states will show a lower incidence rate for persons who have migrated from northern states than for the native population. An attempt was made to test this hypothesis by examining skin-cancer mortality in southern counties known to have a high immigration from 1960 to 1970 and also a high proportion of older residents. The results, however, were inconclusive, perhaps because the data were not entirely suitable: published mortality rates are age-adjusted and are by county, including both migrants and natives. Further, the relative importance of long-time exposure to uv radiation and recent more severe exposure is not known.

As noted earlier, the localities used in our analyses differ in the proportion of persons from northwest Europe. In the Appendix on Proportion of Whites of Foreign Stock from Northwest Europe,* we have tabulated using data from U.S. Bureau of the Census (1973b) these proportions for some of the areas in our study. Fair-skinned people have higher rates for skin cancer. On the other hand, persons who have spent much of their time farther north may have lower rates. Differences in the proportion of fair-skinned population may explain part of the systematic deviations for some localities evident in Figures C.2 and C.7. However, as we have remarked, deviations are not systematic when the interaction model is used.

Not necessarily *

Whenever rates, such as incidence or mortality, are to be estimated, we must suspect the presence of spurious correlation. The predictor variables may be correlated with population size, the denominator in the rate, instead of or as well as with the rate. In order to avoid spurious correlation, we carried out separate analyses for three population size categories when we considered for one study mortality rates among a sample of 94 counties, splitting these counties into subgroups of about 31 counties each, on the basis of population. There was no appreciable difference from one subgroup to another, and so we have assumed that there is no spurious correlation.

I. OTHER CANCERS

It was suggested that cancer of the lip and cancer of the eye might be related to uv radiation. To test these hypotheses we used mortality data of these diseases from the same source and in the same way as malignant melanoma and other skin-cancer mortality data. We found no significant effect of latitude or of uv flux. In order to check against the

* See Note preceding the References.

* Integrated exposure theory vs. acclimatization theory

fortuitous choice of locations, we used mortality data for cancer of the pancreas. For this disease in our simple model, the slope with respect to uv flux is negative. In Fears, Scotto, and Schneiderman (1974), the incidence of all cancers other than nonmelanoma skin cancer is plotted as a function of latitude; there is zero slope, indicating no effect. Also plotted there is a graph of lip-cancer incidence, which shows an increase with decreasing latitude.

V. Research in Progress

1. *Estimates of the actual number of cases and true incidence of nonmelanoma skin cancer.* One use of these estimates would be to express predictions of effects of increased intensities of ultraviolet (uv) radiation in terms of the increased number of cases. Because nonmelanoma skin cancer is grossly underreported, the actual number of cases is much larger than the reported number.

2. *Estimates of today's rates of melanoma mortality and melanoma incidence.* Because of a secular increase, published average annual rates are smaller than today's rates.

3. *Inadequacies of extrapolations.* We are using epidemiological evidence to estimate from an increase in skin cancer from one location to another the increase in skin cancer due to the increase in uv radiation from one location to another. The present dose of uv radiation is not large, and low death rates reflect this fact. If the intensity of uv radiation is appreciably increased, then extrapolations are misleading. It is known for other types of cancer that mortality and incidence increase more sharply than expected from extrapolations.

4. *Estimated percentage increase in skin cancer among the black population due to decreases in stratospheric ozone.*

5. *More realistic models of dependence of skin-cancer rates on characteristics of the locality.* Studies of various factors thought to be of importance, and studies of appropriate measure of uv radiation intensity: total received, maximum received, other measures.

VI. Predictions of Percentage Increases in Skin-Cancer Incidence and Mortality for Various Amounts of Ozone Reduction

Three models are used to predict relative increases in melanoma incidence and mortality and nonmelanoma skin-cancer incidence for various amounts of ozone reduction. Such predictions depend on the wavelength or the weighted average of wavelengths of uv flux used in the models. For the most part, models that use smaller wavelengths give larger predictions. In the interaction model, predictions also depend on the level of solar radiation; except for some inconsistencies, they are larger in areas of greater insolation.

From our models, we can make predictions of the relative increases in incidence and mortality rates of skin cancer among whites for any place in the world. Because there are inconsistencies among predictions obtained from our various models and data bases, we give predictions as an illustration for only two localities, Minneapolis–St. Paul and Dallas–Fort Worth. The legends of the following tables give the models used (simple, interaction, or an interaction model with seven additional predictor variables as well as sex) and the data to which the models were fitted: incidence rates of melanoma and nonmelanoma skin cancer for areas of the Third National Cancer Survey (TNCS), estimates of prevalence rates of nonmelanoma skin cancer and moderate and severe actinic keratosis among 35 areas (all now available of the 65) of the Health and Nutrition Examination Survey (HANES), incidence rates of skin cancer among the 10 cities from the Second National Cancer Survey (SNCS), and mortality rates of melanoma taken from Mason and McKay (1974) (M & M) for the 9 TNCS areas and for 73 areas. Corresponding to a tabulated percentage of ozone reduction in one of the two areas, a table shows an estimate of the percentage increase of the corresponding rate of skin cancer. Under each estimate appears a 90 percent confidence interval for the percentage increase in the rate of skin cancer. The methods of constructing the estimates and confidence intervals are explained in the Appendix on Methodology.*

The estimates for each source of data are shown for one skin-cancer action spectrum midway between erythema and DNA damage. For the TNCS and M & M, the action spectrum was estimated from a weighted average (altitude-adjusted) effective uv flux (weight 0.63 at 296.7 nm, 0.28 at 300.4 nm, 0.07 at 305.5 nm, and 0.02 at 311.4 nm, all corresponding to albedo 0.2). For estimates from the SNCS and the HANES, a weighted average effective uv flux at sea level was used (weight 0.72 at 300.4 nm, 0.24 at 305.5 nm, and 0.04 at 311.4 nm, all corresponding to albedo 0.2). If the appropriate spectrum is shifted to shorter wavelength, the estimated increase in skin cancer for a specified decrease in ozone will be much larger. Theoretically, as suggested by James McDonald (1971), the action spectrum can be estimated from the epidemiological data, but this will require more extensive data and better knowledge of the estimating model. Tables C.1 and C.2 reveal

* See Note preceding the References.

TABLE C.1 Estimated Percentage Increase in Skin Cancer for Various Amounts of Ozone Reduction, Using Simple Model and Action Spectrum Midway between that of Erythema and of DNA Damage. Estimates and 90 Percent Confidence Intervals (Shown Underneath Estimate) of Percentage Increase in Skin-Cancer Mortality and Incidence among Whites in Two Localities[a]

Disease	Rate	Data Source	No. of Areas	Percentage Increase[b] in Skin Cancer Corresponding to Ozone Reduction of			
				5%	10%	20%	50%
Minneapolis–St. Paul							
Skin cancer	Incidence	SNCS	10	20	48	147	4453
		1947–48		15 to 25	36 to 60	106 to 200	2135 to 11321
Nonmelanoma	Incidence	TNCS	4	8	19	50	512
skin cancer		1971–72		2 to 15	4 to 36	11 to 107	68 to 3661
Nonmelanoma	Prevalence	HANES	35	15	34	94	1196
and actinic		1971–72		6 to 24	13 to 60	34 to 198	251 to 10936
keratosis (severe							
or moderate)							
Malignant	Incidence	TNCS	9	10	22	61	947
melanoma		1969–71		8 to 11	18 to 26	48 to 75	609 to 1535
Malignant	Mortality	M & M	9 in TNCS	6	14	37	374
melanoma		1950–69		5 to 8	11 to 18	28 to 47	241 to 584
			73 in HANES	3	6	15	102
			+TNCS	2 to 3	5 to 7	12 to 19	76 to 135

Dallas–Ft. Worth							
Skin cancer	Incidence	SNCS 1947–48	10	42 32 to 52	109 80 to 145	436 284 to 679	65446 21196 to 355164
Nonmelanoma skin cancer	Incidence	TNCS 1971–72	4	17 4 to 32	40 9 to 83	116 23 to 317	2137 159 to 77366
Nonmelanoma and actinic keratosis (severe or moderate)	Prevalence	HANES 1971–72	35	30 12 to 52	73 27 to 143	231 72 to 668	6620 773 to 334776
Malignant melanoma	Incidence	TNCS 1969–71	9	20 16 to 24	48 39 to 59	154 116 to 200	7033 3533 to 16708
Malignant melanoma	Mortality	M & M 1950–69	9 in TNCS	13 10 to 16	30 23 to 38	85 62 to 113	1585 849 to 3301
			73 in HANES + TNCS	6 4 to 7	13 10 to 15	32 25 to 40	263 182 to 379

[a] See text for details of data and of effective wavelengths.
[b] See comment in text on lack of precision.

TABLE C.2 Search for More Realistic Model—Comparison of Estimated Percentage Increase in Skin Cancer Using Different Models and Action Spectrum Midway between that of Erythema and of DNA Damage. Estimates and 90 Percent Confidence Intervals (Shown Underneath Estimate) of Percentage Increase in Skin-Cancer Mortality and Incidence among Whites in Two Localities Corresponding to Ozone Reduction of 10 Percent [a]

					Percentage Increase for Type of Model Used		
Disease	Rate	Data Source	No. of Areas	Locality	Simple [b] (Flux)	Interaction [c] (F, S, F*S)	Interaction [c] (7 predictors)
Skin cancer	Incidence	SNCS 1947–48	10	Minn.–St. Paul	48 36 to 60	26 13 to 41	
				Dallas–Ft. Worth	109 80 to 145	29 −5 to 81	
Nonmelanoma skin cancer	Incidence	TNCS 1971–72	4	Minn.–St. Paul	19 4 to 36	−2 −6 to +3	
				Dallas–Ft. Worth	40 9 to 83	229 133 to 376	

Nonmelanoma and actinic keratosis (severe or moderate)	Prevalence	HANES 1971–72	35	Minn.–St. Paul	34	32	
					13 to 60	−12 to 112	
				Dallas–Ft. Worth	73	69	
					27 to 143	−19 to 346	
Malignant melanoma	Incidence	TNCS 1969–71	9	Minn.–St. Paul	22	13	
					18 to 26	7 to 19	
				Dallas–Ft. Worth	48	44	
					39 to 59	27 to 65	
Malignant melanoma	Mortality	M & M 1950–69	9 in TNCS	Minn.–St. Paul	14	7	
					11 to 18	2 to 12	
				Dallas-Ft. Worth	30	38	
					23 to 38	24 to 55	
			73 in HANES +TNCS	Minn.–St. Paul	6	7	6
					5 to 7	4 to 10	4 to 9
				Dallas–Ft. Worth	13	−2	0
					10 to 15	−7 to +4	−5 to +5

[a] See text for details of data and of effective wavelengths.
[b] See comment in text on lack of precision.
[c] See comments in text on lack of reliability.

several inconsistencies in the estimates from different tentative models. Here, we should emphasize that the estimated confidence intervals are wrong (usually too short) if the model is wrong, especially when there are only a few areas. Also, the 20-year melanoma mortality data should be corrected for the striking rise in melanoma among the young.

Monte Carlo studies reinforce the conclusion that quantitative results using data from few areas are not precise even for the simple model and are not reliable for the more complex models. The predicted percentage increases for the two sexes and the many age groups are in excellent agreement, but they change with the choice of areas fitted by the models, indicating that the models are not complete. Data from many areas are needed; we await the forthcoming data on skin-cancer prevalence for the 65 stands of the Health and Nutrition Examination Survey (HANES), a survey not intended to study cancer so that the sampling proportion is small for our purposes.

Addendum: The Variables

SKIN-CANCER MORTALITY AND INCIDENCE

Average (1950–1969) annual age-adjusted (1960 U.S. standard) rates of reported mortality for each of malignant melanoma of the skin; other skin cancer; and cancer of the pancreas, of the lip, and of the eye for white males and females were obtained from those rates for each county and state (Mason and McKay, 1974). Rates for an area that is a cluster of counties, such as an SMSA, were obtained by an average of rates for constituent counties each weighted by the size of the respective white population in 1960. Mortality rates were compiled for each of the 65 stands of the Health and Nutrition Examination Survey (National Center for Health Statistics, 1973; Murphy, 1974) as well as the areas of the Second and Third National Cancer Surveys. If an area had a 1960 white population of less than 100,000, then contiguous counties with similar altitude and closeness to water were added until this population was near 100,000.

Average (1969–1971) annual age-adjusted (1970 U.S. standard) and age-specific incidence rates of malignant melanoma of the skin for white males and females were computed from the reported age- and sex-specific numbers of cases diagnosed among whites during the three years of the Third National Cancer Survey (Schneiderman, 1974). White populations for each of the nine areas of the survey were obtained from the 1970 census (U.S. Bureau of the Census, 1973a).

Average annual age-adjusted (1970 U.S. standard) and age-specific incidence rates of nonmelanoma skin cancer for white males and females were computed from the reported age- and sex-specific numbers of patients with one or more skin-cancer lesions diagnosed during the six-month supplementary skin-cancer survey (September 1, 1971, to February 29, 1972), which covered four areas of the Third National Cancer Survey (Scotto, 1974). The age-adjusted rates calculated from these numbers appear in the paper by Scotto, Kopf, and Urbach (1974).

Average (1962–1966) annual age-adjusted (1970 U.S. standard) incidence rates of reported nonmelanoma skin cancer among non-Latin white males and females in five regions of Texas (El Paso County and four cluster of counties that form Hill-Burton hospital regions) were computed from age-specific incidence rates provided by Eleanor Macdonald of the M. D. Anderson Hospital and Tumor Institute (Macdonald, 1974a).

X check
El Paso vs
Corpus Christi
on exposure

The definitions of age-specific incidence in the three surveys are not the same. In the Third National Cancer Survey, such incidence of melanoma is the number of reported diagnoses of primary malignant melanoma of the skin in the age group made during the three-year survey divided by three times the corresponding 1970 population size; such incidence of nonmelanoma skin cancer is the number of patients in an age group with one or more reported diagnoses of a nonmelanoma skin cancer lesion during the six-month survey divided by one half of the corresponding 1970 population size. In the M. D. Anderson Hospital and Tumor Institute survey, age-specific incidence is defined to be the reported number of those patients in an age group diagnosed during the six-year survey as having any form of nonmelanoma skin cancer who have had no previous diagnosis of any form of skin cancer divided by the sum of the estimated population sizes for each of the six years in that age group. A comparison of the definitions and methodology of the two nonmelanoma skin-cancer surveys is part of a separate study of the Panel to Review Statistics on Skin Cancer.

Average annual age-adjusted (1950 U.S. standard) reported incidence rates of skin cancer among white males and females were obtained from the Second National Cancer Survey of 10 cities during 1947 and 1948 (Auerbach, 1961; Dorn and Cutler, 1958). This survey does not distinguish between malignant melanoma of the skin and other skin cancer and is based primarily on hospital pathology reports. Such a method of obtaining cases was discovered early in the Third National Cancer Survey to be incomplete; thus, a special supplementary skin-cancer survey was conducted in 1971–1972.

Estimates of prevalence of nonmelanoma skin cancer and of moderate or severe actinic keratosis among the U.S. civilian noninstitutional population at ages 1–74 years and at ages 45–74 years were obtained from the Health and Nutrition Examination Survey of the National Center for Health Statistics (McDowell, 1974). These estimates were based on preliminary data from the first round of 35 areas visited between April 1971 and October 1972 among the 65 stands that were selected for the survey. These 35 areas form a national probability sample.

MEASURES OF ULTRAVIOLET FLUX

Using all available ozone data from well-established stations (measurements at least through 1971 provided directly on data tapes by the Department of Transport, Canada) and guided by upper tropospheric circulation maps, S. V. Venkateswaran and his students have drawn isopleths of mean total ozone on world maps for each month. In addition to the latitudinal variation, each map shows a longitudinal variation, and collectively they display a seasonal variation; neither of the latter is negligible. With interpolations from these maps, Venkateswaran estimated the mean monthly amounts of ozone for areas corresponding to the TNCS and the HANES areas and determined, for each month, the altitude-adjusted direct and diffuse components of uv flux at four wavelengths for each area by integrating over all values of the solar zenith angle in a day (Venkateswaran, 1974a, 1974b, 1974c). The uv fluxes for Fort Collins and Boulder were used and later averaged for Colorado; uv fluxes for Des Moines were used for Iowa.

For the 12 areas at which there are ozone stations with a history of observations, Venkateswaran calculated for each selected wavelength the flux at sea level for each ozone reading and computed from these the daily average fluxes (diffuse and direct) and the standard deviation of the distributions of these fluxes for each month.

The four wavelengths selected for our studies are 296.7, 300.4, 305.5, and 311.4 nm, which are representative values of putative carcinogenic wavelengths. These four wavelengths are among those used to determine the amount of ozone; precise information about the ozone absorption coefficient is known for them.

The calculations of uv flux are based on the assumptions of a clear day and a uniform albedo of 0.2. Additional calculations were also made for albedos of 0 and 0.6. The uv fluxes used in our linear models are adjusted for the approximate altitude of each area. Some uv fluxes at sea level were available for preliminary analyses.

For each area, the mean monthly sky cover (sunrise to sunset) was obtained for each month from the Climatic Atlas of the United States (Environmental Data Service, 1968). In addition, the average noon sky cover for each month was obtained (U.S. Weather Bureau, 1965). Using the linear relation of percent y of uv flux (direct plus diffuse) to cloud cover in tenths x proposed by Büttner (1938),

$$y = 100 - 5.62(x),$$

we multiplied each of the monthly totals of direct plus diffuse components by the appropriate percent y for the reported monthly cloud cover x. Schulze and Gräfe (1969) display this relationship of Büttner with the comment that, "on average this curve has been proved correct." Taking an average of each of the so corrected monthly fluxes weighted by the number of days in the month, we obtained the average annual effective uv flux for each of the four wavelengths.

The measurement of daily average flux has the advantage that it incorporates the temporal variability of ozone for a specified month. Since the log of uv flux at a specific wavelength is nearly a linear function of the amount of ozone, the uv flux for mean values of monthly ozone will underestimate the average uv flux. Such an expectation from theory is confirmed by our observations. In the case of 300.4 nm among twelve ozone stations (excluding the station in Midland because of missing data but including a station in Toronto), a linear regression was used to predict daily average uv flux for clear days (the weighted average of the monthly sums of direct plus diffuse components without any correction for cloud cover) from the clear-day flux for mean monthly ozone; and a similar regression was used for the effective fluxes. In each case, the values of the squares of the correlation coefficients exceeded 0.9998. Since daily average flux is so nearly a linear function of flux for mean monthly ozone, it would make little difference for the purpose of prediction from our linear models which type of flux to use over the range of the selected ozone stations.

Correlations between effective sea-level uv fluxes among the nine areas of the Third National Cancer Survey at albedos 0, 0.2, and 0.6 are all 1.0000 for each of three wavelengths, 300.4, 305.5, and 311.4 nm. Thus, there is, for each wavelength, a linear transformation between uv flux at one albedo and uv flux at another, with the result that, in our linear models, it would make no difference which of the albedos we had assumed: the models would fit the same. These linear relationships are peculiar to the range of wavelengths and intensities that we are using; no such general relationships follow from theory (Venkateswaran, 1974d). What is more, the slopes of the linear relations between sea-level uv fluxes at a specific wavelength for two albedos are nearly the same for each of the different amounts of ozone. Thus, our predictions would not be changed much if we assumed a different uniform albedo. The assumption of a uniform albedo, however, neglects seasonal variation.

The sea-level uv fluxes, for clear days among the nine areas of the Third National Cancer Survey, at different wavelengths are highly correlated (see Table C.3), and the correlations are respectively larger for effective sea-level fluxes (see Table C.4). In each case, the eigenvector corresponding to the largest eigenvalue of the correlation matrix for the three wavelengths in Tables C.3 and C.4 has nearly the same elements and accounts for more than 99 percent of the total variance. With such high correlations, it is impractical to distinguish which of the wave-

TABLE C.3 Correlations of Sea-Level Ultraviolet Flux (0.2 Albedo) for Clear Days at Three Wavelengths, Solar Radiation, and Latitude among the Nine Areas of the Third National Cancer Survey

	λ=300.4	λ=305.5	λ=311.4	Solar Radiation	Latitude
λ=300.4	1	0.992	0.978	0.781	−0.905
λ=305.5	0.992	1	0.996	0.773	−0.949
λ=311.4	0.978	0.996	1	0.755	−0.972
Solar radiation	0.781	0.773	0.755	1	−0.697
Latitude	−0.905	−0.949	−0.972	−0.697	1

TABLE C.4 Correlations of Effective Sea-Level Ultraviolet Flux (0.2 Albedo) at Three Wavelengths, Solar Radiation, and Latitude among the Nine Areas of the Third National Cancer Survey

	λ=300.4	λ=305.5	λ=311.4	Solar Radiation	Latitude
λ=300.4	1	0.996	0.989	0.777	−0.860
λ=305.5	0.996	1	0.998	0.785	−0.898
λ=311.4	0.989	0.998	1	0.787	−0.912
Solar radiation	0.777	0.785	0.787	1	−0.697
Latitude	−0.860	−0.898	−0.912	−0.697	1

lengths in a multiple regression using all of them is a best predictor. We considered instead individual models for each wavelength and for weighted-average wavelengths.

SOLAR RADIATION

A principal concomitant variable that we incorporate in some of our analyses is average annual solar radiation in Langleys per day. Measures of solar radiation were obtained from the Climatic Atlas of the United States (Environmental Data Service, 1968). Helmut Landsberg provided us with assistance in interpolating the values of solar radiation for the areas of the Third National Cancer Survey. The correlations of solar radiation with sea-level uv flux for clear days and effective sea-level uv flux among the nine areas of the Third National Cancer Survey are given in Tables C.3 and C.4.

THE QUALITY OF MEDICAL CARE

The number of active physicians per 1000 resident population in 1967 was used as a measure of the quality of medical care for each area. The numbers of active M.D.'s and Doctors of Osteopathy in 1967 were available for each county and state (National Center for Health Statistics, 1971). For the two states and constituent counties of the seven SMSA's of the Third National Cancer Survey, a geometric interpolation of 1960 and 1970 censuses was used to estimate the 1967 resident population. For other areas, in our sample of counties, we used the estimated populations on July 1, 1966, which were published in the NCHS report.

OTHER VARIABLES

For each area that is a cluster of counties, the density of the 1960 white population per square mile was computed for each constituent county (U.S. Bureau of the Census, 1962). A weighted average of the densities with respect to the 1960 white population was used as a measure of urbanicity. Other predictor variables were effective buying income per person in 1966 (National Center for Health Statistics, 1971) and two comfort indices. The first comfort index is the number of July hours in five years with a temperature of 90° or greater; the second comfort index is the number of July hours in five years with a temperature from 80° to 90° and with a relative humidity of 50 percent or more (U.S. Weather Bureau, 1965).

Acknowledgments

Many gave willingly of their time and effort. We are grateful to Marvin Schneiderman and Joseph Scotto of the National Cancer Institute for permitting the use of the data on skin cancer from the Third National Cancer Survey in advance of publication. Also, we thank Arthur J. McDowell and Jean Roberts of the National Center for Health Statistics for sending us preliminary results on skin-cancer prevalence from the Health and Nutrition Examination Survey. We are also grateful to Eleanor Macdonald of the Department of Epidemiology of the M. D. Anderson Hospital and Tumor Institute for sending compilations from their skin-cancer survey. S. V. Venkateswaran, Dennis St. John, and N. Sundararaman of the Department of Meteorology, University of California, Los Angeles, kindly supplied the particular computations of intensities of ultraviolet radiation and of increases in ultraviolet radiation for various amounts of ozone reduction for the many locations in our study. Harold Johnston and Gary Whitten of the Lawrence Berkeley Laboratory, University of California, Berkeley, were helpful in supplying early information on the relation between ozone and location. The thanks of the Panel go to Bengt Berlin, Steve Casey, Marilyn Hill, Terry Jue, Daphne O'Regan, Josefa Troya, Marcella Wells, and especially to Jeanne Lovasich and Ellen Sherman, all of the Department of Statistics and Statistical Laboratory, University of California, Berkeley, for their part in the analysis. The major computations for our analysis were made on the Control Data 6400 computer of the Computer Center, University of California, Berkeley.

NOTE: The Interim Report of the Panel to Review Statistics on Skin Cancer contains four appendixes. Only the first of these, describing the variables employed in our studies, is included here. The remaining three are more technical. They are entitled Models, Notes on Methodology, and Proportion of Whites of Foreign Stock from Northwest Europe for Selected Areas. They may be obtained from the NRC Committee on National Statistics.

References

Atkin, M., J. A. Fenning Heady, E. L. Kennaway, and N. M. Kennaway. 1949. The mortality from cancer of the skin and lip in certain occupations, *Br. J. Cancer* 3:1–15.

Auerbach, H. 1961. Geographic variation in incidence of skin cancer in the United States, *Public Health Rep.* 76(4):345–348.

Axtell, L. M., S. J. Cutler, and M. H. Myers, eds. 1972. *End Results in Cancer.* Report No. 4. U.S.D.H.E.W. pp. 148–152.

Bain, J. A., H. P. Rusch, and B. E. Kline. 1943. The effect of temperature upon ultraviolet carcinogenesis with wavelengths 2,800–3,400Å, *Cancer Res.* 3:610–612.

Beardmore, G. L. 1972. The epidemiology of malignant melanoma in Australia, *Melanoma and Skin Cancer*. Proc. Int. Cancer Conf., Sydney. V.C.N. Blight, Government Printer, Sydney. March 13–17.

Bener, P. 1972. *Approximate Values of Intensity of Natural Ultraviolet Radiation for Different Amounts of Atmospheric Ozone*. Tech. Rep. Phys.-Meteorol. Obs. Davos, Davos Platz, Switzerland. Document No. AD-752-115. Nat. Tech. Inf. Serv., Springfield, Va.

Black, H. S., and D. R. Douglas. 1973. Formation of a carcinogen of natural origin in the etiology of ultraviolet light-induced carcinogenesis, *Cancer Res.* 33:2094–2096.

Black, H. S., and Wan-bang Lo. 1971. Formation of a carcinogen in human skin irradiated with ultraviolet light, *Nature* 234:306–308.

Blum, H. F. 1959. *Carcinogenesis by Ultraviolet Light*. Princeton U. P., Princeton, N.J.

Blum, H. F. 1974. Uncertainty of growth of cell population in cancer, *J. Theor. Biol.* 46:143–166.

Blum, H. F., and S. W. Lippincott. 1942. Carcinogenic effectiveness of ultraviolet radiation of wavelength 2537Å, *J. Nat. Cancer Inst.* 3:211–216.

Brodkin, R. H., A. W. Kopf, and R. Andrade. 1969. Basal-cell epithelioma and elastosis: a comparison of distribution. *The Biologic Effects of Ultraviolet Radiation: With Emphasis on the Skin*. F. Urbach, ed. Pergamon, New York. pp. 581–618.

Burbank, F. 1971. *Patterns in Cancer Mortality in the United States: 1950–1967*. Nat. Cancer Inst. Monogr. No. 33.

Büttner, K. 1938. *Physikalische Bioklimatologie: Probleme und Methoden*. Akademische Verlagsgesellschaft m.b.H., Leipzig.

Carmichael, G. G. 1962. A survey of skin cancers and solar keratoses in country areas in Queensland, *Med. J. Austral.* 1:395–400.

Carmichael, G. G., and H. Silverstone. 1961. The epidemiology of skin cancer in Queensland: The incidence, *Br. J. Cancer* 15:409–424.

Clark, W. H., Jr., L. From, E. A. Bernadino, and M. C. Mihm. 1969. The histogenesis and biologic behavior of primary human malignant melanomas of the skin, *Cancer Res.* 29:705–727.

Clark, W. H., Jr., and M. C. Mihm. 1969. Lentigo maligna and lentigo-maligna melanoma, *Am. J. Pathol.* 55:39–54.

Cleaver, J. E. 1970. *J. Invest. Dermatol.* 54:181–195.

Coblentz, W. W., and R. Stair. 1934. Data on the spectral erythemic reaction of the untanned human skin to ultraviolet radiation, *J. Res. Nat. Bur. Stand.* 12:13–14.

Davis, N. C., J. J. Herron, and G. R. McLeod. 1966. Malignant melanoma in Queensland—Analysis of 400 skin lesions, *Lancet* pp. 407–410.

Doll, R., C. Muir, and J. Waterhouse, eds. 1970. *Cancer Incidence in Five Continents*. Vol. II. UICC, Geneva.

Dorn, H. F., and S. J. Cutler. 1958. *Morbidity from Cancer in the United States*. Public Health Monogr. No. 56. U.S. Public Health Service.

Dunn, J. E., Jr., E. A. Levin, G. Linden, and L. Harzfeld. 1965. Skin cancer as a cause of death, *Calif. Med.* 102:361–363.

Environmental Data Service, Environmental Science Services Administration. 1968. *Climatic Atlas of the United States*. U.S. Govt. Printing Off., Washington, D.C.

Emmett, E. A. 1973. Ultraviolet radiation as a cause of skin tumors, *CRC Critical Rev. Toxicol.* 2:211–255.

Epstein, J. H., W. L. Epstein, and T. Nakai. 1967. Production of melanomas from DMBA-induced "blue nevi" in hairless mice with ultraviolet light, *J. Nat. Cancer Inst.* 38:19–30.

Everett, M. A., R. L. Olson, and R. M. Sayre. 1965. Ultraviolet erythema, *Arch. Dermatol.* 92:713–719.

Fears, T., J. Scotto, and M. Schneiderman. 1974. The epidemiology of skin cancer, paper given at 45th Aerospace Med. Assoc. Mtg., Washington, D.C. May 6.

Forbes, P. D. 1973. Influence of long wave UV on photocarcinogenesis, *Am. Soc. Photobiol. Abs.* 136. Am. Soc. for Photobiol., Sarasota, Fla.

Freeman, R. G. 1974. The action spectrum for ultraviolet carcinogenesis. Preprint.

Freeman, R. G., H. T. Hudson, and R. Carnes. 1970. Ultraviolet wavelength factors in solar radiation and skin cancer, *Int. J. Dermatol.* 9:232–235.

Freeman, R. G., and J. M. Knox. 1964. Influence of temperature on ultraviolet injury, *Arch. Dermatol.* 89:858–864.

Freeman, R. G., D. W. Owens, J. M. Knox, and H. T. Hudson. 1966. Relative energy requirements for an erythemal response of skin to monochromatic wavelengths of ultraviolet present in the solar spectrum, *J. Invest. Dermatol.* 47:586–592.

Gates, D. M. 1966. Spectral distribution of solar radiation at the earth's surface, *Science* 151:523–529.

Gellin, A. G., A. W. Kopf, and L. Garfinkel. 1969. Malignant melanoma—a controlled study of possibly associated factors, *Arch. Dermatol.* 99:43–48.

Gordon, D., and H. Silverstone. 1973. World-wide epidemiology of premalignant and malignant cutaneous lesions, manuscript.

Gordon, D., H. Silverstone, and B. A. Smithurst. 1972. The epidemiology of skin cancer in Australia, *Melanoma and Skin Cancer,* pp. 23–37 in Proc. Int. Cancer Conf., Sydney. V.C.N. Blight, Government Printer, Sydney, March 13–17.

Herron, J. 1969. The geographical distribution of malignant melanoma in Queensland, *Med. J. Austral.* 2:892–894.

Higginsen, J., and A. G. Oettlé. 1960. Cancer incidence in the Bantu and "Cape Colored" races of South Africa: Report of a cancer survey in the Transvaal (1953–55), *J. Nat. Cancer Inst.* 24:589–671.

Kirby-Smith, J. S., H. F. Blum, and H. G. Grady. 1942. Penetration of ultraviolet radiation into skin, as a factor in carcinogenesis, *J. Nat. Cancer Inst.* 2:403–412.

Koller, L. R. 1965. *Ultraviolet Radiation.* 2nd ed. Wiley, New York.

Lane Brown, M. M., and D. F. Melia, 1973. A genetic diathesis to skin cancer. *J. Invest. Dermatol.* 61:39–41.

Lane Brown, M. M., C. A. B. Sharpe, D. S. Macmillan, and V. J. McGovern. 1971. Genetic predisposition to melanoma and other skin cancers in Australians, *Med. J. Austral.* 1:352–353.

Lee, J. A. H., and J. M. Merrill. 1970. Sunlight and the aetiology of malignant melanoma—a synthesis, *Med. J. Austral.* 2:846–851.

Lee, J. A. H., and S. Yongchaiyudha. 1971. Incidence of and mortality from malignant melanoma by anatomical site, *J. Nat. Cancer Inst.* 47:253–263.

Lund, H. Z. 1965. How often does squamous cell carcinoma of the skin metastasize?, *Arch Dermatol.* 92:635–637.

Lynch, H. T., D. E. Anderson, J. L. Smith, Jr., J. B. Howell, and A. J. Krush. 1967. Xeroderma pigmentosum, malignant melanoma, and congenital ichthyosis: A family study, *Arch. Dermatol.* 96:625–635.

Macdonald, E. J. 1974a. A statement on skin cancer incidence and sun exposure, manuscript.

Macdonald, E. J. 1974b. Cancer of the skin in five regions in Texas: 1962–1966, manuscript.

Macdonald, E. J., M. S. Johnson, and A. Murphy. 1971. Regional patterns in morbidity from melanoma in Texas: 1944–1966, *Cancer Bull.* 23:51–55.

Macdonald, E. J., V. McGuffee, and E. White. 1973. Status of epidemiology of melanoma 1971, *Epidemiology of Melanoma, Pigment Cell.* Vol. 1. Karger, Basel. pp. 222–228.

McDonald, J. E. 1971. Statement submitted at hearings before the House Subcommittee on Transportation Appropriations, March 2.

McDowell, A. 1974. Nat. Center for Health Statistics. Personal communication.

McGovern, V. J. 1972. Melanoma: Growth patterns, multiplicity and regression, p. 95 in *Melanoma and Skin Cancer.* Proc. Int. Cancer Conf., Sydney. V.C.N. Blight, Government Printer, Sydney. March 13–17.

McGovern, V. J., and M. M. Lane Brown. 1969. *The Nature of Melanoma.* Charles C Thomas, Springfield, Ill.

Magnus, I. A. 1964. Studies with a monochromator in the common idiopathic photodermatoses, *Br. J. Dermatol.* 76:245–264.

Magnus, K. 1973. Incidence of malignant melanoma of the skin in Norway, 1955–1970: Variations in time and space and solar radiation, *Cancer* 32:1275–1286.

Mason, T. J., and F. W. McKay. 1974. *U.S. Cancer Mortality by County: 1950–1969.* DHEW Pub. No. (NIH) 74-615. U.S. Govt. Printing Off., Washington, D.C.

Melanoma Clinical Cooperative Group: Harvard University, New York University, Temple University, and the University of California, San Francisco. 1974. Preprint.

Moore, C., and P. C. Iverson. 1954. Xeroderma pigmentosum showing common skin cancer plus melanocarcinoma controlled by surgery, *Cancer* 7:377–382.

Movshovitz, M., and B. Modan. 1973. Role of sun exposure in the etiology of malignant melanoma: Epidemiologic inference, *J. Nat. Cancer Inst.* 51:777–779.

Murphy, R. S. 1974. Nat. Center for Health Statistics. Personal communication.

National Cancer Institute. 1974. The Third National Cancer Survey Advanced Three Year Report: 1969–1971 Incidence. Preprint.

National Center for Health Statistics, U.S. Dept. H.E.W. 1971. *Health Manpower: A County and Metropolitan Area Data Book.* Public Health Serv. Pub. No. 2044. U.S. Govt. Printing Off., Washington, D.C.

National Center for Health Statistics, U.S. Dept. H.E.W. 1973. *Plan and Operation of the Health and Nutrition Survey: United States 1971–1973.* Vital and Health Statistics, Ser. 1, No. 10a. U.S. Govt. Printing Off., Washington, D.C.

Oettlé, A. G. 1963. *Skin Cancer in Africa.* Nat. Cancer Inst. Monogr. No. 10. U.S. Dept. H.E.W., pp. 197–214.

Oettlé, A. G. 1964. Cancer in Africa, especially in regions south of the Sahara, *J. Nat. Cancer Inst.* 33:383–439.

Owen, L. G., M. T. Fliegelman, R. L. Jetton, and J. W. Musgrave. 1974. Seasonal variation of basal cell epitheliomas in Kentucky, *Arch. Dermatol.* 109:205–206.

Owens, D. W., J. M. Knox, H. T. Hudson, and D. Troll. 1974. Influence of wind on ultraviolet injury, *Arch. Dermatol.* 109:200–201.

Reed, W. B., B. Landing, G. Sugarman, J. E. Cleaver, and J. Melnyk. 1969. Xeroderma pigmentosum: Clinical and laboratory investigation of its basic defect, *J. Am. Med. Assoc.* 207:2073–2079.

Ringertz, N. 1970. Cancer incidence in Sweden 1962–1965, *Cancer Incidence in Five Continents,* Vol. 2. R. Doll, C. Muir, and J. Waterhouse, eds. UICC, Geneva.

Robbins, J. H., K. H. Kraemer, M. A. Lutzner, B. W. Festoff, and H. G. Coon. 1974. Xeroderma pigmentosum: An inherited disease with sun sensitivity, multiple cutaneous neoplasms, and abnormal DNA repair, *Ann. Intern. Med.* 80:221–248.

Robertson, D. F. 1968. Solar ultraviolet radiation in relation to sunburn and skin cancer, *Med. J. Austral.* 2:1123–1132.

Robertson, D. F. 1974. U. of Queensland, Australia. Personal communication.

Schneiderman, M. 1974. Personal communication of data from the Third National Cancer Survey Advanced Three Year Report.

Schulze, R., and K. Gräfe. 1969. Consideration of sky ultraviolet in the measurement of solar ultraviolet radiation. *The Biologic Effects of Ultraviolet Radiation: With Emphasis on the Skin.* F. Urbach, ed. Pergamon, New York, pp. 359–373.

Scotto, J. 1974. Nat. Cancer Inst. Personal communication.

Scotto, J., A. W. Kopf, and F. Urbach. 1974. Nonmelanoma skin cancer among Caucasians in four areas of the United States, *Cancer* 34:1333–1338.

Setlow, R. B. 1974. The wavelengths in sunlight effective in producing skin cancer: A theoretical analysis, *Proc. Nat. Acad. Sci.* 71:3363–3366.

Silverstone, H., C. B. Campbell, C. S. Hosking, L. P. Lang, and R. G. Richardson. 1963. Regional studies in skin cancer—first report: North-western Queensland. *Med. J. Austral.* 1:312–315.

Silverstone, H., and D. Gordon. 1966. Regional sutdies in skin cancer—second report: Wet tropical and subtropical coasts of Queensland, *Med. J. Austral.* 2:733–740.

Silverstone, H., and J. A. H. Searle. 1970. The epidemiology of skin cancer in Queensland: The influence of phenotype and environment. *Br. J. Cancer.* 24:235–252.

Swanbeck, G., and L. Hillström. 1971. Analysis of etiological factors of squamous cell skin cancer of different locations. 4: Concluding remarks, *Acta Dermato-vener* (Stockholm) 51:151–156.

Szymanski, F. J. 1963. Keratoacanthoma and squamous cell carcinoma, *South. Med. J.* 56:609–612.

U.S. Bureau of the Census, Dept. of Commerce. 1962. *City and County Data Book 1962.* U.S. Govt. Printing Off., Washington, D.C.

U.S. Bureau of the Census, Dept. of Commerce. 1973a. *1970 Census of Population: Characteristics of the Population.* U.S. Govt. Printing Off., Washington, D.C.

U.S. Bureau of the Census, Dept. of Commerce. 1973b. *1970 Census of Population, Subject Reports, National Origin and Language.* PC(2)-1A. U.S. Govt. Printing Off., Washington, D.C.

U.S. Weather Bureau, Dept. of Commerce. 1965. Summary of hourly observations, *Climatography of the United States*. Series No. 30. U.S. Govt. Printing Off., Washington, D.C.

Urbach, F. 1969. Geographic pathology of skin cancer, *The Biologic Effects of Ultraviolet Radiation: With Emphasis on the Skin*. F. Urbach, ed. Pergamon, New York, pp. 635–650.

Urbach, F. 1971. Geographic distribution of skin cancer, *J. Surg. Oncol.* 3:219–234.

Urbach, F., D. Berger, D. F. Robertson, and R. E. Davies. 1974. Field measurements of biologically effective UV radiation and its relation to skin cancer in man, manuscript.

Urbach, F., D. B. Rose, and M. Bonnem. 1972. Genetic and environmental interactions in skin carcinogenesis, Symp. on Fundamental Cancer Res., the 24th, M.D. Anderson Hosp. and Tumor Inst., Houston, 1971. *Environment and Cancer: A Collection of Papers*. Williams and Wilkins, Baltimore, Md., pp. 355–371.

Venkateswaran, S. V. (UCLA) 1974a. The solar ultraviolet radiation at the earth's surface, preprint.

Venkateswaran, S. V. 1974b. Radiation in the natural and perturbed troposphere. CIAP Monograph 4, Chap. 5.

Venkateswaran, S. V. 1974c. Solar ultraviolet radiation reaching the earth's surface, *Proc. of the Third Conf. on the CIAP,* Cambridge, Mass. DOT-TSC-OST-74-15, pp. 428–435.

Venkateswaran, S. V. (UCLA) 1974d. Personal communication.

Ward, W. H. 1967. Melanoma: Carcinoma of the skin and sunlight, *Austral. J. Dermatol.* 9:70–75.

Yates, F. 1955. The use of transformations and maximum likelihood in the analysis of quantal experiments involving two treatments, *Biometrika* 42:382–403.

Zagula-Mally, Z. W., E. W. Rosenberg, and M. Kashgarian. 1974. Frequency of skin cancer and solar keratoses in a rural southern county as determined by population sampling. *Cancer* 34:345–349.

Appendix D

Effects of Ultraviolet on Plants

The references and tables that follow are some of the foundations for the conclusions reached in Chapter 3 concerning the effects of uv-B on plants. The tables are abstracted from CIAP Monograph 5, Chapter 3, and describe experiments in which the ultraviolet irradiance in the uv-B range was increased by an amount corresponding to an approximate 50 percent decrease in ozone. The ultraviolet supplementation was achieved by use of sunlamps whose light was filtered through thin cellulose acetate sheets so as to cut out wavelengths <290 nm. Control plants were exposed through thin Mylar films that cut out wavelengths below 310 nm. Thus the Mylar transmitted supplementary uv-A and the cellulose acetate uv-B plus uv-A. Supplementation was carried out on plants grown in growth chambers, in greenhouses, and in the field.

This Appendix presents the findings of the Panel on Biological and Medical Effects: R. Setlow, *Chairman;* W. Baier, W. Butler, R. Clayton, E. Deevey, T. Fitzpatrick, A. Giese, D. Gordon, F. Haxo, E. Runge, E. Scott, and F. Urbach. Consultant to the Panel: M. Caldwell.

I. Effects of Supplemental uv on Plant Growth

Caldwell, M. M. 1968. Solar ultraviolet radiation as an ecological factor for alpine plants, *Ecol. Monogr.* 38:243–268.

Caldwell, M. M. 1971. Solar uv irradiation and the growth and development of higher plants, *Photophysiology*. Vol. 6. A. C. Giese, ed. Academic, New York. pp. 131–177.

Lockhart, J. A., and U. Brodfuhrer-Franzgrote. 1961. The effect of ultraviolet radiation on plants, *Encyclopedia of Plant Physiol.* Vol. 16. W. Ruhland, ed. Springer-Verlag, Berlin. pp. 532–554.

A. IN GROWTH CHAMBERS

The data are from CIAP Monograph 5, Chapter 3, Tables 3.36 to 3.40.

	Dry Weight per Plant (g) Exposed to Supplementary uv-radiation	
Plant	(uv-A and uv-B)	(uv-A only)
Peas, "little marvel"	0.74	1.10
Peas, "early Alaska"	0.17	0.39
Cabbage	1.10	2.07
Collard	1.62	2.45
Corn	0.32	0.40

B. IN GREENHOUSES

Figure D.1 shows the effects of supplemental uv-B on the growth of *Rumex mexicanus*.

II. Effects of Supplemental uv on Photosynthesis and Chlorophyll Content of Leaves

Mantai, K. E., J. Wong, and N. I. Bishop. 1970. Comparison studies on the effects of ultraviolet irradiation on photosyntheses, *Biochim. Biophys. Acta* 197:257–266.

This material is excerpted from CIAP Monograph 5, Section 3.2.5, pp. 3-61 to 3-64.

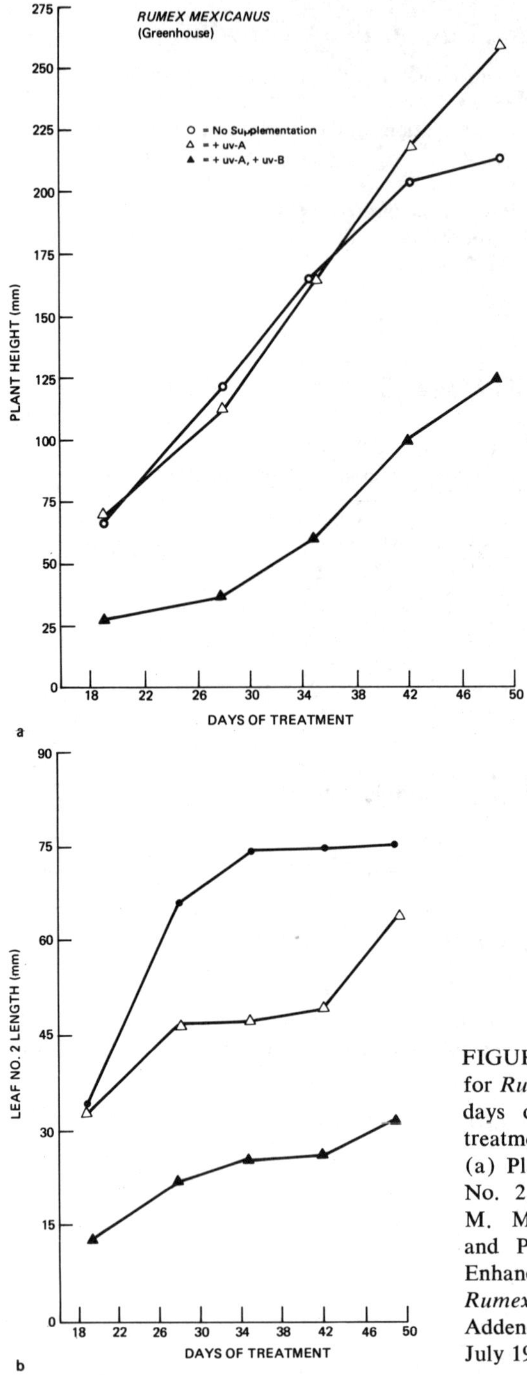

FIGURE D.1 Plant measurements for *Rumex mexicanus* from 19 to 48 days of uv-enhanced and control treatments in greenhouse studies. (a) Plant height; (b) length of leaf No. 2. (From W. B. Sisson and M. M. Caldwell, "Morphological and Photosynthetic Studies of an Enhanced U.V. Radiation Regime on *Rumex mexicanus* and Soybean," Addendum to CIAP Monograph 5, July 1974).

Net Photosynthesis of Collards Exposed to a uv-B-Enhanced Light Regime *

Supplementary Light	Experiment I	Experiment II		
	mg $CO_2/dm^2/h$	mg CO_2/g dm^2/h	mg CO_2/g fr. wt/h	mg CO_2/mg total chlorophyll/h
uv-A	26.14	19.62	7.77	3.49
	22.21	19.63	6.88	4.28
	16.40	25.54	7.98	4.30
	22.70	21.11	8.26	4.37
	23.87	23.16	8.13	4.35
	22.58	—	—	—
AVERAGE	22.32	21.81	7.80	4.16
uv-A plus uv-B	16.10	15.21	5.92	4.44
	12.94	18.87	8.07	4.43
	14.72	17.85	6.76	4.72
	18.97	19.41	6.73	3.38
	12.01	—	—	—
AVERAGE	14.95 †	17.83 ‡	6.87 N.S.	4.24 N.S.

Chlorophyll Content of Beans and Cabbage Exposed to a uv-B-Enhanced Light Regime *

Supple-mentary Light	Beans		Cabbage	
	mg chlorophyll a+b/g fresh wt	chlorophyll a / chlorophyll b	mg chlorophyll a+b/g fresh wt	chlorophyll a / chlorophyll b
uv-A	2.48	2.14	1.69	2.33
	2.71	2.15	1.53	2.89
	2.45	2.18	1.64	2.49
	1.98	2.36	—	—
AVERAGE	2.40	2.21	1.62	2.57
uv-A plus uv-B	1.56	2.06	1.40	2.81
	1.50	2.28	1.31	2.61
	2.00	2.24	1.48	2.72
	2.02	2.07	—	—
AVERAGE	1.77 ‡	2.16 N.S.	1.40 ‡	2.71 N.S.

* uv-B treatment equivalent to that present with 50% ozone depletion.
† Significantly different from control according to F test, $p = 0.01$.
‡ Significantly different from control according to F test, $p = 0.05$.

III. Effects of Supplemental uv on Cell Structure in Plants

Barton, R. 1966. Fine structure of mesophyll cells in senescing leaves of Phaseolus, *Planta* 71:314–325.

Butler, R. S. 1967. The fine structure of senescing cotyledons of cucumber, *J. Exp. Bot.* 18:535–543.

Cline, M. G., and F. B. Salisbury. 1966a. Effects of ultraviolet radiation on the leaves of higher plants, *Radiat. Bot.* 6:151–163.

Cline, M. G. and F. B. Salisbury. 1966b. Effects of ultraviolet alone and simulated solar ultraviolet radiation on the leaves of higher plants, *Nature* 211:484–485.

Mantai, K. E. 1970. Electron transport and degradation of chloroplasts by hydrolytic enzymes and ultraviolet irradiation, *Carnegie Inst. Washington Yearbook 1968–1969*, pp. 593–603.

This material is excerpted from CIAP Monograph 5, Section 3.2.3, pp. 3-45 to 3-52.

Since cellular damage may be evident long before morphological changes are apparent, experiments were carried out on the effects of supplementary uv-B on the ultrastructure (detected by electron microscopy) of soybeans grown in the field.

Cells from plants grown under supplementary uv-A for 56 hours exhibited no detectable changes from those of the nontreated controls.

However, many mesophyll cells from plants exposed to uv-A plus uv-B exhibited a considerable amount of vesiculation, presumably arising from disruption of the tonoplast, an increased amount of RER, or both. Also, there was a slight disorientation of the grana within some of the chloroplasts, there was a slight aggregation of the plastoglobuli within some chloroplasts, and the chloroplast appeared slightly misshapen.

In any biological system, there will be a certain amount of degradation of organelles concomitantly with growth. Moreover, a certain amount of disruption of normal organellar structure may be attributed to the harsh treatment of preparative procedures for electron microscopy. The recorded alterations also may have developed secondarily, as a result of the uv treatments, if the radiation rendered the membranes more susceptible to damage during preparation of the samples for electron microscopy.

Although statistical analyses are not practical at this time, it was felt that the injury could be attributed to uv radiation.

IV. Effect of Supplemental uv on Plant Cell Mutations

Ichikawa, S., and A. H. Sparrow. 1967. Radiation-induced loss of reproductive integrity in the stamen hairs of *Tradescantia blossfeldiana* Mildbr., a twelve-ploid species, *Radiat. Bot.* 7:333–345.

This material is abstracted from CIAP Monograph 5, Section 3.2.2, pp. 3-108 to 3-112.

The following describes the effects of supplementary uv in stamen hair mutations in *Tradescantia* grown in the field.

Uniformly vigorous plants of *Tradescantia,* clone 4430 (obtained from A. H. Sparrow, Brookhaven National Laboratory) were given 7 hours of supplemental uv radiation per day under field conditions. Non-treated and treated plants were grown under these conditions for one week to equilibrate them, and then somatic aberrations (color changes in the petals and stamen hairs) were scored for 30 days. Somatic aberrations are expressed as the frequency of events of each type per hair; an event is defined as one cell or any sequence of contiguous cells exhibiting the same aberration or mutation.

Effect of uv Radiation on the Frequency of Aberrant Events per Stamen Hair and Total Aberrant Events in Stamen Hairs of *Tradescantia* Clone 4430, Grown under Field Conditions; Plants were equilibrated for 1 Week and Then Scored for 20 days *

Treatment	Aberrant Events per Hair	Total Aberrant Events	Total Number of Stamen Hairs Scored
No supplementation	0.0003 a	20 a	58,800
Plus uv-A	0.0020 b	156 b	79,800
uv-A plus uv-B	0.0069 c	577 c	84,750

* Those values within a column followed by the same letter are not significantly different at the 5 percent level.

V. Effects of Ultraviolet on Algae

A. 3 HOURS OF SUNLIGHT KILLS 90 PERCENT OF THE POPULATION OF THE ALGA *Chlamydomonas*

This material is excerpted from CIAP Monograph 5, Section 3.2.7, supplemental.

Sunlight irradiations were performed on cells on agar in petri dishes surrounded by water that was kept cool (20–27°C) by evaporation or by adding ice cubes. Control plates covered with black paper or ¼-in. Lucite showed no effects of a 4-hour exposure.

Simulated solar uv-B was obtained from FS 20 sunlamps filtered with cellulose triacetate films. The filters are important in blocking wavelengths below 290 nm in order to simulate solar uv-B: at comparable total irradiances below 315 nm, the wavelengths below 290 nm are roughly 20 times more effective in killing *Chlamydomonas* than those between 290 and 315 nm. Without the filter, 50 percent of the cells are killed with a 6-min exposure; with the filter, about 120-min exposure is needed.

Dosimetry of uv-B was accomplished with a Robertson-Berger Sunburning Ultraviolet Meter. This instrument is described and discussed fully by Urbach in CIAP Monograph 5. Doses and dose-rates are presented as SBU or $SBU \cdot h^{-1}$, where SBU refers to Sun Burn Unit and corresponds to 400 counts on the R-B meter. Typical results are shown in Figure D.2.

B. COMPARATIVE SENSITIVITIES OF VARIOUS ALGAE TO UV

Hipkiss, A. R. 1967. Immediate loss of flagella from *Chlamydomonas reinhardi* after ultraviolet irradiation, *Radiat. Bot.* 7:347.

Schiff, J. A., and H. T. Epstein. 1965. The continuity of the chloroplast in *Euglena, Reproduction: Molecular, Subcellular, and Cellular.* M. Locke, ed. Academic, New York, pp. 131–189.

Srivastava, B. S., and H. D. Kumar. 1969. Inactivation and killing of the blue-green alga *Anabaena doliolum* grown on ultraviolet-irradiated culture medium, *Radiat. Bot.* 9:291.

Van Baalen, C. 1968. The effects of ultraviolet irradiation on a colloidal blue-green alga: Survival, photosynthesis, and photoreactivation, *Plant Physiol.* 43: 1689–1695.

This material is excerpted from CIAP Monograph 5, Section 3.2.7, pp. 3-89 and 3-90.

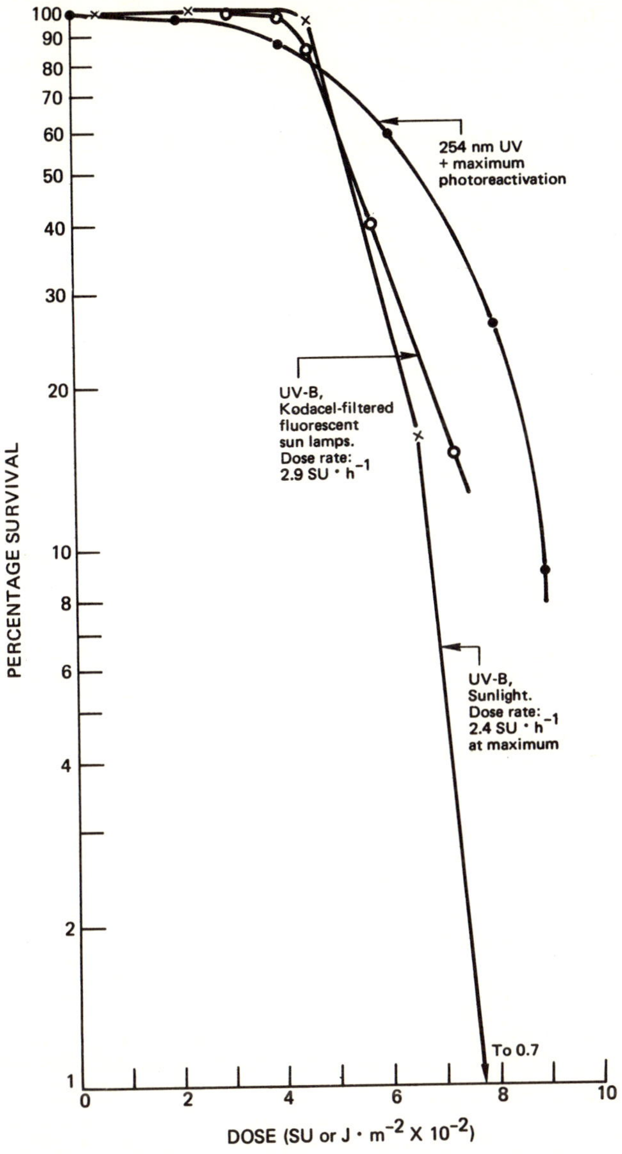

FIGURE D.2 Percentage survival of *Chlamydomonas* versus uv dose. (From D. S. Nachtwey, "Linking Photobiological Studies at 254 nm with uv-B," Addendum to CIAP Monograph 5, July 1974).

The Doses at 254 nm to Inhibit 90 Percent of Green Colony Formation (1) and Colony Formation (2)

	Organism	Dose ($J \cdot m^{-2}$)
(1)	*Euglena gracilis* (dark grown cells)	21.5
	Euglena gracilis (light grown cells)	37
	Hitzschia linearis	54
	Chlamydomonas reinhardi	100
(2)	*Agmenellum quadroplicatum*	110
	Chlorella vulgaris	130
	Chlamydomonas reinhardi	156
	Euglena gracilis	156
	Chlamydomonas reinhardi	200
	Anabaena doliolum (spores)	435
	Chlamydomonas reinhardi	580
	Netrium digitus	1200

C. EFFECTIVE DOSES OF UV-B ON PLANT CELLS WILL BE APPROXIMATELY DOUBLED BY A 50 PERCENT DECREASE OF OZONE

Caldwell, M. M. 1971. Solar uv irradiation and the growth and development of higher plants, *Photophysiology*. Vol. 6. A. C. Giese, ed. Academic, New York, pp. 131–177.

Luckiesh, M. 1946. *Applications of Germicidal, Erythemal and Infrared Energy*. Van Nostrand, New York, p. 70.

Stadler, L. J., and F. M. Uber. 1942. Genetic effects of ultraviolet radiation in maize, *Genetics* 27:84–118.

This material is excerpted from CIAP Monograph 5, Section 3.2.7, pp. 3-72 to 3-81.

Caldwell's method allows us to evaluate the increase in *biocidal qualities* of an increase in intensity of solar uv-B. The method takes into account the fact that the biological effects of uv depend not only on dose (intensity × time) but also on wavelength. Some wavelengths of uv are more effective than others. In Caldwell's method, the biologically effective irradiance of a mixture of uv-B photons of different energies is assessed by weighting the number of photons/cm^2 at each wavelength by a factor that reflects the probability of having an effect and then summing the weighted intensities. Caldwell assigned photons of 280-nm wavelength a relative weight of one and photons of wavelength greater than 280 nm decreasing relative effectivenesses. (See Appendix B.) To obtain the weighting values, Caldwell constructed a generalized action spectrum from portions of published action spectra, which represent

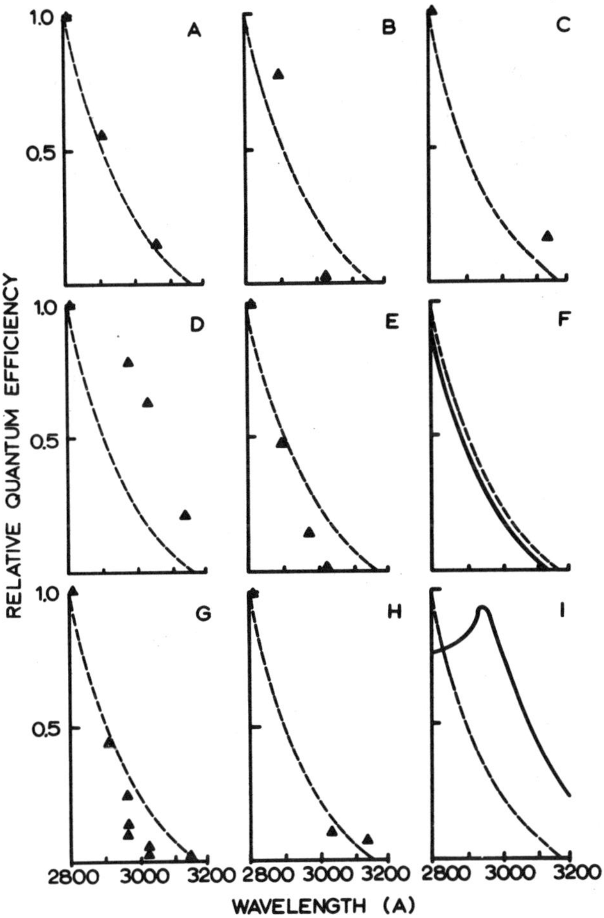

FIGURE D.3 Plant uv-B action spectra relative to 2800 Å (280 nm). A, Mutation of liverwort spores; B, mutation in the fungus *Trichophyton mentagrophytes;* C, inhibition of photosynthesis in *Chlorella pyrenoidosa;* D, cessation of cytoplasmic streaming in epidermal cells of *Allium cepa;* E, frequency of "endosperm deficiencies" in corn due to chromosomal deletion or mutation; F, germicidal action spectrum; G, epidermal cell damage in *Oxyria digyna* leaves; H, induction of chromosomal aberrations in *Tradescantia paludosa* pollen; I, base curvature of *Avena* coleoptiles. Intermittent lines represent the spectral sensitivity of a uv phototube chosen to approximate the biological action spectra for uv-B effects on plant cells (Caldwell, 1971).

experimentally determined relative effectivenesses of different wavelengths of uv in producing given effects. Figure D.3 (Caldwell, 1971) shows his generalized plant uv-B action spectrum and the data points upon which it is based. Using this spectrum and calculated irradiances, it was shown by standard methods that a 25 percent decrease in ozone will result in a 50 percent enhancement of the biocidal power.

Appendix E

Climatic Change (Natural and Man-Made)

Over the last half billion years or so, the climate of the world has been characterized by long, mild, and even epochs interspersed by much shorter intervals of much greater variability. The length of the mild epochs has been of the order of 100 million years; that of the variable intervals, a few million. Currently one of the variable intervals prevails when polar ice sheets have expanded and withdrawn with periods of the order of 100,000 years. Global average rainfall and temperature have varied with the same frequency. At present, the climate is intermediate between coldest and warmest; the global average temperature at the peak of the ice ages was perhaps 5°C below the present temperature.

The causes of these large climatic fluctuations are uncertain. Many theories have been proposed, some with considerable imagination. Among those proposed are the following: postulated changes in the solar-energy output; changes in earth–sun geometry; changes in the earth's atmosphere, its gases and particulate loading; changes in the earth's topography (Mitchell, 1968); and the theory of self-induced (unstable) oscillation of the atmosphere–ocean–cryosphere system.

It is, of course, not necessary that any one mechanism be operating by itself; several may have contributed simultaneously. Further, it is quite possible that climatic changes on different time scales have differ-

---

This Appendix presents the findings of the Panel on Atmospheric Physics and Chemistry: F. Johnson, *Chairman;* J. Friend, *Vice Chairman;* J. Chamberlain, D. Hunten, H. Johnston, F. Kaufman, H. Landsberg, and H. Panofsky. Consultants to the Panel: R. Hudson, C. Leith, E. Reiter, and G. Robinson.

ent causes. For example, drifting of the continents is too slow to be responsible for the advances and recessions of glaciers within one of the "short" and variable intervals. But drifting of the continents almost certainly had an impact on the very long-term changes and is responsible for the fact that glaciation had completely different distributions during the different ice ages.

Changes on the sun, visible or otherwise, could have produced climatic changes on all time scales; but we have no evidence for or against any long-period solar variation. In that connection, most of the theories attempting to explain climatic change on the basis of solar variation postulate that an ice age follows a cooling of the sun; another theory, however, suggests just the opposite.

Several theories have been proposed specifically to account for climatic fluctuations with a time scale of the order of 100,000 years. Only one of these can be checked in a reasonably quantitative fashion. This is the theory by Milankovich (1941), which is based on changes in the characteristics of earth–sun geometry. These can be computed by astronomical techniques and predict the timing of advance and recession of ice sheets.

Of particular interest for this report are two other theories dealing with changes on the same time scale: the CO_2 theory and the volcanic-dust theory. If CO_2 were to increase, because of volcanic eruptions or other causes, the surface temperature would rise as a result of the increased greenhouse effect. Repetitive warming and cooling is explained by periodic exchanges of CO_2 between atmosphere and oceans. Volcanic dust certainly decreases the solar radiation reaching the ground, suggesting that volcanic eruptions should be followed by surface cooling. The particles themselves absorb solar radiation, so that stratospheric warming might be expected.

It is difficult to find statistical evidence for a relation between mean surface temperature anomalies and volcano eruptions in the northern tropical and temperate zones in the last century (Schwarzbach, 1974). Regional cooling has been observed after some eruptions, but such is of doubtful statistical significance. Stratospheric warming by 6°C following the Mt. Agung eruption in 1963 has been found by Newell (Wilson and Matthews, 1972), but the interpretation is still somewhat controversial. No tropospheric effect was observed, suggesting that even relatively large stratospheric contamination may have no observable consequences in the troposphere. In summary, then, it is difficult to show a clear relationship between stratospheric particles and surface climate; but inferences favor climatic cooling with increased stratospheric particle contamination.

Since the recession of the last ice sheet about 11,000 years ago, there have been many less spectacular changes in temperature and precipitation—not only in their magnitudes but also in their patterns. Global average temperature changes have been typically of order 2°C. The warmest period was about 5000 B.C. Again about A.D. 1000, the climate was warmer than it is today. Since that time, there was the "little ice age" from the sixteenth to the nineteenth century. These climatic changes over the past 5000 years have become better and better documented. Weather events, crop conditions, and glacier positions have entered chronicles, diaries, and, later, pictorial representations. Freezing dates of rivers and lakes as well as blooming dates of trees have, in many areas, been faithfully recorded (Chu, 1973; LeRoy Ladurie, 1971; Yamamoto, 1974). For nearly three centuries, instrumental records have supplied reliable climatic information, which, in the past century, has reached global coverage (Lamb, 1966).

All of these data document a substantial range of natural variability. Most of this is adequately dealt with in the pertinent literature and the CIAP treatment (U.S. Department of Transportation, 1974). The evidence shows that during the last century there has been, in the northern hemisphere, a temperature rise to the mid-1940's of about 0.6°C, and since that time a cooling of about 0.3°C (Budyko, 1974). These changes seem to have had natural causes. However, it needs to be emphasized that any global changes have been unevenly distributed over the globe. They are small in equatorial areas and the southern hemisphere. In the northern hemisphere, the changes have been largest on the northern margins, north of 40–50° latitude. There, especially in the inhabited zones near the Arctic Circle, some rather spectacular rises and later falls of temperature have taken place. Such changes have had corresponding consequences on ice fields and plant growth. These high-latitude regions are the climatically vulnerable areas of the globe.

No statements on precipitation changes can be made because of the high variability of this element. The length of record is insufficient to discover trends, but it is clear that intervals of decade length in various localities have been drier or wetter than the average for the respective localities. This applies particularly to marginal rainfall zones near deserts and in the monsoonal regions.

It should be realized that many climate-producing factors have their origin at or near the surface of the earth. One may enumerate here land and water contrasts, land forms (particularly mountains and lakes), vegetation, and soil conditions. Each type of surface has different physical characteristics such as albedo (reflectivity for incoming radiation), heat conductivity, and heat capacity. These are readily changed and disturbed. Every such alteration causes a change in climate, albeit only locally in the layer of air near the surface. This is often referred to as a microclimatic change. When larger areas are affected, the change will result in mesoclimatic change.

Such changes, which are measurable and have been well documented, are caused by deforestation, swamp drainage, conversion to agriculture, and urbanization. Some of these changes can lead to secondary alterations with further local climatic consequences. Examples are deforestation or overgrazing with subsequent soil erosion, leading in areas near arid zones to encroachment of the deserts (Matthews *et al.,* 1971).

In urban areas far-reaching atmospheric changes have taken place (Landsberg, 1974). These include the formation of so-called heat islands with well-marked temperature rises. These coupled with air contaminants cause added rainfall downwind of urban complexes, and the smoke plumes can drift hundreds of kilometers downwind with as yet unascertained effects on cloud and precipitation formation (Hobbs *et al.,* 1974). Although, fortunately for world climate, these unnatural admixtures are eliminated by a number of processes within a few days, they have acidified the rain in industrial areas. This acid rain not only falls in the vicinity of the sources but even hundreds of kilometers downwind to the detriment of agricultural activities there (Likens and Bormann, 1974).

However, there is no evidence whatsoever that these changes brought about by man's alteration of the landscape have caused any global climatic change. The most radical ones so far, namely, the conversion of aboriginal forest land to agricultural pursuits, has certainly changed elements of the water balance, but because this land use covers only about 11 percent of the land surface of the earth it has not had an appreciable effect on the general circulation of the atmosphere. This is not to say that there are no profound ecological changes in the biosphere caused by expansions of the arable land.

Agriculture is also a major producer of "dust," because of wind erosion taking place in fallow areas and off-season. Furthermore, slash-and-burn methods used in native tropical agriculture add to the particulate load of the lower atmosphere. Industry and traffic run a close second in production of particles. The best current estimates place particulate production caused, directly or indirectly, by human activity at 15 percent of natural emission of particulates into the atmosphere (Peterson and Junge, 1971). Hence, the lower troposphere is naturally a turbid medium, and the effects of this turbidity have affected climate throughout history. There is no evidence that, except for the already mentioned precipitation effects, anthropogenic aerosols have had any appreciable climatic effects up to the present time (Elsaesser, 1973). There is also no convincing evidence that the atmosphere on a global scale has become discernably more turbid in recent decades due to human activities; of course, big increases have occurred on a local scale.

A more difficult problem arises with man-made additions of carbon dioxide (CO_2) to the atmosphere. There have been massive injections of this gas from combustion processes using fossil fuels. These went beyond the absorbing capacity of the great sink for CO_2, namely, the ocean water. Thus the concentration of this atmospheric admixture has risen from about 290 parts per million to 330 parts per million during the past century. The reason for concern about this increase stems from the fact that CO_2 is an absorber of infrared radiation. As such it intercepts radiation from earth to space, and hence it has been claimed that this change in the planetary radiation balance should lead to a global temperature rise. In fact, when a rise in the mean temperature of the planet was noted in the 1920 to 1940 interval it was promptly associated with the rapid increase in CO_2. Some rather simplified simulations of the process would lead one to calculate a 0.3°C global temperature rise for the 10 percent increase in CO_2. But feedback mechanisms would bring about increased cloudiness, thus causing a higher albedo, and compensate, at least in part, for the CO_2 effect. Even more notable is the fact that an increase of CO_2 of over 10 parts per million has taken place in the past 15 years, during which all indications have pointed to a global cooling. So again, there is no observational evidence of a man-induced climatic change caused by this factor (Keeling, 1973; Schneider and Kellogg, 1973).

Effects on climate by human interference are more likely to be expected from changes brought about in the higher layers of the atmosphere. There has been considerable discussion of the possible effects of persistent condensation trails produced by high-flying aircraft. The first effect of such man-made clouds could be an increase in global albedo, which is usually associated with cooling. The second effect could be the dropping of ice crystals from these clouds into lower cloud layers, where they might induce precipitation. There is some local evidence for an increase in cirrus cloud cover during the last decade in several areas of the United States. This has been linked to the increase in jet fuel consumption, which is a rough measure of substratospheric air traffic. This increase in high clouds over the prior periods has been about 10 percent (Machta, 1971). Inspection of satellite pictures does not indicate, up to the present, that such condensation trails cause an effective increase in total global cloudiness or albedo. This aspect could, however, be a concern for the future. There are no observations to support the contention that falling crystals from such man-made cirrus clouds have initiated precipitation in lower clouds. If such cases have occurred, present knowledge of cloud physics suggests that this precipitation would have started naturally anyway but perhaps a little later in time.

There is a potential for climatic change by man-made stratospheric injections of pollutants. Many of these (NO_x, SO_x) are treated elsewhere in this report, but a few words on stratospheric aerosols are in order in this context. Because of their long residence time and possible steady accumulation (at least until a new steady state is reached), they have an impact on global climate. A probably minor effect would be their influence on the vertical temperature distribution. More important would be the effect of a stratospheric aerosol layer on an increase of the global albedo. The best available evidence suggests that this increase would be small and hence diminish the energy flux to the earth's surface only very slightly. Yet a small cooling effect has been suggested as a consequence. Transitory massive dust injections into the stratosphere by volcanic eruptions during historical times have been adduced as analogies, but they have had only minor, if any, effect on surface temperatures.The climatic effect of such injections may have been small because of their transitory nature, but effects from a continuing source, such as stratospheric aircraft, may be larger.

In summary, man's activities have had measurable and significant influences on local climatic conditions. There is no evidence that they have had any observable influence on climate on a larger, particularly global, scale.

References

Budyko, M. I. 1974. *Climate and Life*. Academic Press, New York. 508 pp.

Chu, K. C. 1973. A preliminary study on the climatic fluctuations during the last 5000 years in China, *Scientia Sinica* 16(2):226–256.

Elsaesser, H. W. 1973. The upward trend in airborne particulates that isn't, *EOS* 54:1096–1097.

Hobbs, P. V., H. Harrison, and E. Robinson. 1974. Atmospheric effects of pollutants, *Science* 183:909–915.

Keeling, C. D. 1973. The carbon dioxide cycle, *Chemistry of the Lower Atmosphere*. S. I. Rasool, ed. Plenum, New York, pp. 251–329.

LeRoy Ladurie, E. 1971. *Times of Feast, Times of Famine*. Doubleday, Garden City. 426 pp.

Lamb, H. H. 1966. *The Changing Climate*. Methuen, London. 236 pp.

Landsberg, H. E. 1974. Inadvertent atmospheric modification through urbanization, *Weather Modification*. Chap. 20. W. Hess, ed. Wiley, New York, pp. 726–763.

Likens, G. E., and F. H. Bormann. 1974. Acid rain: A serious regional environmental problem, *Science* 184:1176–1179.

Machta, L. 1971. Global effects of contaminants in the upper atmosphere. Paper presented at 64th Ann. Mtg. Am. Inst. of Chem. Eng. San Francisco.

Matthews, W. H., W. W. Kellogg, and G. D. Robinson, eds. 1971. *Man's Impact on Climate*. MIT Press, Cambridge, Mass. 594 pp.

Milankovich, M. 1941. Kanon der Erdbestrahlung und seine Anwendung auf des eigeit Problem, *Kgl. Serb. Akad. Spec. Pub.* Vol. 132. Belgrad.

Mitchell, J. M., Jr., ed. 1968. Causes of climatic change, *Meteorol. Monogr. 8(30)*. Am. Meteorol. Soc., Boston. 159 pp.

Peterson, J. T., and C. E. Junge. 1971. Sources of particulate matter in the atmosphere, *Man's Impact on Climate*. W. H. Matthews *et al.*, eds. MIT Press, Cambridge, Mass., pp. 310–320.

Schneider, S. H., and W. W. Kellogg. 1973. The chemical basis for climatic change, *Chemistry of the Lower Atmosphere*. S. I. Rasool, ed. Plenum, New York, pp. 203–249.

Schwarzbach, M. 1974. Vulkane, vom paläoklimatologen betrachtet, *Bonner Meteorol. Abhandl.* 17:575–593.

U.S. Department of Transportation. 1974. The natural and radiatively perturbed troposphere, CIAP Monograph 4 (draft).

Wilson, C. L., and W. H. Matthews, eds. 1972. Man's impact on the global environment, *Study of Critical Environmental Problems (SCEP)*. MIT Press, Cambridge, Mass. 319 pp.

Yamamoto, T. 1974. The changing climate through the period of meteorological records and the historical past in Japan and some considerations about the cold climate in the alluvial epoch, *Quaternary Rec.* 12(4):145–159.

Appendix F

Perturbation of Solar and Terrestrial Radiation by Changes in Stratospheric Composition

Laboratory experiments in which sulfur dioxides and water vapor in air were exposed to radiation in the wavelength region above 250 nm have shown conclusively that diluted sulfuric acid aerosol is produced. There is also ample laboratory evidence that the presence of ammonia gas greatly enhances the oxidation rate of SO_2 to sulfate, producing particles of ammonium sulfate. Laboratory studies of the kinetics of SO_2 reactions show that O, OH, and HO_2 radicals are capable of initiating the oxidation of SO_2 (Davis, 1974). On the basis of measured reaction rate constants and profiles of O, OH, and H_2O concentrations computed from one-dimensional models of the stratospheric chemical system, it has been shown that the reaction of SO_2 with OH is much more rapid than with any of the other radical or molecular species present in the lower stratosphere. The exact mechanism of the formation of H_2SO_4 from the reaction of $SO_2 + OH$ is not known. That H_2SO_4 is formed has been well demonstrated in laboratories. The major question about the formation of stratospheric H_2SO_4 is whether there is some species of which we are currently unaware that could oxidize SO_2 at an even faster rate than the OH radical. Studies using one-dimensional models incorporating the SO_2 and OH reaction at least indicate the reasonableness of the idea that the natural sulfate aerosol layer originates

This Appendix presents the findings of the Panel on Atmospheric Physics and Chemistry: F. Johnson, *Chairman;* J. Friend, *Vice Chairman;* J. Chamberlain, D. Hunten, H. Johnston, F. Kaufman, H. Landsberg, and H. Panofsky. Consultants to the Panel: R. Hudson, C. Leith, E. Reiter, and G. Robinson.

from gas-phase reactions in the lower stratosphere. There can be hardly any doubt that SO_2 emitted by aircraft flying in the stratosphere will to some extent be converted to diluted sulfuric acid aerosol.

If material that absorbs solar radiation is added to the stratosphere, the solar energy reaching the troposphere and the earth's surface is correspondingly reduced, as is the proportion of incident solar radiation reflected back to space from the atmosphere and surface. There is, therefore, a tendency to warm the stratosphere, to warm the planet by reduction of the planetary albedo, and to cool the troposphere. If material absorbing solar energy is removed from the stratosphere, the opposite effects occur—there is a tendency to cool the stratosphere, to cool the planet by increase of the planetary albedo, and to warm the troposphere. If material that scatters solar radiation—a proportion of it back to space—without absorption is introduced into the stratosphere, energy reaching the troposphere and surface is reduced, as is the energy reflected from troposphere and surface. There is a tendency to cool the troposphere, but the effect on the stratosphere and the planet as a whole depends on details of the scattering process and is not obvious without careful calculation. If the added material both scatters and absorbs solar radiation, there will be a tendency to warm the stratosphere and to cool the troposphere: again the net effect on the planet is not obvious *a priori*.

There is also local heating or cooling of the surface and atmosphere by exchange of terrestrial radiation: this depends on both the composition and temperature structure of the atmosphere. Composition and temperature structure are not independent; estimation of the actual change of the heating rate and temperature structure following a change of composition requires consideration of the dynamics and thermodynamics of the whole atmosphere. The only computation that can be made using the radiative transfer equations alone is of the initial perturbation of heating rate that would follow a sudden change of composition with no change of temperature structure. Any changes of heating rates would, in fact, be followed by changes in temperature structure, and such changes would have climatic implications. These implications are the concern of climatic models and are not considered here.

We examine three major perturbations of the composition of the stratosphere.

I. Addition of NO

Of the gases into which emitted NO is transformed, one—NO_2—has a very high cross section for absorption of solar radiation, extending through the near ultraviolet and peaking in the blue, where the incident radiation is of high intensity. In the terrestrial radiation, the NO_x (i.e., NO and NO_2) emissions are masked by the much stronger emissions of H_2O and CO_2 and may be ignored in the present state of precision of either computation or measurement.

II. Reduction of O_3

Ozone has an extremely high cross section in the solar ultraviolet ($\lambda < 300$ nm) and also absorbs in the visible with a peak in the yellow-orange region (Chappuis band). When computed in terms of absorption of total solar radiation, the O_3 cross section, although large by absolute standards, is about 1/100 that of NO_2.

Ozone makes a significant contribution to the field of terrestrial radiation, and perturbation of this, following ozone reduction, cannot be ignored.

III. Addition of Particles

Little is known of the optical properties, which depend critically on size as well as composition, of the sulfuric acid/water particles whose addition to the stratosphere is postulated. If they are uncontaminated and of a size similar to naturally occurring particles, they can be expected to scatter solar radiation throughout the spectrum and absorb in the solar infrared. Their influence on the terrestrial radiation field can be computed and shown to be a smaller perturbation than that of solar radiation. Details of these perturbations follow.

A. THE NO_2–O_3 PERTURBATION AND SOLAR RADIATION

As is clear from the body of this report and Appendix A, there remain areas of uncertainty among the various techniques (models) available for study of O_3 reduction following injection of NO_x in the stratosphere. Computation of the net heating rate tendency following this perturbation requires details of the perturbed concentrations of NO_2 (not NO_x) and O_3 at all heights. Different models produce different heating rate perturbations. Here we will present two illustrations, the first an approximate approach, which is perhaps the most elaborate that the situation warrants, the other an illustration of the consistent use of one of the models.

Example 1:

NO_2 perturbation $+5 \times 10^{15}$ molecule cm^{-2} in a vertical column.

O_3 perturbation -7.5×10^{17} molecule cm^{-2} in a vertical column.

Solar elevation 30° for 12 h in 24 (which produces the correct global annual mean power).

Troposphere–surface albedo 0.3 in green-red (ozone Chappuis band) and 0.35 in near uv and blue (NO_2 band).

Computation shows:

1. Decrease of downward solar radiation at the tropopause due to increase in NO_2, 0.11% (0.37 W/m^2 global mean).
2. Increase in the same due to decrease in O_3, 0.15% (0.5 W/m^2).
3. Increase in stratospheric absorption due to NO_2 increase, 0.15% (0.5 W/m^2).
4. Decrease in the same due to O_3 decrease, 0.16% (0.54 W/m^2).

Within the accuracy of this computation (the albedo assumptions are particularly doubtful), it is reasonable to say that the net heating effect of the addition of NO_x on the solar radiation field is negligible in both stratosphere and troposphere. There would, of course, be changes in the spectral composition of solar radiation within the atmosphere.

Example 2:

F. M. Luther has made detailed calculations of heating rate changes following injection of 1.42×10^{12} g per year of NO_x (as NO_2) in a 1-km layer centered on 20 km, using the one-dimensional atmospheric model of J. Chang. Figure F.1 illustrates his results for the mean solar elevation, 30°. This computation shows that over most of the stratosphere the decrease of heating rate due to ozone depletion is about twice the increase due to addition of NO_2. The major difference between this computation and the approximate computation of Example 1 is the magnitude of the NO_2 perturbation, i.e., the increase over "natural" concentrations. The changes in integrated column densities given by the Chang model for the stated injection are

$O_3 - 8.9 \times 10^{17}$ molecule/cm^2	(−11.2%)
$NO_2 + 2.74 \times 10^{15}$ molecule/cm^2	(+67%)

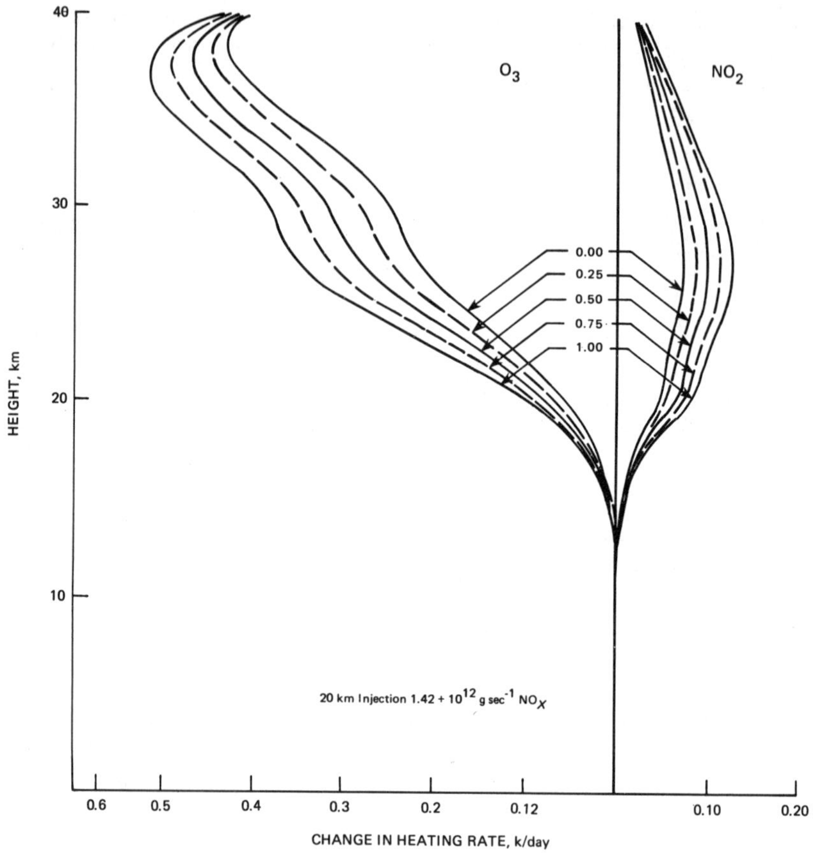

FIGURE F.1 Change in stratospheric heating rates (after F. Luther). Solar elevation, 30°.

With current levels of uncertainty concerning both the natural and perturbed populations of O_3 and NO_2, even the sign of the resultant perturbation of solar radiation must be considered uncertain. The magnitude is likely to be small.

B. THE NO_2–O_3 PERTURBATION AND TERRESTRIAL RADIATION

The NO and NO_2 perturbations are not expected to have any appreciable effect on the terrestrial radiation. O_3 makes a significant contribution to this radiation. The effect of reduced stratospheric O_3 content may be estimated by linear perturbation of the heating rates computed by standard methods. Using the diagram that appears as Figure 5.45 in CIAP Monograph 4 (and there attributed to an unpublished report by Newell and Gray), the perturbation shown in Figure F.2 is obtained. This perturbation in heating rate corresponds to a decreased absorption of terrestrial radiation in the stratosphere of about 0.5 W/m^2.

C. PERTURBATION OF SOLAR RADIATION FOLLOWING AN INCREASE OF STRATOSPHERIC PARTICLE CONTENT

In CIAP Monograph 3, two different models were used to estimate stratospheric burdens of sulfate aerosol. In one model (a box model), it was crudely estimated that about 70 percent of the sulfur dioxide emitted in the stratosphere was converted into sulfate aerosol. In a one-dimensional model with eddy diffusion and SO_2+OH chemistry, the

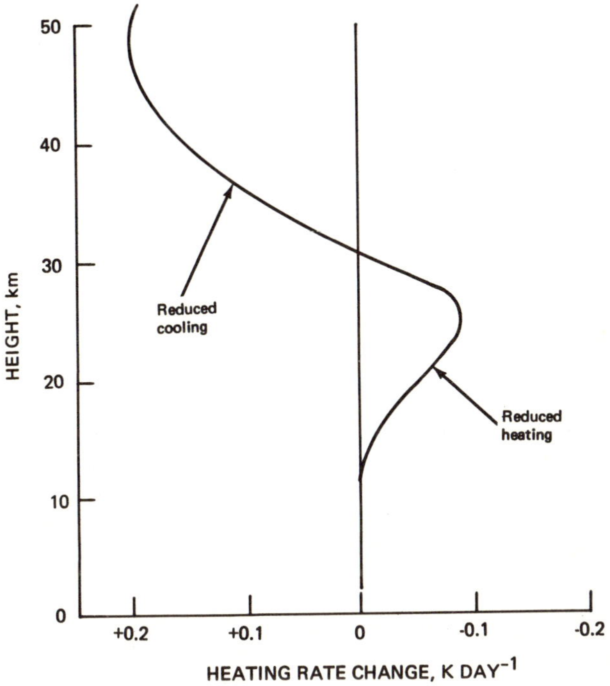

FIGURE F.2 *Perturbation* of stratospheric heating rates following 10 percent decrease of O_3 and consequent changes in the terrestrial radiation.

calculated fractional conversions ranged from 0.8 to 0.96. For the purpose of estimating a maximum effect of SO_2 emissions on global average climate, we may reasonably assume that all the SO_2 is oxidized to sulfate. With these considerations and using either model, it is estimated that the atmospheric average column density of diluted sulfuric acid aerosol (calculated as 75% H_2SO_4 by mass) resulting from aircraft traffic of the CIAP 1990 upperbound fleet would be less than 300 μg m^{-2}. This would correspond to an average mass concentration of 0.03 μg m^{-3} in a 10-km-thick layer of the stratosphere. The optical effect of this sulfate burden is expressed as a change in optical thickness of the atmosphere, and according to model computations by Herman using the scattering and radiation transfer theory for the atmosphere would be less than 0.0015. In effect then, the augmentation of the stratospheric aerosol layer would diminish solar radiation reaching the ground by less than 0.2 percent.

Calculation of effects due to particles in the stratosphere is subject to much more uncertainty than that due to a perturbation of ozone amount because of lack of knowledge of certain optical characteristics of the particles and of the albedo of the underlying system—atmosphere, cloud, and surface—in the solar infrared. We require to know, for each particle, and for each wavelength, the angular distribution of scattered light $S(\theta,\lambda)$ and the albedo for single scatter, $\omega_0(\lambda)$, which is the ratio of total scattered to total incident energy. $S(\theta,\lambda)$ and $\omega_0(\lambda)$ are calculable if we know the size of the particle and the complex refractive index $m=n-ia$ for all wavelengths. We do not, of course, know the size distribution of the sulfate particles that might result from the burning of sulfur-containing fuel, but it seems reasonable to assume that it will be the same as that of naturally occurring stratospheric sulfate particles. The refractive index poses a more difficult question: if the particles are pure hydrated sulfuric acid with water content controlled by the stratospheric humidity, we have some information but not in great spectral detail. If the particles are nucleated, we have no information but can make limit estimates. The main problems lie in the solar infrared near and between the water vapor absorption bands. Through the visible, that is, for about half of the solar energy, the particle absorption must bc low.

Computations have been made for a variety of refractive indices. Addition of particles, which certainly absorb in the solar infrared if not in the visible, must result in a local heating in the stratosphere and a reduced solar input to the troposphere, but even the sign of the perturbation of the planetary albedo has not been determined unequivocally.

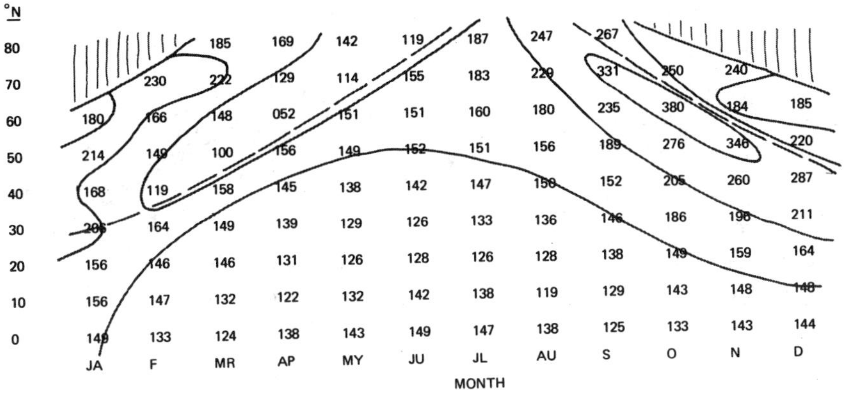

FIGURE F.3 Albedo *increase* by latitude and month following addition of a 10-km layer of 0.08 μg m^{-3} of particles $\omega_0 = 1$ (no absorption). Units are $10^5 \times$ fractional albedo change. ‖ ‖‖‖‖ ‖‖, no sun; — — limit of snow and ice. (After B. Herman.)

Figure F.3 is an example of one type of computation. It shows the change in the local albedo of the planet for a 500-nm wavelength caused by addition of 0.08 μg m^{-3} of particles in a 10-km stratospheric layer (corresponding change in optical depth $\tau = 0.003$), with $\omega_0 = 1$; i.e., no absorption. At 500 nm, mean local values of planetary albedo are reasonably well known from satellite observations. There is an increase of albedo overall but with considerable latitudinal and seasonal variations. Figure F.4 shows the same computation but with $\omega_0 = 0.9$. There is now an overall decrease of albedo.

It seems unlikely that the single scatter albedo of the pollutant particles can be as low as 0.9 in the visible. Calculations have, therefore, been made for three solar angles and a range of surface albedos for particles with a realistic size distribution and refractive index

$$m = 1.45 - 0.005i,$$

which have $\omega_0 = 0.973$ at 500 nm and 0.88 at 2450 nm. These calculations are still unrealistic in the neighborhood of the water absorption bands but are probably more realistic than those illustrated in Figures F.3 and F.4 for the visible and for the total energy integrated through the spectrum. Table F.1 (after F. Luther) contains the results of some of these computations and refers to a perturbation of optical depth $\Delta\tau$ of 0.005, which corresponds to a loading of about 0.1 μg m^{-3} over a 10-km layer. For realistic perturbations, the effects scale linearly with the optical depth change. The numbers in Table F.1 have been averaged

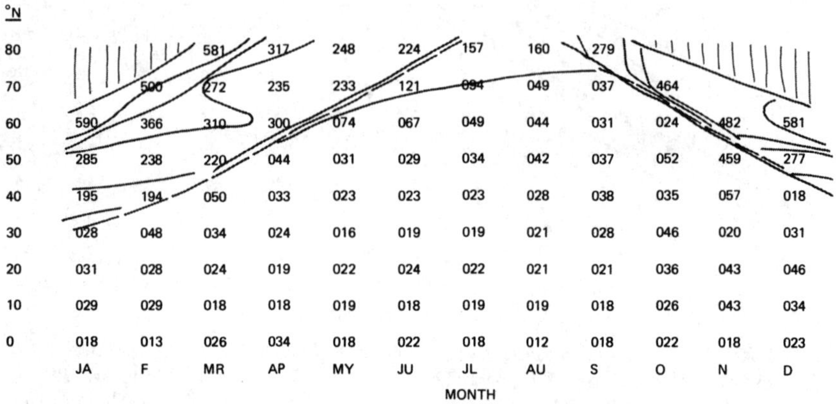

FIGURE F.4 Albedo *decrease* by latitude and month following addition of a 10-km layer of 0.08 μg m^{-3} of particles $\omega_0 = 0.9$. Units are $10^5 \times$ fractional albedo change. \| | | \ | | | /, no sun; —— limit of snow and ice. (After B. Herman.)

TABLE F.1 Perturbation of Solar Radiation Due to a Change in Particle Content of the Stratosphere [a]

	Perturbation
Planet	
Planetary albedo	+0.0005
Net downward flux at top of atmosphere	−0.15 W/m^2
Lower Stratosphere	
Energy absorbed in stratosphere	+0.15 W/m^2
Troposphere	
Downward flux at tropopause	−0.3 W/m^2
Energy absorbed by troposphere	−0.05 W/m^2
Earth Surface	
Energy absorbed at the surface	−0.25 W/m^2

[a] $\Delta\tau = 0.005$, particle refractive index $(1.45 - 0.005i)$. Particle size distribution, Deirmendjian, Haze H. Solar altitude 30°.

over the day and night hemispheres and represent the perturbation of the global energy budget. The computed perturbations depend on assumed fixed albedos for the troposphere–surface system.

D. PERTURBATION OF TERRESTRIAL RADIATION FOLLOWING A CHANGE OF PARTICLE CONTENT

Computations of this perturbation have been carried out in considerable detail for a model tropical atmosphere (by Luther using methods due to Ellingson). These show that a particle layer similar to that considered in Section III.C, above, with no temperature or other change, would decrease the net outward ir radiative flux at its upper boundary by 0.15 W/m^2 and would decrease the net upward radiative flux at its lower boundary by 0.08 W/m^2, corresponding to a tendency to heat both the particle-containing layer and the troposphere–surface system.

References

Davis, D. D. 1974. A kinetics review of atmospheric reactions involving H_xO_y compounds. *Can. J. Chem.* 52:1405–1414.

Friend, J. P., R. Leifer, and M. Trichon. 1973. On the formation of stratospheric aerosols. *J. Atmos. Sci.* 30:465–479.

Appendix G

Some Limitations on Ecological Models

In contrast to effects of ozone on ultraviolet radiation, climatic effects in the troposphere resulting from stratospheric chemistry changes cannot easily be predicted because of the inadequacy of physical models for nonlinear interaction variables in the complex atmosphere–ocean system. It is probable that little or no long-term effects would be observable (see Chapter 5); but complacency on this point does not necessarily lead to complacency about small climatic effects on the biosphere because of the nonlinearity of innumerable interactions within it.

Modeling of large natural ecosystems, technically called "biomes," has progressed under the International Biological Program (IBP) to an extensive degree. In the CIAP program, explicit analysis of several of the biome models has yielded some encouragingly firm conclusions and some unexpected examples of nonlinear interaction for forest, grassland, and the tundra. The main conclusion is that primary production is significantly diminished by small decreases in temperature, as would be expected in well-watered regions where runoff is linearly related to precipitation. The effects of diminutions of rainfall are compensated for by decreasing temperatures. For deserts, where production is closely attuned to precipitation, the effect of increasing temperature is diminish-

This Appendix presents the findings of the Panel on Biological and Medical Effects: R. Setlow, *Chairman;* W. Baier, W. Butler, R. Clayton, E. Deevey, T. Fitzpatrick, A. Giese, D. Gordon, F. Haxo, E. Runge, E. Scott, and F. Urbach. Consultant to the Panel: M. Caldwell.

ing production. Although increasing precipitation when coupled with decreasing temperature was not a factor included in the IBP models, the overall conclusions are in good accord with those of the agricultural climatologist for production of various crops (see Chapter 5).

Nonlinear interaction effects in the models were of most importance in consideration of animal production and diversity and in mineral cycling between forest, soil, and runoff. For example, decomposition and mineral cycling, being in large part a function of a healthy soil microflora and fauna, proved to be exceptionally sensitive to seasonal *distributions* of temperature. Both in deciduous forest and in grassland, changes in the patterns of mineralization of plant litter may alter long-term successional trends in complex ways. The coniferous forest model made no allowance for mineral cycling. Unfortunately, mineral cycling was considered in all the other models solely as a simulated output, and the idea that most of the nutrient minerals reaching ecosystems in eastern North America are airborne and are only loosely coupled to the quantity of precipitation could not be tested. This idea is explicit in several ongoing ecosystem studies.

As pointed out in Chapter 5, evolutionary strategies have been used over geologic time to adapt to changing averages; however, we may overgeneralize the physiological and biogeochemical consequences by saying that the strategies adopted by organisms and ecosystems are designed to respond sensitively to the variance of climatic parameters and to pay little attention to their averages; for example, (a) a small change in the length of frost-free seasons if continued for a very few years has the effect of shifting the range of some important organisms for biotic communities by several degrees of latitude or longitude or by several hundred meters of altitude; (b) a run of cool cloudy summers in northern Europe has the effect of expanding valley glaciers and blanket bog with relatively large effects on remote ecosystems and on human culture; and (c) a short run of exceptionally warm summers may raise soil temperatures by at least 1 °C in July, accelerate litter decomposition and mineralization, and favor a soil-generated pathogen that removes a predominant species from the forest stand.

What climatologists regard as small effects (i.e., negligible effects on long-term averages and changes in variance by a few percent), can be amplified. Ecologists therefore react with unease, if not alarm, to the likelihood that a stratospheric fleet will make "small" changes in the heat balance of the troposphere.

Appendix H

Modeling Techniques for Crops

This Appendix is abstracted from the paper by R. A. Leeper, E. C. A. Runge, and W. M. Walker. 1974. Effect of plant-available stored soil moisture on corn yield, Part II, under variable climatic conditions. *Agron. J.* 66:728–733.

Agronomists have long been interested in how variation in soils and climate are related to crop yields. An experiment was conducted to quantify the effect of weekly plant-available stored soil moisture (PASSM) during 10 weeks of the growing season on corn (*Zea mays* L.) yield under constant and variable climatic conditions. The objective of this study was to determine how much of the variation in corn yield was due to variations in soils and climate and to develop a model relating corn yield to soil, rainfall, and temperature that could be used to predict corn yields for other locations not studied. Plot yields from four Illinois locations (farms) for three years (1969–1971) were correlated with weekly PASSM, rooting depth, or available water-holding capacity in the rooting zone and climate data (rainfall and temperature). Experimental plots had high fertility and management. Weekly PASSM, rainfall, and temperature values were summed by a method similar to Fisher's polynomial technique as previously modified.

Differences in corn yields due to soils, locations, and among years within locations are due to differences in climate (rainfall and tempera-

This Appendix presents the findings of the Panel on Biological and Medical Effects: R. Setlow, *Chairman;* W. Baier, W. Butler, R. Clayton, E. Deevey, T. Fitzpatrick, A. Giese, D. Gordon, F. Haxo, E. Runge, E. Scott, and F. Urbach. Consultant to the Panel: M. Caldwell.

ture) and differences in the ability of the soil to supply water. When correlated with rainfall and temperature data, PASSM determined on a weekly basis was slightly less efficient ($R^2=0.80$) than was rooting depth ($R^2=0.83$) or amount of preseason water available in the rooting zone ($R^2=0.81$) in explaining variations in corn yield. Equations fitted to the data should be useful models for predicting corn yields for other locations and soil conditions not studied. The model utilizing the amount of preseason water available in the rooting zone is particularly easy to utilize for predicting corn yield for other locations from soils and climatic data. The model determined was the following:

$$\begin{aligned}\text{Yield (bu/acre)} &= 1566.37 - 83.0688W - 1.0697W^2 + 42.9392 \\ &\Sigma(R_it_i) - 8.1130\,\Sigma(R_it_i^2) + 0.3654\,\Sigma(Th_it_i) - 0.1013\,\Sigma(Th_it_i^2) \\ &- 0.5014\,\Sigma(RTh_it_i) + 0.0974\,\Sigma(RTh_it_i^2) - 3.9802W\,\Sigma(R_it_i) \\ &+ 0.7907W\,\Sigma(R_it_i^2) - 0.061W\,\Sigma(Th_it_i) + 0.0121W\,\Sigma(Th_it_i^2) \\ &+ 0.0482W\,\Sigma(RTh_it_i) - 0.0097W\,\Sigma(RTh_it_i^2),\end{aligned}$$

where R is weekly precipitation in inches, Th is weekly mean maximum temperature in degrees Fahrenheit, W is available water-holding capacity within the rooting zone in inches, and t_i is the time identification of weekly weather data, taking on the value of 1 to 10.

Appendix I

Climate Effects on Crop Yields and Water Conservation

TABLE I.1 Estimated Percent Change in Corn Yield [a] as a Result of Changes in Temperature and Precipitation [b]

Change in Temperature (°C)	Change in Precipitation (%)				
	−20	−10	0	+10	+20
−2	19.8	21.2	22.7	24.2	25.6
−1.0	8.4	9.8	11.3	12.8	14.2
0	−2.9	−1.5	0	1.5	2.9
+1.0	−14.2	−12.8	−11.3	−9.8	−8.4
+2.0	−25.6	−24.2	−22.7	−21.2	−19.8

[a] Normal = 85 ± 16 bu/acre, 1901–1972 average for selected stations in Missouri, Illinois, Indiana, Nebraska, Iowa, and Kansas, where 65 percent of U.S. corn production is located.
[b] From J. F. Benci and E. C. A. Runge. 1974. Effects of hypothetical climatic changes on production and yields of corn. Report for CIAP study, DOT, Dec. 1974.

This Appendix presents the findings of the Panel on Biological and Medical Effects: R. Setlow, *Chairman;* W. Baier, W. Butler, R. Clayton, E. Deevey, T. Fitzpatrick, A. Giese, D. Gordon, F. Haxo, E. Runge, E. Scott, and F. Urbach. Consultant to the Panel: M. Caldwell.

TABLE I.2 Effect of Various Levels of Temperature and Precipitation Changes on State Average Wheat Yields Based on L. Thompson's Model [a]

Climatic Change: Temp. (°C)	Climatic Change: Precip. (%)	North Dakota	South Dakota	Kansas	Ill.	Ind.
		Average Wheat Yields by State (bu/acre)				
		25.0	21.1	25.9	36.3	36.9
		Changes in Wheat Yields (bu/acre)[b]				
−2	−20	−3.3	−1.1	−0.4	+4.6	+3.9
	−10	−0.1	0	+0.6	+3.6	+2.8
	0	+1.2	+0.5	+1.4	+2.4	+1.7
	+10	+2.4	+0.9	+2.3	+0.9	+0.6
	+20	+4.4	+1.1	+2.9	−0.7	−0.6
−1	−20	−1.8	−0.7	−1.1	+3.4	+3.1
	−10	−0.6	+0.3	−0.1	+2.4	+2.0
	0	+0.7	+0.9	+0.7	+1.1	+0.9
	+10	+1.9	+2.3	+1.5	−0.8	−0.3
	+20	+3.1	+1.5	+2.2	−1.9	−1.4
0	−20	−2.5	−1.6	−1.8	+2.3	+2.2
	−10	−1.1	−0.7	−0.9	+1.2	+1.1
	0	0	0	0	0	0
	+10	+1.2	+0.4	+0.8	−1.4	−1.1
	+20	+2.4	+0.6	+1.4	−3.0	−2.3
+1	−20	−3.4	−2.8	−2.6	+1.0	+1.3
	−10	−1.9	−4.7	−1.6	0	+0.2
	0	−0.9	−1.2	−0.8	−1.2	−0.9
	+10	+0.2	−0.7	0	−2.7	−2.1
	+20	+1.5	−0.2	+0.7	−4.3	−3.2
+2	−20	−4.4	−3.2	−3.4	−0.4	+0.3
	−10	−3.1	−2.4	−2.4	−1.5	−1.8
	0	−1.9	−1.6	−1.6	−2.7	−1.9
	+10	−0.7	−1.1	−0.8	−4.1	−3.1
	+20	+0.5	−0.7	−0.2	−5.7	−4.2

[a] Compiled by W. Baier. 1974. Canadian Dept. of Agriculture, based on data in Supplement to report on response of wheat yields to induced climatic changes. Addendum to CIAP Monograph 5 (draft), Sec. 4.3.1.B. May.

[b] Climatic changes applied to weather during the growing season. Data for Oklahoma excluded. Yield depression boxed in.

TABLE I.3 Annual Values of Precipitation, Runoff, Evapotranspiration (ET) and Drainage for a Watershed at Riesel, Texas, and the Expected Changes in ET and Drainage Resulting from the Assumed Climatic Changes [a]

					Parametric Climate Changes			
					1	2	3	4
	Precipitation	Runoff	ET	Drainage				
Year	(mm)				Change in ET (%)			
1969	759	51	640	0	−3.9	−2.0	−0.8	−0.2
1970	698	98	526	22	−2.2	−1.1	−0.4	0.0
1971	1037	191	734	105	−3.3	−1.6	−0.8	+0.4
					Change in Drainage (%)			
1969	759	51	640	0	—	—	—	—
1970	698	98	526	22	−50.5	−25.7	−7.8	−2.3
1971	1037	191	734	105	+0.2	−2.2	−0.2	−0.7

[a] From CIAP Monograph 5 (draft), pp. 4–28.

Appendix J

Detailed Report on Engines and Fuels

I. Introduction

As background for Chapters 6 and 7, which address aircraft engines and fuels problems related to possible climatic impact problems arising from the operation of high-flying aircraft in the stratosphere, the Panel prepared this Appendix concerning jet aircraft engines, jet aircraft fuels, and effluents from jet aircraft engines. Using this information as a basis, the Panel studied the specific problems of nitrogen oxides in the effluents from jet aircraft engines (Section V) and of sulfur oxides in the effluents from jet aircraft engines (Section VI). The Panel's conclusions are presented in Section VII.

II. Jet Aircraft Engines

The engine used in present aircraft is the gas turbine engine. Presently projected aircraft will also use this engine or its derivatives. The simplest aircraft gas turbine (turbojet) has a compressor that forces sufficient high pressure air into a burner so that the fuel can be burned efficiently. During compression the temperature of the air increases. The air is heated by the combustion with fuel; then the gases flow through a tur-

---

This Appendix presents the findings of the Panel on Engines and Fuels: F. D. Rossini, *Chairman;* L. Brewer, A. Ferri, F. Johnson, and R. Miller. Consultants to the Panel: J. English, I. Glassman, S. Penner, and R. Rudey.

bine, which drives the compressor. Although some work is extracted from the burned gases to drive the turbines, the gases have sufficiently high total enthalpy (specific plus kinetic) to be expanded through a nozzle to achieve high exhaust velocities. Because of its self-accelerating characteristics and the efficiencies it can achieve, the turbojet and its derivatives have become the power plant for present-day modern aircraft (Zucrow, 1958).

The two most common derivatives of the basic turbojet engines are the turbofan and the turbojet with afterburner. In order to obtain greater thrust for a given engine airflow, it is possible to inject additional fuel into the turbine exhaust stream to be burned with the excess air present. A schematic diagram of this afterburner-type engine is given in Figure J.1. The higher exhaust velocities obtained give greater thrust.

A different derivative of the turbojet is the turbofan engine. Here the turbine does more work at the expense of the exhaust velocity and drives a fan that compresses air in excess to the air entering the main compressor (see Figure J.2). The fan moves a large quantity of air. Only a part of the air enters the main engine; the rest is discharged through a short duct surrounding the front of the engine. Thus, the turbofan at a given thrust level moves a larger mass of gas at an average exhaust velocity lower than the basic turbojet. The turbofan engine improves efficiency, especially at low flight speed, and decreases noise; therefore, it is the preferred configuration for subsonic cruising airplanes.

A derivative of the turbofan has been considered for flights in the high-speed regime. Simply, the idea is to inject and burn fuel in the turbofan duct. A schematic of a duct-burning turbofan is given in Figure J.3.

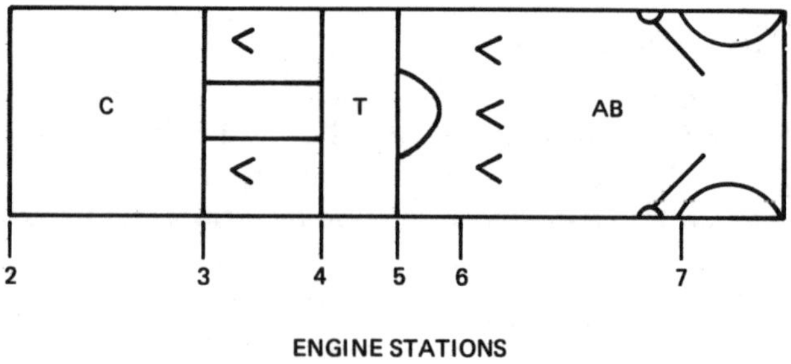

FIGURE J.1 Schematic diagram of turbojet engine with afterburner: C, compressor; T, turbine; AB, afterburner.

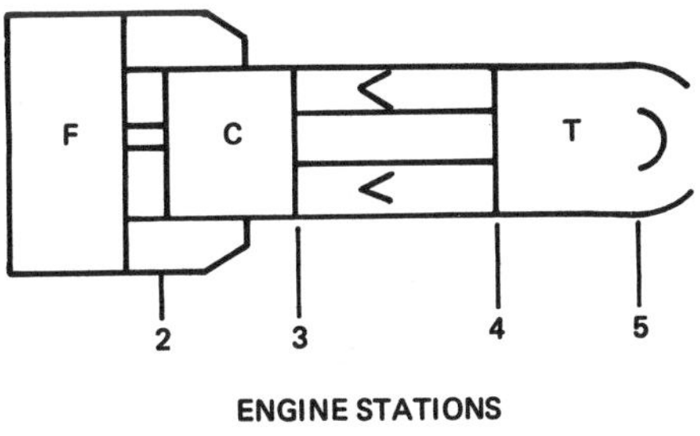

FIGURE J.2 Schematic diagram of turbofan jet engine: F, fan; C, compressor; T, turbine.

As in many propulsion devices, the limits on the performance of the basic turbojet are associated with the ability to develop high-temperature materials and to keep the materials that are in contact with the high-temperature gases within limits acceptable by the materials. The performance of the engine can be shown to be strongly dependent on the turbine inlet temperature. Thus turbine materials, cooling methods, and blade design limit the performance of the turbojet and its derivatives. The fuel–air ratio in the combustor is chosen so that the temperature at the exhaust of the combustor is compatible with turbine performance. This temperature is presently of the order of 1400–1500 K at maximum power and below such values at cruise conditions. Goals for the future are as high as 1700 K at cruise conditions. The fuel–air ratios associated

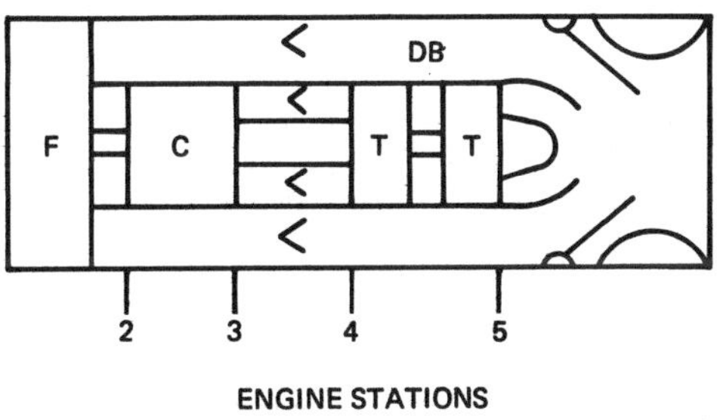

FIGURE J.3 Schematic diagram of duct-burning turbofan jet engine: F, fan; C, compressor; T, turbine; DB, duct burner.

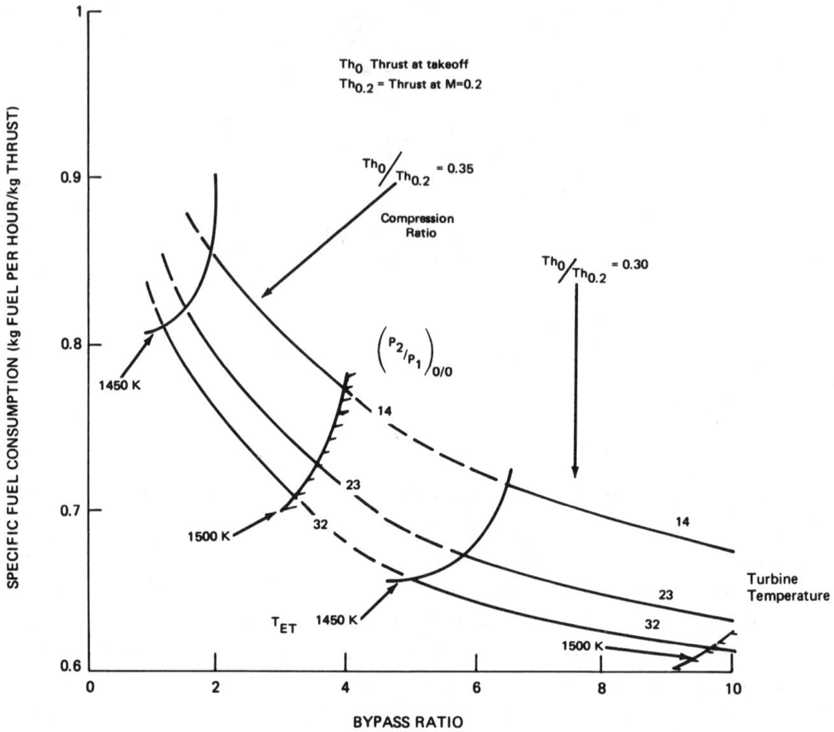

FIGURE J.4 Specific fuel consumption at flight Mach 0.8 and 9-km altitude. Maximum turbine temperature 1600 K.

with these temperatures are very much on the lean (excess air) side of the stoichiometric mixture ratio.*

Increase in turbine temperature increases engine performances. Figures J.4 and J.5 give typical plots of the variation of specific fuel

* The stoichiometric mixture ratio is that ratio at which there is exactly sufficient air (oxygen) present to burn the hydrocarbon fuel completely to carbon dioxide and water vapor. At about this ratio, one achieves the maximum adiabatic equilibrium temperature obtainable with the fuel. When there is excess air, the equilibrium temperature drops because the excess air acts as a diluent. When there is excess fuel, there is not sufficient air to burn all the fuel to carbon dioxide and water; all the chemical energy is not converted to thermal energy; and thus the temperature drops again. It has become a convenience to define the mixture ratio in terms of another ratio—the equivalence ratio. The equivalence ratio, ϕ, is 1 when the mixture is stoichiometric; the equivalence ratio is above 1 when the air present is less than that required for stoichiometric mixture; and the equivalence ratio is below 1 if excess air is present. The equivalence ratio is given by the ratio of the two quantities of air.

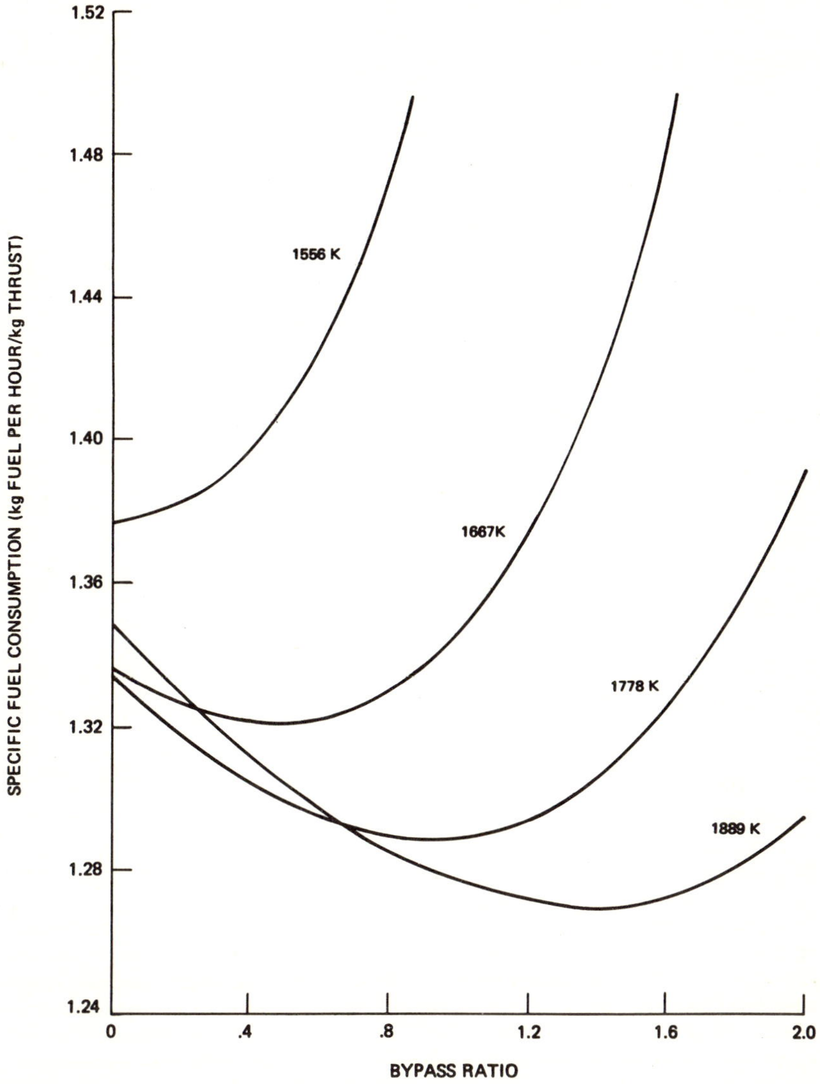

FIGURE J.5 Specific fuel consumption as a function of the bypass ratio, at various temperatures, at Mach 2.7 to 3.

consumption (kilograms of fuel burnt per kilogram of thrust per hour) as a function of the bypass ratio (air used in the fan outside the burner/ air used in the burner and turbine) for subsonic and supersonic flight conditions. Figure J.4 shows that, at subsonic cruising speed, efficient bypass engines require large bypass ratio and high compression ratio and turbine inlet temperature. High compression ratio produces high entrance temperature in the burner. The entrance burner temperature is an important parameter for the NO_x production in present burner design. Figure J.5 indicates the importance of high turbine temperature for the performance at high speed.

Figure J.6 gives typical entrance burner temperature as a function of compression ratio of the engine at subsonic velocity (Mach 0.85).

Actual engine cruise operating conditions for many of the systems described above and so important to this study are given in Table J.1. Specifically, two classes of engine to power subsonic aircraft (P&WA JT9D-7 and future turbofans) and two classes to power supersonic aircraft (Concorde afterburning turbojet, Olympus 593, and future duct-burning turbofans) are reviewed in this table.

Additional details relating to jet aircraft engines, the combustion process, flame stability, and spontaneous ignition may be found in Anderson (1973), Longwell (1953), Mullins (1955), National Science Foundation (1974), and Palmer and Cullis (1965).

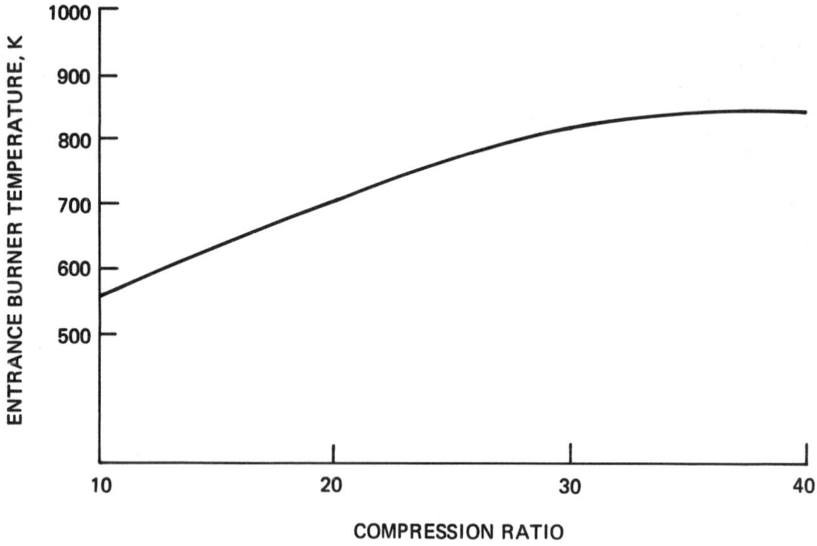

FIGURE J.6 Entrance burner temperature as a function of compression ratio, at Mach 0.85.

TABLE J.1 Cruise Operating Conditions (from Grobman and Ingebo, 1974)

(a) Subsonic

Condition	Production Turbofan (P&WA JT9D-7)	Future Turbofans (ATT)		
Fuel	JP	JP	LNG	Hydrogen
Cruise Mach number	0.85	0.85	0.85	0.85
Cruise altitude, m	10 700	12 200	12 200	12 200
Total airflow rate, kg/sec	304	151	136	90.6
Bypass ratio	4.9	7.8	7.8	7.8
Compressor discharge airflow rate, kg/sec	51.5	17.2	15.5	10.3
Combustor inlet temperature, K	710	661	661	661
Combustor inlet pressure, atm	9.7	7.2	7.2	7.2
Combustor exit temperature, K	1410	1540	1540	1540
Fuel–air ratio	0.018	0.026	0.023	0.0096
Fuel flow rate (per engine), kg/h	2800	1610	1280	356

(b) Supersonic

Condition	Concorde After-burning Turbojet (Olympus 593) [a]	Future Duct-Burning Turbofans (ASST)		
Fuel	JP	JP	LNG	Hydrogen
Cruise Mach number	2.0	2.7	2.7	2.7
Cruise altitude, m	17 700	19 800	19 800	19 800
Total airflow rate, kg/sec	83	336	306	202
Bypass ratio	0	2.36	2.36	2.36
Compressor discharge airflow rate, kg/sec	83	100	91	61
Combustor inlet total temperature, T_3, K	824	810	810	810
Combustor inlet total pressure, P_3, atm	6.5	4.7	4.7	4.7
Combustor exit average temperature, K	1320	1770	1770	1770
Combustor fuel–air ratio	0.0141	0.0299	0.0265	0.01187
Combustor fuel flow rate, kg/h	4200	10 750	8700	2560
Augmentor airflow rate, kg/sec	—	236	215	141
Augmentor inlet total temperature, K	—	635	635	635
Augmentor inlet total pressure, atm	—	2.6	2.6	2.6
Augmentor inlet velocity, m/sec	—	90	90	90
Augmentor exit average temperature, K	—	835	835	835
Augmentor fuel–air ratio	—	0.00495	0.0045	0.00187
Augmentor fuel flow rate, kg/h	—	4200	3500	940
Total fuel flow rate (per engine), kg/h	4200	14 950	12 200	3500

[a] Afterburner is not used for steady-state cruise.

III. Jet Aircraft Fuels

A. NATURAL PETROLEUM

Although it contains small amounts of sulfur, nitrogen, and oxygen compounds, natural petroleum is principally hydrocarbon material, with molecules ranging in size from 1 to 50 or more carbon atoms per molecule and comprising hydrocarbons of several molecular types, including paraffins, cycloparaffins, and aromatics (Rossini, 1960). Conventionally, petroleum can be thought of as initially being separated by the process of distillation into several broad fractions of decreasing volatility: gas, gasoline, kerosine, gas oil, lubricating oil, and residue (Rossini, 1960). These materials, and variations of them, are subjected to the many different processes developed by scientists and engineers to produce the multitude of specialized products provided by the petroleum industry.

B. JET AIRCRAFT FUEL

The jet aircraft fuel currently in use is essentially made from the kerosine fraction of petroleum. This material is subjected to appropriate treatment to improve its stability with respect to temperature, oxidation, and corrosivity. Appropriate other treatment, which may include blending with selected other hydrocarbon material, is used to control the following properties of the fuel (Hurn, 1974): the amount of olefin hydrocarbons (to several percent or less by volume); the amount of aromatic hydrocarbons (to 20 percent or less); the flash point, minimum of 40.6°C and maximum of 65.6°C; the freezing point maximum, −37.8°C for Jet A-1 and −47.8°C for Jet A; test distillation temperatures, maximum, 10 percent, 204°C, 90 percent, 232°C, and final, 288°C. In addition, the jet aircraft fuel must satisfy requirements pertaining to the content of total sulfur, "mercaptan" sulfur, and naphthalene hydrocarbons, as well as requirements for the "smoke point" and "existent gum" (Hurn, 1974). The processing and handling of jet aircraft fuels are carefully controlled to make sure that the material is essentially clean, dry, and free of contamination on delivery into the tanks of the aircraft (Hurn, 1974).

The foregoing specifications result in jet aircraft fuels having a composition with respect to types of hydrocarbons as follows, in percent by volume (Hurn, 1974): paraffins, about 80; aromatics, about 18; olefins, about 2. The sulfur content ranges from 0.02 to 0.15 percent by weight, with the worldwide average roughly estimated to be about 0.05 percent by weight (Hurn, 1974).

The jet aircraft fuels currently in use in Canada and Europe have a broader distilling range, being more volatile and having a higher end point, although it is expected that, in time, such fuels will approach those of the United States in properties (Hurn, 1974). The properties of the jet aircraft fuel used in the Soviet Union generally fall between the properties of those used in the United States and those used in Canada and Europe (Hurn, 1974).

C. IMPURITIES IN JET AIRCRAFT FUEL

Naturally occurring petroleum contains minute quantities of metals, in amounts of the order of 10 to 50 parts per billion (Hurn, 1974). These metals, along with small amounts of nitrogen and sulfur atoms, are largely bound within large complex porphyrin molecules, so that the process of hydrogenation simultaneously releases sulfur, nitrogen, and free metals, with the sulfur and nitrogen being removed as gases, in uncombined or combined states, and with the metals remaining in the catalyst bed used in the process of hydrogenation (Hurn, 1974). A variety of commercial processes are used by the petroleum industry to remove such contaminants, and it appears that the increasing need to lower the sulfur content for many fractions of petroleum is leading generally to the desulfurization of broad fractions of petroleum (Hurn, 1974). Further processing of separate fractions, such as the gasoline or kerosine fractions can lead to extremely low levels of sulfur in such materials (Hurn, 1974). It is estimated that the cost of lowering the sulfur content in jet aircraft fuel to 20 parts per million is about 1 cent per gallon (Axelrod, 1974; Davis *et al.,* 1974; Hurn, 1974).

Present-day jet aircraft fuel, made from natural petroleum, is relatively low in content of combined nitrogen. It is estimated that about half of the combined nitrogen in the fuel ends up in nitrogen oxides, but the amount of nitrogen in this fuel is so low that no significant amounts of nitrogen oxides are produced from this source. "Synthetic" petroleum from tar sands is also relatively low in nitrogen content. But "synthetic" petroleum from oil shale and from coal is relatively high in combined nitrogen, and therefore jet aircraft fuel from these sources would need to have the nitrogen content specified to tolerable low levels. Such removal of nitrogen can be accomplished by well-established refining processes in much the same manner as in the removal of sulfur from the fuel.

D. COSTS OF JET AIRCRAFT FUEL

In late 1973, the cost of jet aircraft fuel was about 13 cents per gallon, which can be calculated as $4.00 per barrel for the crude petroleum plus processing costs of $1.50 per barrel, or a total of $5.50 per barrel of 42 gallons (Hurn, 1974). Similarly, one can calculate that, with a processing cost of $1.50 per barrel, crude petroleum at $10.00 per barrel would yield jet aircraft fuel at 27.5 cents per gallon. An increase in the processing cost of $0.85, to $2.35 per barrel, would increase each of these costs by 2 cents per gallon.

In 1975, the use of jet aircraft fuel in the United States is estimated to be near 0.5 billion barrels per year; while by 1990, the amount is projected to be somewhat more than one billion barrels per year (Hurn, 1974). Referred to the total amount of petroleum used in the United States, the percentage needed for jet aircraft fuel is estimated as follows (Longwell, 1974): 1975, about 6 percent; 1990, about 9 percent.

E. SUPPLIES OF JET AIRCRAFT FUEL

It is reported that supplies of petroleum will be fully available to meet the projected needs for jet aircraft fuel through 1990 (Hurn, 1974; National Petroleum Council, 1972). However, should there develop significant shortages of jet aircraft fuel from petroleum, such an insufficiency can be alleviated with this petroleum material in several ways (Longwell, 1974): (a) by modifying the specifications for jet aircraft fuel to encompass wider boiling materials, (b) by hydrogenation of aromatics to the acceptable cycloparaffins, and (c) by hydrocracking (combination of hydrogenation and cracking to smaller molecules) higher boiling material. The latter two conversion processes will, of course, add to the cost of the jet aircraft fuel so produced, in an amount of the order of 5 to 10 cents per gallon of aromatic material (Longwell, 1974).

It is estimated that, in the inventory of ultimate resources of fossil fuels in the United States expressed in units of energy, oil shale plus tar sands will yield five times as much as petroleum (including natural gas) and that coal will yield about five times as much as oil shale plus tar sands (National Petroleum Council, 1972). It should be borne in mind that such estimates have large uncertainties. Also, it should be noted that, while the United States has large deposits of oil shale (in Utah, Colorado, and Wyoming), Canada (in Alberta) has very large deposits of tar sands (or bituminous sands).

F. "SYNTHETIC" PETROLEUM FROM TAR (BITUMINOUS) SANDS

"Synthetic" petroleum is already being produced commercially from tar (bituminous) sands in Alberta, Canada, on a large scale, in an amount of the order of 100,000 barrels per day (Camp, 1974). The bitumen associated with these sands is largely hydrocarbon material, with significant amounts of sulfur (about 4 percent by weight) and nitrogen (about 0.5 percent by weight) (Camp, 1974). The "synthetic" petroleum refined from the bitumen can have properties about as follows, in percent by weight (Camp, 1974): "paraffin" and "cycloparaffin" hydrocarbons, about 80; aromatic hydrocarbons, about 20; sulfur, 0.03; nitrogen, 0.02. The content of sulfur and nitrogen in the refined product can be reduced down to a level of a few parts per million (Camp, 1974). Counting the existing production, along with the applications submitted to the Alberta government, the production of "synthetic" petroleum from tar (bituminous) sands may, in a few years, reach 0.25 billion barrels per year.

G. "SYNTHETIC" PETROLEUM FROM OIL SHALE

Raw oil produced from oil shale is a highly viscous oil that contains appreciable quantities of nitrogen and sulfur and does not pour below about 32°C (Hartley and Brinegar, 1957). One representative shale oil has the following properties (Hartley and Brinegar, 1957; Huffman, 1974): sulfur, percent by weight, 0.7 in the raw oil, 0.0016 in the refined oil; nitrogen, percent by weight, 1.8 in the raw oil, 0.046 in the refined oil; aromatics in the refined oil, 38 percent. To upgrade the raw shale oil to the equivalent of a good petroleum, two processes may be used: (a) "coking," the use of heat to convert a small part of the oil to coke, leaving the remainder as a lighter, freer-flowing oil; and (b) "hydrogenation," at high temperatures and pressures, with a catalyst (Hartley and Brinegar, 1957). The "synthetic" petroleum so produced can be refined to produce high-octane gasoline, jet aircraft fuel, diesel fuel, and other top-grade products, all indistinguishable from similar products made from the best natural petroleum (Hartley and Brinegar, 1957). The content of aromatic hydrocarbons in the material from "synthetic" petroleum from oil shale that would correspond to jet aircraft fuel is near 20 percent, about the same as similar fuel from natural petroleum and from tar (bituminous) sands (Longwell, 1974). Commercial production of "synthetic" petroleum from oil shale is expected to begin by 1980 (Huffman, 1974) and may reach a level of about 0.33 billion barrels per year by about 1985 (Longwell, 1974).

H. "SYNTHETIC" PETROLEUM FROM COAL

Raw oil produced from coal is somewhat similar to natural petroleum except that the content of sulfur and nitrogen is high and the amount of aromatics is very high, being about 80 percent (Hurn, 1974; Longwell, 1974). Regular refining of raw coal oil reduces the content of sulfur and nitrogen to low levels but the aromatic content remains high (Hurn, 1974; Longwell, 1974). However, the aromatic content can be reduced to specification levels for jet aircraft fuel by the process of hydrogenation, at a cost estimated to be of the order of 5 to 10 cents per gallon of aromatic material (Longwell, 1974).

I. AVAILABILITY OF JET AIRCRAFT FUEL FROM OIL SHALE, TAR SANDS, AND COAL

The cost of producing specification grade jet aircraft fuel from "synthetic" petroleum from tar (bituminous) sands and from oil shale appears to be competitive with the cost of producing the same material from natural petroleum, assuming petroleum to cost about $10.00 per barrel. On the same basis, jet aircraft fuel produced from "synthetic" petroleum from coal may cost somewhat more.

With regard to the time when commercial production exists, we can say that "synthetic" petroleum from tar (bituminous) sands already exists, "synthetic" petroleum from oil shale is near (1980), while "synthetic" petroleum from coal may require several more years to reach significant commercial production.

J. HYDROGEN AS FUEL FOR JET AIRCRAFT

As an aircraft fuel, hydrogen has an advantage in that the thermal energy available is about 2.75 times the energy available from conventional hydrocarbons, per unit mass. Hydrogen is also a clean-burning fuel, producing no carbon monoxide, hydrocarbons, sulfur oxides, or smoke. By proper design, emissions of nitrogen oxides may also be kept low. Because the reaction rates for the formation of nitrogen oxides are much slower than the energy-producing reactions, the residence time in the combustion region can be reduced and the formation of nitrogen oxides inhibited (Ferri, 1973a; Glassman, 1970). Because of its wider flammable range, hydrogen fuel can also be burned leaner, further reducing emissions of nitrogen oxides.

Another possible use of hydrogen as fuel is for hypersonic aircraft engines. The type of engine required for this application is the Scramjet, which uses combustion generated in a stream moving during combustion at supersonic velocity, with high local temperatures being obtained in the combustion region. Then the only mechanism available to limit the formation of nitrogen oxides is the residence time.

The disadvantages of hydrogen as an aircraft fuel are its low density (approximately 15 percent of hydrocarbon fuels) and the fact that its low boiling point (−253°C) requires cryogenic handling, with insulated tanks and venting. Hydrogen is also relatively easily ignited over wide composition limits and thus requires greater care in handling.

IV. Effluents from Jet Aircraft Engines

As reported in the preceding sections, the currently used jet aircraft fuel is largely hydrocarbon material, the combustion of which in oxygen in an ideal process produces simply carbon dioxide and water as the products:

$$C_aH_b + (a + \tfrac{1}{4}b)\ \mathrm{O_2} = a\mathrm{CO_2} + \tfrac{1}{2}b\mathrm{H_2O}. \qquad (1)$$

The amounts of carbon dioxide and water formed in the combustion are stoichiometrically proportional to the amount of fuel burned and hence are fixed in quantity.

Since the oxygen for the combustion comes from air, which contains about four times as much nitrogen as oxygen, there is opportunity for nitrogen and oxygen to react, and, if the temperature produced in the combustion is high enough, significant amounts of nitrogen oxides will be formed.

If the hydrocarbon fuel contains some sulfur, the sulfur will react with oxygen during the combustion and sulfur oxides will be formed in an amount proportional to the amount of sulfur in the fuel.

If the engine, with hydrocarbon fuel, operates in such a way that the combustion process is not fully efficient, there will be incomplete combustion, with resultant formation of carbon monoxide and unburned and partially burned hydrocarbons.

With a jet aircraft engine operating with hydrogen as fuel, the principal product is water, according to the reaction

$$\mathrm{H_2} + \tfrac{1}{2}\mathrm{O_2} = \mathrm{H_2O}. \qquad (2)$$

The amount of water formed is proportional to the amount of hydrogen burned and hence is fixed in quantity.

Since air is used to provide oxygen for the combustion of hydrogen, nitrogen oxides may also be formed during the combustion by reactions involving nitrogen and oxygen. But with hydrogen as fuel, there will be no sulfur oxides nor any products of incomplete combustion involving carbon compounds.

Thus, the effluents that are discharged by a gas turbine engine during operation are composed of constituents related to the air used in the combustion process, including constituents resulting from complete and incomplete combustion, constituents produced from heating the air, and constituents related to the fuel composition used in the combustion process (Boccio *et al.,* 1973; Goodman, 1974). Apart from the principal products of combustion, carbon dioxide and water, the effluents of concern here are the contaminants, including nitrogen oxides, sulfur oxides, and carbon monoxide and unburned and partially burned hydrocarbons. As indicated here, there are two types of contaminant that possibly can be emitted by a turbojet engine. One type is formed by the combustion of the carbon and hydrogen in the fuel, and the other arises from impurities in the fuel. The impurities in the fuel are metals and sulfur, bound in large molecules with carbon and hydrogen. It is generally accepted that the best manner to eliminate contaminants arising from impurities in the fuel is simply to remove the impurities from the fuel prior to use. The contaminants that would possibly form in the combustion process are carbon monoxide, unburned hydrocarbons, carbon soot, and the nitrogen oxides (NO_x). At cruise conditions, the combustion process in the turbojet is nearly 100 percent efficient, and thus there is essentially no problem from carbon monoxide and unburned hydrocarbons. It has been shown recently (Dryer *et al.,* 1971) that the rate of conversion of carbon monoxide to carbon dioxide is very rapid above 1000 K. In present engines at cruise conditions, there are no regions in a turbojet combustor in which temperatures drop as low as 1000 K. Specifically then, for aircraft operating in the stratosphere, the major problem concerns the formation of nitrogen oxides, NO_x.

Nitrogen oxides are produced by several rate-controlling reactions. The rate at which nitrogen oxides are formed in a fuel-lean combustion system can be described approximately by the following equation (Thompson *et al.,* 1972):

$$\frac{dx_{NO}}{dt} = 10^{15.563} T^{-1} \exp \frac{-135500}{RT} x_{N_2} x_{O_2}{}^{\frac{1}{2}} P^{\frac{1}{2}} \ (\mathrm{sec}^{-1}), \qquad (3)$$

where T is in kelvins, P is in atmospheres, and x represents mole fraction. In normal combustion processes, the nitrogen oxides produced are composed primarily of NO and NO_2, where the NO varies from 90–95 percent (see Glassman, 1973; Marteney, 1970; Zeldovitch, 1946).

Equation (3) shows the strong dependence of both time and temperature, with temperature being the predominant factor because of its exponential nature. Since the formation of NO is exponentially dependent on local flame temperature (function of ϕ), large quantities of NO are formed only when high temperatures are produced, as in the case of combustion in a locally rich (near stoichiometric) region.

Simple estimates of residence times within burners and actual rates lead to a rule of thumb that very little NO will form until 1800 K is reached. Since the initiating step is one of high activation energy, it is apparent that the amounts of NO that will form will increase rapidly above 1800 K. Typical values obtained for residence time corresponding to practical burners are shown in Figure J.7 for hydrocarbons as a function of the maximum temperature reached in the flame (Ferri, 1974; Ferri and Agnone, 1974).

Since permissible turbine inlet temperatures are below 1800 K, one questions why the NO_x problem arises in the turbojet engine. The principal causes for the formation of this contaminant in an engine are related to the techniques currently employed in the design of combustors (Ferri, 1973a). Most present-day state-of-the-art combustors consist of a primary zone, where fuel is injected as a liquid using pressure atomizing nozzles, which provides a combustion region with an average equivalence ratio ϕ (see Section II) of slightly less than unity, and a secondary zone where diluent air is mixed with the combustion products to provide the proper temperature for the turbine (see Figure 6, p. 69).

The first stage, referred to as the primary, operates at an equivalence ratio close to 1 (stoichiometric) in order to stabilize the process. The excess air is then added to the burner in a secondary zone to cool the burner and reduce the temperature of the primary zone gases to permissible turbine inlet temperatures.

In order to understand the characteristics of the crucial primary zone, it is important to distinguish between homogeneous premixed flames and diffusion flames. Droplet (or spray) burning, the burning of separate jets of fuel and air, and the secondary burning of various gaseous fuel–air pockets of different composition make up the realm of diffusion flames. The burning rate of these flames is controlled by the rate at which the fuel and air can diffuse together. The burning rate of premixed flames is dominated by the much faster chemical kinetic reaction rates.

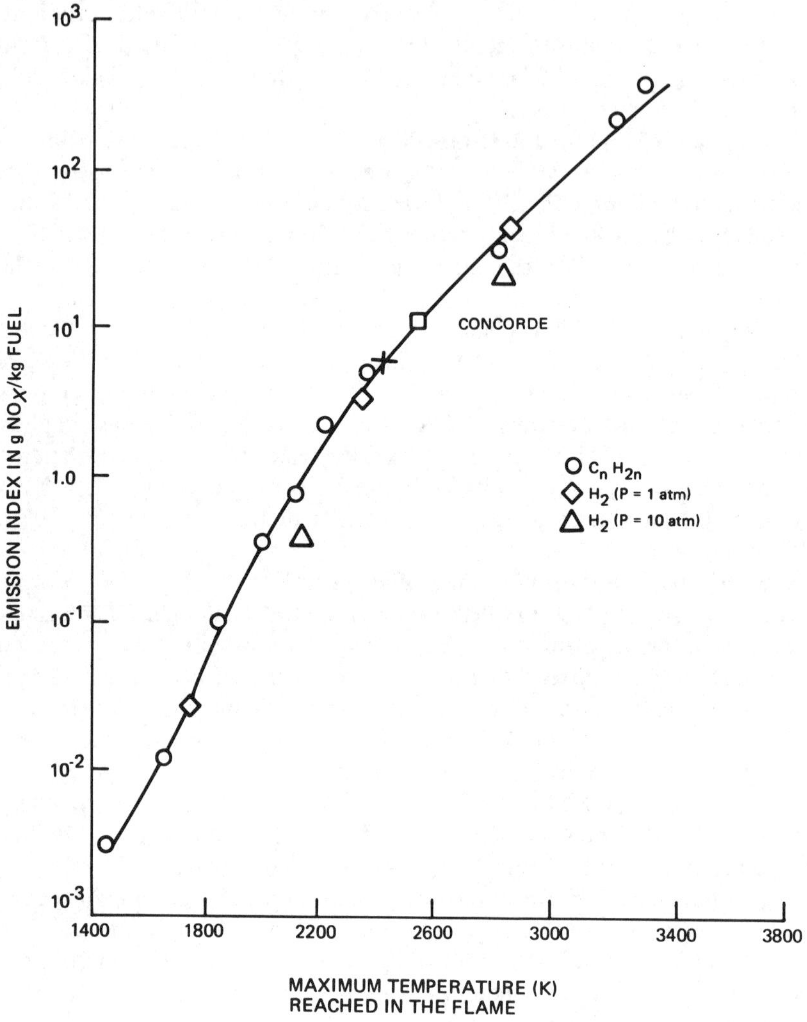

FIGURE J.7 Emission index as a function of maximum temperature reached in the flame.

Consider, for example, fuel droplets burning in air with the overall equivalence ratio being very lean. The final temperature of this burning stage will correspond to the overall lean composition. However, each droplet is surrounded by a flame that burns at the stoichiometric ($\phi=1$) temperature because the droplet fuel vapor and air diffuse together at rates that are in stoichiometric proportion, as in Figure J.8.

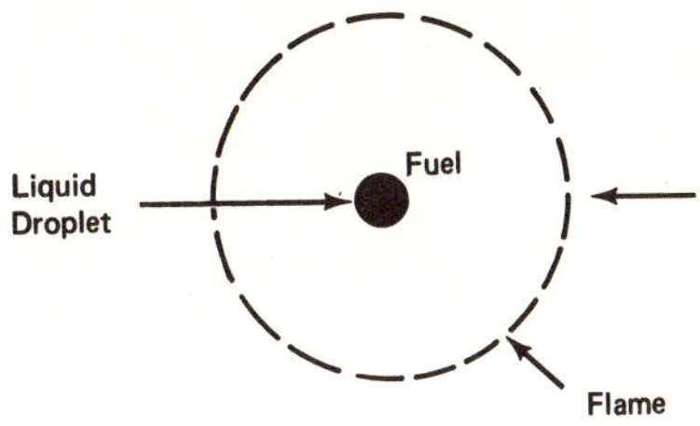

FIGURE J.8 Diagram of droplet flame in air.

The high-temperature products of the droplet flame are then diluted by the excess constituent to reach the fuel temperature corresponding to the overall composition. Typical temperatures reached locally at $\phi=1$ in the flame region are shown in Figure J.9 as a function of combustor inlet temperature.

Present-day combustors, in which fuel is injected as a droplet spray and combustion occurs at conditions close to stoichiometric, produce substantial amounts of NO_x as shown in Figure J.7 for temperatures corresponding to values shown in Figure J.9. (The figure has been constructed on the basis of short residence time at high temperature.)

In present design combustors, the maximum temperature reached increases when the air temperature increases. Typical experimental results are shown in Figure J.10.

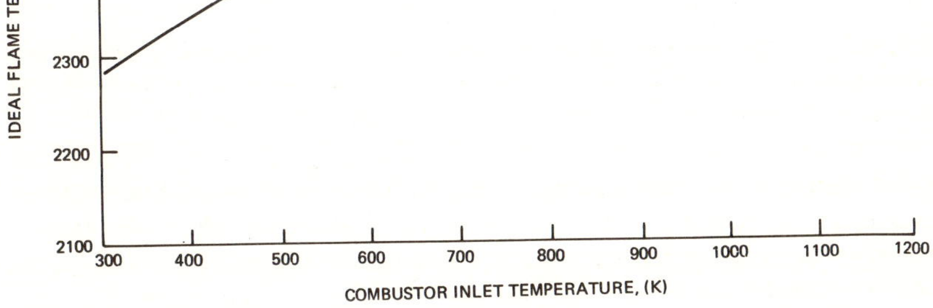

FIGURE J.9 Typical temperatures released locally in the flame region, at an equivalence ratio of one, as a function of combustor inlet temperature.

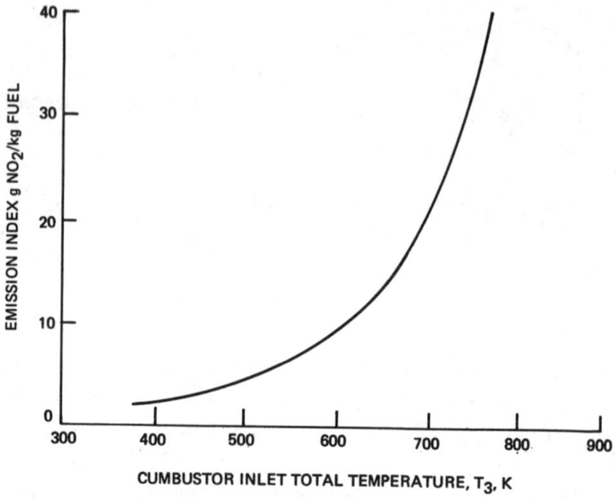

FIGURE J.10 Variation of NO_x emissions with combustor inlet temperature for current production engine combustors.

The NO_x formation starts after the heat release has occurred. Thus, it is possible to quench the NO_x formation reactions and still have complete energy release.

Figure J.11 gives the time history of variation of species for a combustion process (Glassman, 1970). Such a figure indicates that the NO formation starts only when water and carbon dioxide have already been formed in large quantities.

The secondary injection in present-day combustor cans achieves the quenching to a degree. The faster the burning rate, the shorter the required stay time in the primary zone and the greater the chance to quench NO_x formation. It is for this reason that early approaches to reduce NO_x emissions seek to increase mixing during combustion. The quenching can effect substantial reduction of NO produced, as shown in Figure J.12, which shows that the increase of fuel injected in the primary zone decreases quenching and increases production of NO (Anderson, 1973). It has been shown recently (Bracco, 1973) that, for a given mass, the smaller the droplet sizes the less the amount of NO formed; that is, a large droplet burning will give more NO than a group of small droplets adding to the same mass. However, the decrease of NO_x obtainable from this method is small because the NO formation occurs a few microseconds after the temperature has risen, and

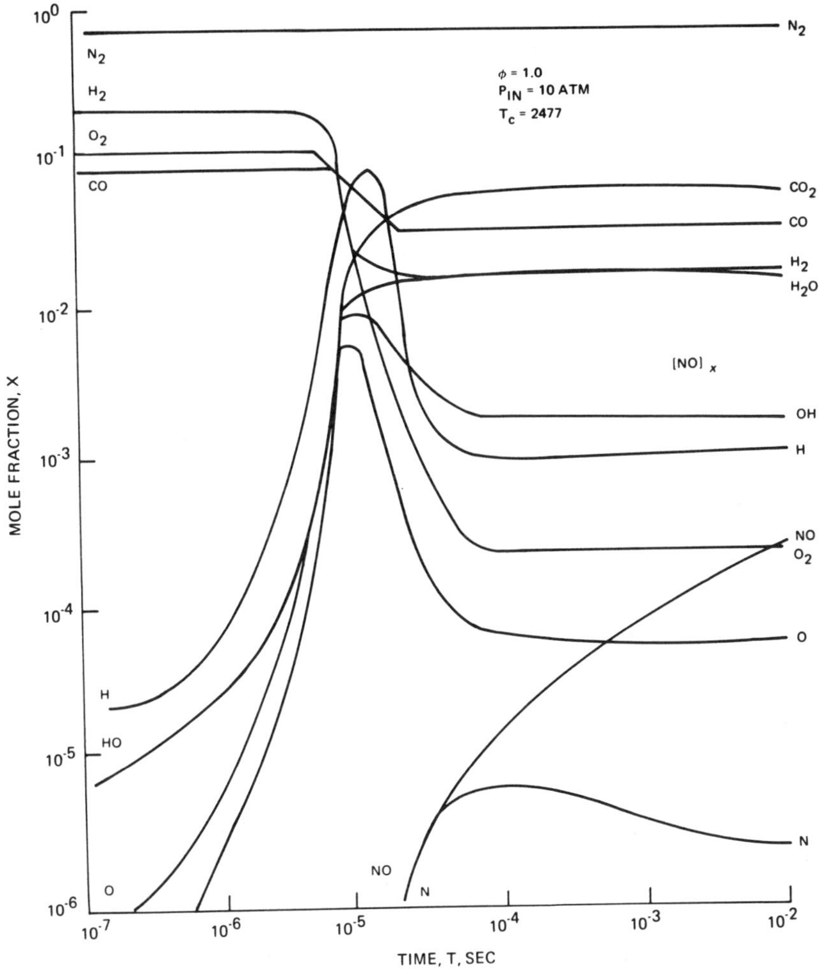

FIGURE J.11 Variation of species with time during the combustion process.

this time is too small to permit cooling substantially the products of combustion unless the size of the droplet is of molecular size.

The ultimate in increasing the homogeneity is to have a completely homogeneous prevaporized fuel–air mixture (Anderson, 1973; Boccio *et al.*, 1973; Ferri, 1973a, 1973b; Glassman, 1970; Goodman, 1974; Huffman, 1974; Thompson *et al.*, 1972). The question then arises as to whether a stable primary zone can be achieved with a lean homogeneous

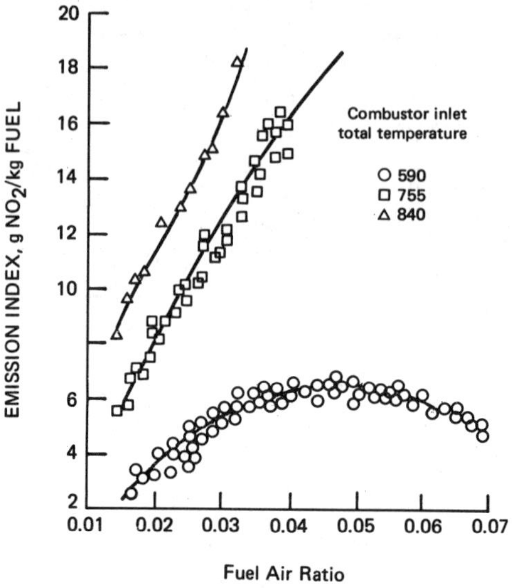

FIGURE J.12 Variation of NO_x with fuel–air ratio for an experimental combustor at 6 atm pressure.

premixed fuel–air mixture. Analyses (Glassman, 1967) and experiments (Anderson, 1973; Ferri, 1973a; Roffe and Ferri, 1974) indicate that the flammability limit is not a problem provided that ϕ does not go below 0.3. This approach, if implemented, will permit limiting the amount of NO produced to very low values, which are related to the inlet temperature of the turbine.

V. The Specific Problem of Nitrogen Oxides in Effluents from Jet Aircraft Engines

A. INTRODUCTION

A schematic diagram of a conventional turbojet combustor is shown in Figure J.13. The current generation of SST combustors (Concorde, Tupolev) emit 16 to 20 g of NO_x (calculated as NO_2) per kg of fuel during cruise. As noted previously, NO_x formation is largely determined by the maximum operating temperatures reached during burning in the primary combustion zone, and these maximum temperatures are

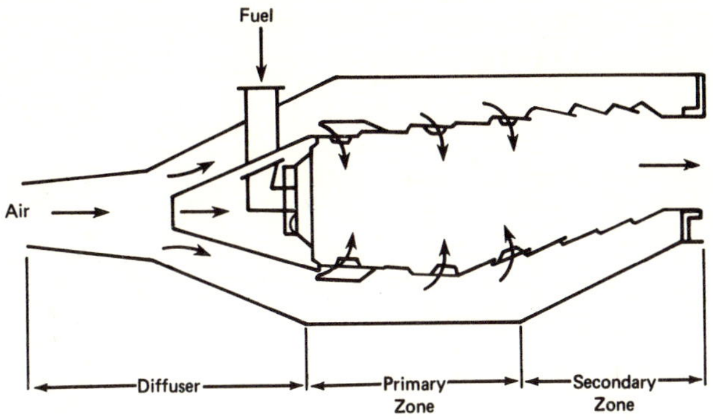

FIGURE J.13 Schematic diagram of a conventional turbojet combustor.

appreciably higher than the temperatures required for optimal engine performance. The turbine-inlet temperatures, and hence the turbine efficiencies, are limited by material considerations relating to turbine-blade construction and long turbine life. An obvious approach for reducing NO_x formation is, therefore, the design of combustors in which the local peak temperatures are reduced to levels that are compatible with the maximum inlet temperatures that the turbines can accept. A useful operational procedure for implementing this desirable objective is to burn fuel-lean mixtures during cruise operation. The fuel used in present airplanes is a liquid and is injected in the air as a spray. Then small pockets of high-concentration fuel are produced where high temperature is generated by the combustion process. Therefore in order to obtain uniformly fuel-lean mixtures, a mixture of vaporized fuel and air is required. The vaporization can be obtained easily during the mixing process, because of the sufficiently high temperature of the air downstream of the compressor.

We shall now consider two engine-modification schemes that have been suggested for fuel-lean combustion and indicate the reductions in NO_x concentration that have been achieved in prototype experiments.

B. PREMIXING AND PREVAPORIZING JET FUEL AND AIR

1. *Experimental Studies at the Advanced Technology Laboratories*

A schematic diagram of a system proposed by Ferri (1974) is reproduced in Figure J.14. In prototype experiments (Bracco, 1973; Roffe

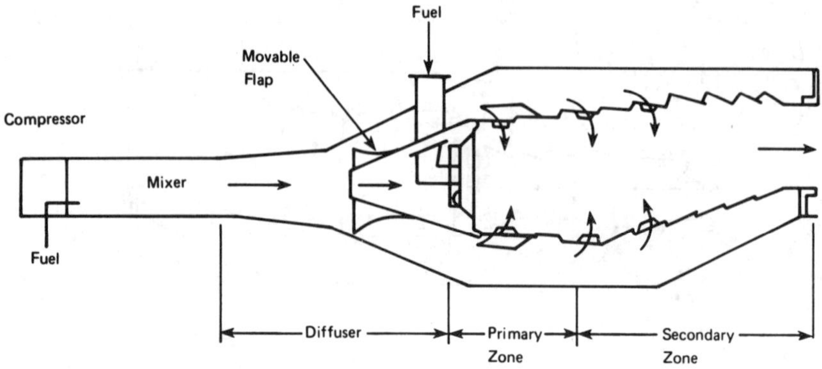

FIGURE J.14 Ferri's proposed modified turbojet combustor for reduction of NO_x.

and Ferri, 1974), liquid turbojet engine fuel (JP5) was preheated and injected into air in a test apparatus (see Figure J.15) at 833 K and 4 atm, which approximates SST engine cruise conditions. Measurements of combustor efficiency and of NO_x output indicated an NO_x emission index of about 0.38 g/kg of fuel with a measured combustion efficiency of 0.993 at an equivalence ratio, ϕ, of 0.25 and 0.3, and an emission index of 0.8 g/kg at $\phi=0.5$ with a combustion efficiency of 0.99. These values were measured with residence times in the combustor of the

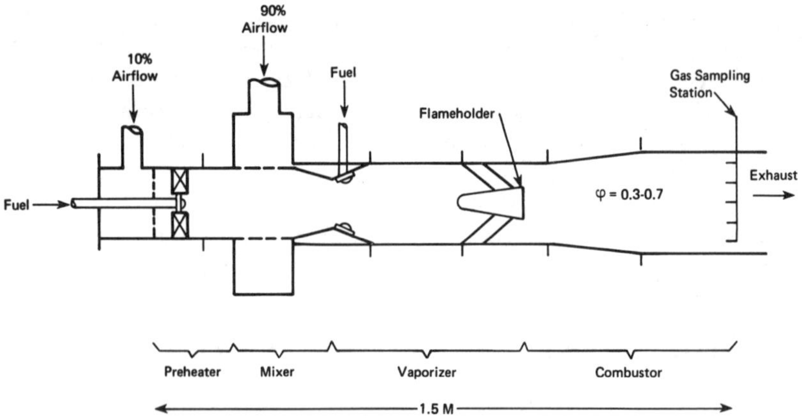

FIGURE J.15 Schematic diagram of the premixed, prevaporized combustor used by Ferri (1974) and Roffe and Ferri (1974).

same order as in actual engines. Thus a reduction in NO_x production of about a factor of 50 was achieved below that observed with conventional combustors, even though Roffe and Ferri (1974) noted that the air–fuel mixtures were not uniform. With further development of this technique, efficient operation may be achievable at lower emission indices.

The scheme proposed by Ferri would use two fuel control systems: one to be used for acceleration and flame-out conditions, which is the normal fuel system; the second to be used only at high altitude and high speed where the fuel is injected immediately downstream of the last stator of the compressor in a region of sufficiently high velocity to avoid flashback.

The scheme proposed by Ferri will require careful control of fuel/air ratio in the mixing section of Figure J.14 and involves also the important problem of prevention of flashback in high-speed flows. It is possible that practical implementation of this scheme will be facilitated by the introduction of flame-quenching surfaces, provided the corresponding pressure drops do not become excessively large.

2. *Experimental Studies at NASA-Lewis*

The NASA-Lewis Research Center has also been conducting fundamental combustion experiments to try to define the lower limit of NO_x that can be achieved by using premixed fuel-lean mixtures. A 10-cm-diameter flame tube (see Figure J.16) was used to simulate the primary-combustor zone. Premixed, prevaporized fuel was burned, and a perforated plate flame-holder was employed. The fuel used in these experiments was propane, which has somewhat similar combustion characteristics of kerosine-type jet fuels. Test conditions of air inlet

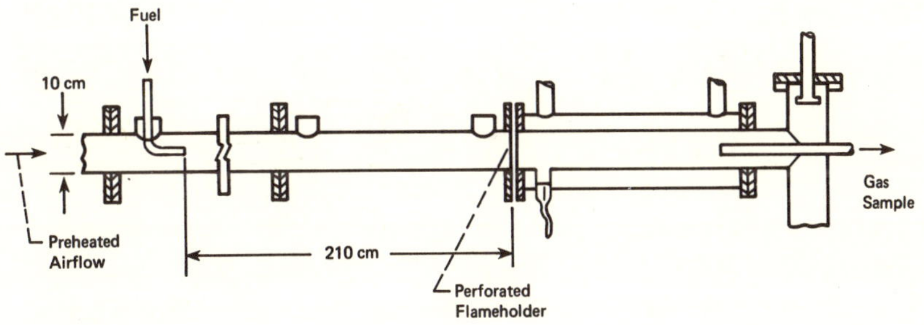

FIGURE J.16 Schematic diagrams of the NASA-Lewis premixed primary-zone test section used for NO_x-production measurements (Anderson, 1973).

temperature, pressure, and reference velocity were set to simulate high-altitude supersonic flight conditions; the effects of variations in equivalence ratio and residence time were investigated.

The results obtained in a wide variety of tests are cross-plotted in Figure J.17 to show the relationship between NO_x production (measured as NO_2), equivalence ratio, residence time, and combustion efficiency. Reference to Figure J.17 shows that NO_x emissions below

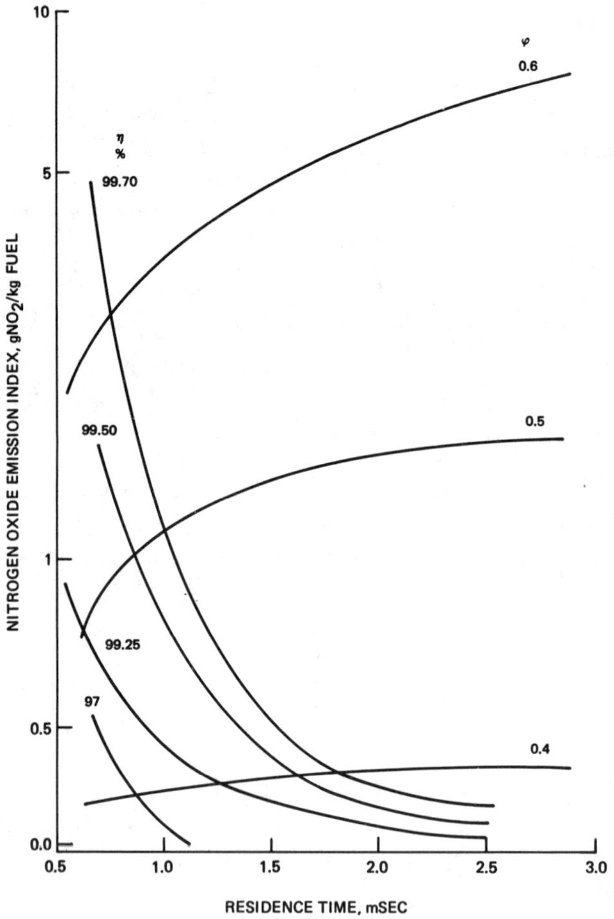

FIGURE J.17 The NASA-Lewis test results on fuel-lean propane–air mixtures have been cross-plotted to show the effect of residence time on NO_x emission index for various equivalence ratios ϕ; the combustion efficiencies, η, are also indicated. $T_3 = 800$ K; $P_3 = 5.5$ atm.

1 g/kg of fuel were achieved with greater than 99 percent combustion efficiency. A lower limit of 0.3 g/kg of fuel was obtained when operating near the lean flammability limit at simulated supersonic cruise conditions. This was accomplished with very lean mixtures ($\phi \approx 0.4$) and relatively long residence time (≈ 2.0 msec). Figure J.17 illustrates the tradeoffs that are available between equivalence ratio and residence time for low NO_x emissions, while maintaining high combustion efficiency.

C. CATALYTIC JET-FUEL COMBUSTORS

Another procedure for achieving high combustion efficiency with fuel-lean mixtures at relatively low peak temperatures involves the use of catalytic combustors (Blazowski and Bresowar, 1974; Roessler *et al.*, 1973; Roffe and Ferri, 1974). In preliminary experiments at Wright-Patterson Air Force Base (Blazowski and Bresowar, 1974) with combustor inlet temperatures of 566 to 648 K, exhaust temperatures of 1200 to 1589 K, pressures of 4 to 10.5 atm, equivalence ratios of 0.26 to 0.45, and combustor reference velocities of 13.7 to 20.3 m/sec, a decrease in nitric oxide production to 0.1 g of NO_x per kg of fuel was achieved. The upper factor exhaust temperatures of 1589 K may presumably be raised still higher to 1750 K at pressure drops of only 1 percent across the catalytic reactor. The combustion efficiency was nearly 100 percent. The catalytic reactor showed a strong decrease in efficiency with decreasing temperature.

The test section shown in Figure J.16 has also been used at the NASA-Lewis Research Center for tests of a catalytic combustor in place of the flame reaction zone (Blazowski and Bresowar, 1974). Experimental experience with this type of device using premixed prevaporized propane fuel at supersonic aircraft engine operating conditions indicates that the emission levels of NO_x could be reduced below the limits of the instrument-measuring capability (0.3 g/kg of fuel burned).

Premixing and prevaporizing problems were encountered with the fuel, with flashback and pre-ignition upstream of the catalyst bed being particular problems. Catalyst life may be short and a highly sensitive function of mixture ratio in off-design operation, particularly if excessively high temperatures are generated temporarily.

D. UNTRIED PROCEDURES FOR REDUCING NO_x EMISSION BY FUEL-LEAN COMBUSTION OF JET FUEL

A number of as yet untried procedures for efficient operation of fuel-lean turbojet combustors have been proposed to solve the problem and, in due course, should be investigated. These include a heat-exchange reformer unit to produce hydrogen and fuel fragments, lean-burning fuel with fuel droplets exploded by water, staged combustors, post-combustion cleanup of NO_x (Environics, 1974), and use of leaner fuel–air mixtures (Johnson, 1974; Roessler *et al.*, 1973).

E. HYDROGEN FOR USE AS A FUEL IN FUEL-LEAN MIXTURES

We expect that the use of hydrogen as fuel will greatly facilitate low-temperature operation in fuel-lean mixtures. Either catalytic or conventional combustors may be used.

Because of its higher heat content per kilogram, compared with conventional hydrocarbon fuel for jet aircraft engines, the emissions of nitrogen oxide from hydrogen combustion would be reduced by a factor of about 2.75, on the basis of energy output, all other factors remaining the same. In fuel-lean hydrogen–air mixtures, it should be possible to achieve significantly lower levels of emissions than with hydrocarbon jet fuel (Anderson, 1973; Ferri and Agnone, 1974; Grobman *et al.*, 1973).

However, it should be noted that no experimental studies of this type under simulated flight conditions have as yet been performed. The amount of NO_x per unit energy release can be made very low in fuel-lean hydrogen–air mixtures because the extraordinarily wide flammability limits and high flame speed for hydrogen–air mixtures permit operation very far from stoichiometric conditions. The lean-limit flame temperature is considered to be about 1100 K for hydrogen and about 1900 K for kerosine. While it is unreasonable to contemplate engine operation at such low average temperatures, we note nevertheless that the combustor designer has available a variety of fuel-control options that should allow significant reduction of NO_x as a combustion product with hydrogen.

F. REQUIRED DEVELOPMENT PROGRAMS

The fundamental nature of the laboratory experiments described in the preceding sections must be stressed when trying to assess the practical implications of the results obtained. The controlled laboratory experiments refer essentially to discrete operating points. No attempt has as yet been made to adapt the laboratory approaches to the wide range of operating conditions that an aircraft engine combustion chamber must be capable of meeting efficiently. Nevertheless, we must appropriately emphasize the experimentally established fact that a reduction of about two orders of magnitude in NO_x emission index for JP fuel has been achieved in laboratory experiments.

A number of developmental problems can be envisaged that must be solved before the new combustion-control concepts can be incorporated into flight-worthy engines. Such engines must be capable of operating over a wide range of temperature, from idle to full power, must be easily restarted, and must not be subject to flame-out. It may, therefore, be necessary to use either a variable-geometry configuration or a multiple combustor, either in series or in parallel. For such combustors, two modes of operation are contemplated. One follows present methods, with a primary region of combustion followed by a dilution zone. In this mode, only a small part of the air is introduced in the primary region and the remainder is injected to dilute the combustion gases downstream. In the second mode, a larger part or all the air is mixed with fuel and then injected in the primary combustion zone.

There are other design concepts that should permit the desired stable operations over wide limits. Implementation of any of these will, however, require considerable time and effort before the engines can be certificated for airline use.

It is reasonable to expect that the following development schedule could be followed:

(a) Existing subsonic engines could be modified to reduce NO_x emissions by a factor of 2 to 5 by using methods already under development by NASA. Such retrofits could probably be incorporated at costs and over a time frame similar to that experienced with the retrofit of smokeless cans in the past (the development time was about 2 years and the cost about $20 million). However, it is possible that larger improvements may be required. Then new concepts using the approaches mentioned above will be necessary.

(b) It is expected that 2 to 3 years of laboratory effort will have to be directed to the selection and design of optimal operational combustors. Then application of fuel-lean combustion to operational engines should be demonstrated. The present results indicate that the scheme proposed can be utilized for SST cruise conditions. However, a practical engine must operate in a variety of conditions. Substantial additional work is required to develop an actual burner. Special attention should be directed toward investigation of the transient characteristics of the engine. At the present time, it is difficult to predict whether some modification of the engine (for example, lengthening the shaft a few inches) will be required, in which case an additional cost would be encountered, or whether the revised combustor can be incorporated within the geometry of present engines. In either case, several years will be needed to develop and demonstrate the revised engine to certification status. On the basis of past experience, it may be expected that 5 to 8 years will be required for the engine design and development program. It thus appears that a minimum of 7 to 10 years will be required in all before a suitable program producing a low NO_x engine is ready for service. Research and development expenditures for such a program are difficult to estimate and will depend on the complexity and degree of engine design modification.

G. CONCLUSIONS AND RECOMMENDATIONS

The results that have already been obtained (see Sections V. B and V. C) indicate that a 10- to 20-fold reduction in NO_x concentration is practically achievable, and more is possible. Before embarking on a development program of new engine types, it would appear appropriate to conduct more extensive research on the schemes presently successfully tested and to make a preliminary evaluation of as yet untried techniques, such as those mentioned in Section V.D. A program should be definable within 2 or 3 years, after research and development expenditures of perhaps $50 million, and should be concluded successfully within 8 to 11 years (5 to 8 years after prototype-design selection). The *incremental cost* for NO_x reduction in building an entirely new engine should be relatively small, perhaps a few percent (less than 5 percent) of the development and production costs, of about $0.5 billion to $1.5 billion.

VI. The Specific Problem of Sulfur Oxides in Effluents from Jet Aircraft Engines

As sulfur is a common constituent of petroleum products, the question of the impact of sulfur oxides from the exhaust of stratospheric aircraft must be examined seriously. Reduction in sulfur content of fuel can be of economic importance in reducing jet turbine corrosion. The possible decrease in the lubricity of the fuel, with decrease in sulfur content, may need to be investigated. But, importantly, the question of substantial alterations of the fragile stratospheric environment by the oxidation products of sulfur must be carefully considered. The seriousness of the problem of changes in the stratosphere is enhanced by the very slow rate of recovery from any damage that might be done to the stratosphere.

Sections III.C and III.D of Appendix F give a detailed account of the manner in which aerosols may be formed in the stratosphere, starting with the sulfur oxides emitted by jet aircraft engines operating in the stratosphere with jet aircraft fuel containing sulfur. The sulfur oxides may be completely oxidized to sulfur trioxide, which, with water, can form sulfuric acid aerosols, which can affect the radiation to the earth and its climate.

The calculations presented in Sections III.C and III.D of Appendix F indicate that there exists a potential climatic effect arising from the presence of such aerosols in the stratosphere. Since this constitutes a potentially serious risk, it is important to take any economically feasible steps to reduce the risk to acceptable limits by lowering the amount of sulfur in the fuel.

The present allowable limit for sulfur in Jet A and A-1 fuels is 0.3 percent by weight. Hurn (1974) presents analyses of representative samples ranging from 0.01 to 0.167 percent by weight, along with an estimate that the worldwide average might be about 0.05 percent. He ascribes the low value to the fact that the hydrogenation process necessary to remove metal and nitrogen impurities also removes sulfur. Hurn refers to a study by Bonner & Moore Associates (1974) on the cost of reducing sulfur in fuels to lower levels. On the assumption that (1) the petroleum contains sulfur compounds that are of the easily removed type, (2) hydrogen gas production can be adequately expanded, (3) capital investment for equipment can be depreciated in 16 yr, and (4) energy costs are $0.60 per million BTU, they calculate that the sulfur level can be reduced to 0.01 weight percent at a cost of $0.15 to $0.25 per barrel or reduced to 0.0025 percent at a cost of $0.20 to $0.30 per barrel by a conventional catalytic hydrodesulfurization treatment. Oliver (1974) has obtained estimates from the Union Oil Company on the cost of producing jet fuel with sulfur contents of 1000, 500, and 10 ppm by weight from petroleum containing 0.4 percent sulfur. The main cost is the capital plant investment. The cost of about ½ cent per gallon for a reduction to 10 to 20 ppm, or about 2 percent of total cost, agreed with the values of Bonner & Moore Associates. More recent information from the M. W. Kellogg Company (Axelrod, 1974) indicates a cost, in terms of 1974 dollars, of about 1.0 cent per gallon for a reduction of sulfur to 10 to 20 ppm. This cost may be lowered if the reduction in the content of sulfur is carried out on broader fractions with part of the costs properly allocated to materials other than jet aircraft fuel alone.

VII. Conclusions

On the basis of the information summarized in this report, the Panel on Engines and Fuels is led to the following conclusions:

A. REGARDING THE MAIN PRODUCTS OF COMBUSTION

1. The main products of the reaction of combustion in jet aircraft engines, using hydrocarbon fuel of specification grade from natural petroleum or from "synthetic" petroleum are carbon dioxide and water, the amounts of both of which are directly proportional to the amount of fuel consumed and are thus fixed in quantity.

2. The main product of the reaction of combustion in jet aircraft engines, using hydrogen as the fuel, is water, the amount of which is directly proportional to the amount of fuel consumed and is thus fixed in quantity.

B. REGARDING PRODUCTS OF INCOMPLETE COMBUSTION

1. In jet aircraft engines operating at cruise conditions, using hydrocarbon fuel of specification grade from natural petroleum or from "synthetic" petroleum, the amounts of the products arising from possible incomplete combustion of the fuel, such as carbon monoxide and hydrocarbons, are negligible because the jet engine operates at high efficiency, with essentially complete combustion.

C. REGARDING NITROGEN OXIDES IN THE EFFLUENTS FROM JET AIRCRAFT ENGINES

1. One of the two principal minor products of the reaction of combustion in jet aircraft engines, using hydrocarbon fuel of specification grade from natural petroleum or from "synthetic" petroleum, are nitrogen oxides, arising in the combustor from reactions of the nitrogen and oxygen of the air. The amount of these nitrogen oxides in the engine effluents can be reduced to relatively low values by appropriate changes in the design and operation of the combustor in jet aircraft engines.

2. Within a time frame of 10 to 15 years, new jet aircraft engines, using hydrocarbon fuel of specification grade from natural petroleum or from "synthetic" petroleum, can be designed, developed, tested, and put into use that will produce, in the engine effluents, amounts of nitrogen oxides that are less than the amounts produced by the supersonic aircraft (Concorde and Tupolev) flying in 1974 by factors ranging from 10 to 20, possibly more, assuming that adequate resources and high priority are put on the task. The factor of 10 is a fully realizable target, while the factor of 20 has a high probability of success in achieving the goal, both assuming adequate resources and high priority, taking into

account our present knowledge of completed and proved laboratory experimentation and recognizing the ability of man to create new knowledge and produce new devices in response to a given challenge. It is difficult at the present state of research to estimate the added complexity required in the burner design, because only steady-state conditions have been considered. At present, on the basis of present knowledge, no indication exists that the complexity will be large. The time frame of 10 to 15 years is reasonable in terms of the rate of scientific, technical, and industrial productivity on such problems. It is important to note that, with a time frame of 10 to 15 years, the necessary research, development, and other work should be started in 1975, if the new engines are to be in use by 1985 to 1990.

3. It is estimated that the cost of bringing into being new jet aircraft engines, using hydrocarbon fuel of specification grade from natural petroleum or from "synthetic" petroleum and embodying changes in design and operation that will produce, in the engine effluents, amounts of nitrogen oxides that are less than the amounts produced by the supersonic aircraft flying in 1974 by factors ranging from 10 to 20, is small in comparison with the possible consequences of not performing such work. The cost for completing the laboratory investigations is of the order of $5 million to $15 million, while the complete research program can be several times larger, for example, $100 million. This estimate assumes that, with a time frame of 10 to 15 years, engine manufacturers will be developing new jet aircraft engines in course, and that the incorporation of the changes in design and operation of the combustor to reduce nitrogen oxides in the engine effluents will be a minor part of the overall development cost of a new engine (which overall development cost can be of the order of $1 billion). It should be emphasized that significant increase in the complexity of the combustor system in the new engine may increase the cost of each engine by as much as about 5 percent of the cost of the engine. Also, there may be an increase in operating cost.

4. The principal minor product of the reaction of combustion in jet aircraft engines, using hydrogen as the fuel, are nitrogen oxides, arising in the combustor from reactions of the nitrogen and oxygen in the air. The amount of these nitrogen oxides in the effluents from turboengines, using hydrogen as fuel, can be reduced to relatively low values by appropriate design and operation of the combustor in such engines. For jet engines in hypersonic flight, using hydrogen as the fuel, the problem of inhibiting the formation of nitrogen oxides in the engine effluents is more difficult, but believed solvable.

D. REGARDING SULFUR OXIDES IN THE EFFLUENTS FROM JET AIRCRAFT ENGINES

1. The second of the two principal minor products of the reaction of combustion in jet aircraft engines, using hydrocarbon fuel of specification grade from natural petroleum or from "synthetic" petroleum, are sulfur oxides, arising from the reaction of sulfur in the fuel with oxygen of the air. The amount of these sulfur oxides in the engine effluents can be reduced to relatively low values by appropriate reduction of the amount of sulfur in the jet fuel.

2. The amount of sulfur oxides in the engine effluents from jet aircraft engines, using hydrocarbon fuel of specification grade from natural petroleum or from "synthetic" petroleum, can be reduced by applying established technology, already available in the petroleum industry, to lower the amount of sulfur in the hydrocarbon fuel itself to an amount of the order of 10 to 20 parts per million, by weight.

3. It is estimated that the cost of reducing the amount of sulfur in the hydrocarbon fuel itself to the order of 10 to 20 parts per million, by weight, is about 1 cent per gallon of jet aircraft fuel. This cost may be lowered if the reduction in the content of sulfur is carried out on broader fractions, that is, on more than the kerosine fraction of petroleum in the one process, so that part of the cost is properly allocated to material other than the jet aircraft fuel alone.

References

Anderson, D. 1973. *Effect of Premixing on Nitric Oxide Formation*. NASA-Lewis Report. NASA TMX-68220.

Avery, W. H., and R. W. Hart. 1953. Combustor performance with instantaneous mixing, *Ind. Eng. Chem.* 45:1634.

Axelrod, L. 1974. M. W. Kellogg Co. Private communication.

Blazowski, W. S., and G. E. Bresowar. 1974. Preliminary study of the catalytic combustor concept as applied to aircraft gas turbines, *Air Force Aero Propulsion Lab. Tech. Rep.* AFAPL-TR-74-32. May.

Boccio, J. L., *et al.* 1973. *A Mathematical Model for Jet Engine Combustor Pollutant Emission*. NASA CR-12108. GASL TR-781.

Bonner & Moore Associates, Houston, Tex. 1974. 1973 presented as Appendix C, Economic and social measures of biologic and climatic change. CIAP Monogr. 6 (draft), pp. 35–41.

Bracco, F. V. 1973. Nitric oxide formation in droplet diffusion flames, *14th Symp. on Combustion*. Combustion Inst., Pittsburgh.

Camp, F. W. 1974. *The Tar Sands of Alberta, Canada*. Cameron Eng., Inc., Denver, Colo.

Davis, W. O., G. Daly, R. C. Oliver, J. M. English, and G. L. Kernan. 1974. Costs and technology of control. CIAP Monogr. 6 (draft), Chap. 4, pp. 4-1–4-42.
Dryer, F. L., D. W. Naegeli, and I. Glassman. 1971. Temperature dependence of the reaction $CO+OH=CO_2+H$, *Combustion and Flame* 17:270.
Environics. 1974.
Ferri, A. 1973a. Better marks on pollution for the SST, *Astron. Aeron.* 11:37–41.
Ferri, A. 1973b. *Reduction of NO Formation by Premixing.* NYU-AA-73-02.
Ferri, A. 1974. The problem of pollution for the SST, *9th Congr. Int. Council Aeron. Sci.* ICAS 74-29. August.
Ferri, A., and A. Agnone. 1974. The jet engine design that can drastically reduce oxides of nitrogen. Aerospace Sciences Meeting Paper 74-160, AIAA 12.
Glassman, I. 1967. *Special Topics in Combustion and Chemical Rocket.* Dept. of Aero. Eng., Technion, Haifa, Israel.
Glassman, I. 1970. *Combustion Science and Technology.* Guggenheim Labs., Princeton U.
Glassman, I. 1973. Current understanding of NO_x formation in mobile and stationary combustion sources, *Proc. Symp. Current Status NO_x Problem and Its Control.* Middle Atlantic Consortium on Air Pollution.
Goodman, I. 1974. *Jet Engine Exhaust Emission of High-Altitude Commercial Aircraft Projected to 1990.* NASA TMX-3007.
Grobman, J., and R. D. Ingebo. 1974. *Jet Engine Exhaust Emissions of High-Altitude Commercial Aircraft Projected to 1990.* NASA TMX-3007.
Grobman, J., C. Norgren, and D. Anderson. 1973. *Turbojet Emissions, Hydrogen versus JP.* NASA TMX-68258. IASA Tech. Memo. May.
Hartley, F. W., and C. S. Brinegar. 1957. *Sci. Mon.* 84:275–289.
Huffman, H. C. 1974. Union Oil Co., Res. Center, Brea, Calif. Private communication.
Hurn, R. W. 1974. Fuels. CIAP Monogr. 2 (draft), Chap. 2, pp. 2-1–2-17.
Johnson, F. S. 1974. Atmospheric Physics and Chemistry Panel, Woods Hole. July.
Longwell, J. P. 1974. EXXON Res. and Eng. Co. Private communication.
Longwell, J. P., E. E. Frost, and M. A. Weiss. 1953. Flame stability in bluff body recirculation zones, *Ind. Eng. Chem.* 45:1627.
Marteney, P. J. 1970. Analytical study of the kinetics of formation of nitrogen oxide in hydrocarbon-air combustion, *Comb. Sci. Technol.* 1:461.
Mullins, B. P. 1955. Spontaneous ignition of liquid fuels. AGARD No. 4. Butterworths, London.
National Petroleum Council. 1972. Report of the Committee on U.S. Energy Outlook.
National Science Foundation Workshop. 1974. *Combustion for Future Energy Needs.* Dept. Aerospace and Mech. Sci., Princeton U., Princeton, N.J. June.
Oliver, R. 1974. CIC summer study. Woods Hole, Mass.
Palmer, R. F., and C. F. Cullis. 1965. The formation of carbon from gases, *Chemistry and Physics of Carbon.* Marcel Dekker, New York.
Roessler, W. V., *et al.* 1973. *Investigation of Surface-Combustion Concepts for NO_x Control in Utility Boilers and Stationary Gas Turbines.* EPA-650/2-73-014.
Roffe, G., and A. Ferri. 1974. *Prevaporization and Premixing for Low NO_x Gas Turbine Combustion.* ATL TR 203. August.
Rossini, F. D. 1960. *J. Chem. Educ.* 37:554.

Thompson, D., *et al.* 1972. NO_x formation in combustion, *Combustion and Flame* 19:69–79.

Zeldovich, V. R. 1946. The oxidation of nitrogen in combustion and explosions, *Acta Physicochem. URSS* 21:577.

Zucrow, J. M. 1958. *Aircraft and Missile Propulsion.* Vol. II, Chap. 8. Wiley & Sons, New York.

Appendix K

Uncertainty Analysis

I. Introduction

Despite the advances made by the Climatic Impact Assessment Program (CIAP) over the past several years, it cannot be denied that we remain in a state of considerable uncertainty about the effects of aircraft effluents in the stratosphere. There is uncertainty about each link in the chain connecting nitrogen oxide (NO_x) emissions to ozone reduction, to increased ultraviolet radiation, to final effects such as skin cancer, although some links are much more uncertain than others. There is even more uncertainty about the effects of effluents [sulfur oxides (SO_x), water (H_2O), and NO_x] and of the ozone reduction resulting from NO_x on climate (temperature and precipitation). Sophisticated simulation models of global climate provide some insights, but their outputs are only as valid as the assumptions that go into them. And these assumptions—including the very specification of the models in terms of omitted variables and functional relationships—are by no means claimed to be true reflections of reality.

The methodology that is best equipped to structure these uncertainties systematically is *decision analysis* (H. Raiffa, *Decision Analysis: Introductory Lectures on Choices under Uncertainty,* Addison-Wesley,

---

This Appendix presents the findings of the Panel on Decision Analysis—Monitoring and Surveillance: K. Arrow, *Chairman;* D. Hunten, H. Jacoby, H. Odishaw, R. Solow, J. Tukey, and R. Zeckhauser. Consultants to the Panel: R. d'Arge, J. English, S. Shavell, and M. Weinstein.

Reading, Mass., 1968). Since the present discussion focuses on the uncertainty regarding the *consequences* of alternative flight and emission scenarios, and not on the decisions among them (the latter requiring an analysis including economic benefits as well as risks), we might refer to the present mode of analysis as *consequence analysis*.

The aim of consequence analysis is to break down the components of uncertainty systematically and then to derive quantitatively the probability distributions for final outcomes from the quantitative specifications of the component probabilities. It is this breaking-down and putting-back-together character of the analysis that is its greatest asset, since it allows experts in different areas to assess their own uncertainties separately.

The method used also allows one to include uncertainties of very different varieties in the overall assessment of probabilities. Regarding climatic impact, there are uncertainties *within* any given model of climatic change due to parameter errors and to randomness in the natural system. But there is also variation *between* different models with different assumptions and specifications.* While the present analysis does not attempt to separate the two kinds of uncertainty, they are both included implicitly in the overall probability assessments.

Thus, although models provide us with useful insights, they cannot claim to reduce all uncertainties to parametric uncertainties within the models themselves. One must turn to expert opinion to assess the relative credence placed in different models and, specifically, to assess their own uncertainties. In the present analysis, the opinions of several experts were assessed in order to capture the range of uncertainty among individuals who attach differing degrees of credence to the validity of the various models.

The consequence analysis presented here is flexible. The structure permits one to input one's own assumptions as to (a) fuel flow, (b) regulations in effect, (c) emission indices, and, especially, (d) probability assessments of the component uncertainties. Thus, one can use this structure to examine the implications of various policy options under various probability distributions.

The importance of this kind of model is that it treats the uncertainty explicitly and, in principle, as completely as one wishes. No source of

* The disparity between the Budyko-Sellers and MacCracken models of the effects of decreased radiation on temperature and the disparity between the Hunten and Chang models of the effect of NO_x on ozone reduction are examples of this phenomenon.

uncertainty need be omitted, even one as elusive as expert hunch or skepticism about models. One can input one's own probabilities and derive the consequences—in probabilistic terms, of course.

The present approach has its strengths and limitations. One strength is that it is not wedded to any particular probability distribution given the great uncertainty in the extremes. This is a valuable asset since the commonly used probability distributions (e.g., normal or lognormal) may not exhibit sufficient "spread" at the extremes (kurtosis) to describe one's sense of vast uncertainty.

Another strength is that the approach allows one to input one's own judgmental probabilities. This allows the decisions made by a policy maker to be consistent with his own best judgment of uncertainty based on contact with experts. But this strength is also a limitation. Psychologists have shown that people tend not to be very good assessors of probability, at least people without much experience at it. In particular, they tend to be overconfident, placing too much weight at the mean and not enough at the extremes. This has been demonstrated in tests by which it was shown that in estimating the probability distribution for a known, but obscure, number (such as the number of ice cream cones consumed in New York City in 1963), the true value lay outside the subjectively assessed 98 percent confidence limits more than 25 percent of the time (M. Alpert and H. Raiffa, "Progress Report on the Training of Probability Assessors," Harvard Business School, 1967). For ideal assessors, this would have occurred only 2 percent of the time.

Climatologists, especially those with meteorological experience, can be expected to be better than average probability assessors. Although their assessments must be viewed as possibly reflecting overconfidence in climatic models that may be grossly misspecified, one can use their judgments with quite a bit of confidence.

The consequence analyses are presented as follows. First, the particular chains of consequences are laid out in terms of flowcharts and in terms of parameterized quantitative relationships. Next, some alternative scenarios of altitudes, fuel flows, and emissions are given, reflecting alternative options on fleet size and emission indices. Then, the method of obtaining probability estimates for the links in the chains of consequences is given, involving assessment and aggregation of expert opinions. Next, the method for combining these component probability estimates into probability estimates of final impacts on skin cancer and mean surface-temperature change are given. Finally, the results are presented in the form of probability distributions for these ultimate outcome measures.

II. Chains of Consequences

Table K.1 shows the relationship between the present mass in th stratosphere of several chemicals emitted by aircraft and the level o emissions expected from 10^{11} kg of fuel under current technology. It i clear that NO_x and SO_x are major potential problems, with 50 percen and 20 percent, respectively, of the present stratospheric content emitte each year by combustion of 10^{11} kg of fuel. Carbon monoxide (CO) and water (H_2O) are also potential problems, with 10 percent and 7 percent, respectively, of existing loadings produced each year. When Space Shuttle technology is available, hydrogen chloride (HCl) and aluminum oxide (Al_2O_3) will also need to be reckoned with. However, this report will concentrate on the effects of NO_x and SO_x, as well as H_2O. The reason for including H_2O is twofold: first, it appears that the climatic effect of H_2O may be comparable with that of SO_x if sulfur is reduced to very low levels in fuels; and, second, water is a major by-product of the hypersonic aircraft envisaged for the future and may be substantially more important a problem under that technology than it is with subsonic or supersonic technology.

Five chains of consequences are considered in this report. These are summarized in Figure K.1.

Chain A stems from the emission of oxides of nitrogen. By chemical reaction, NO_x resident in the stratosphere causes some degree of reduction of ozone (O_3). This linkage is displayed in Figure K.2. Notice that

TABLE K.1 Direct Alterability of Stratospheric Composition (Stratospheric Mass $= 1 \times 10^{18}$ kg [a])

Component	~Mass in Stratosphere, kg	Mass from 10^{11} kg of fuel [b]	Ratio	Examine Further (?)
CO_2	5×10^{14}	3.2×10^{11}	0.0006	No
H_2O	2×10^{12}	1.3×10^{11}	0.07	Yes
CH_4	5×10^{11}	1.0×10^{9}	0.002	No
CO	3×10^{10}	3.0×10^{9}	0.10	Yes
NO_x (as HNO_3)	3×10^{9}	1.6×10^{9}	0.5	Yes
SO_2 (as H_2SO_4)	$\sim 10^{9}$	2×10^{8}	0.2	Yes
		Mass from 60 Shuttles, kg		
HCl	$< 8 \times 10^{8}$	6×10^{6}	> 0.008	Yes
Al_2O_3	? (LOW)	8×10^{6}	HIGH [c]	Yes

[a] Source: Institute for Defense Analyses, 1974.
[b] Assumes duct-burning turbofan; "EPA" technology.
[c] Small compared with total particulates.

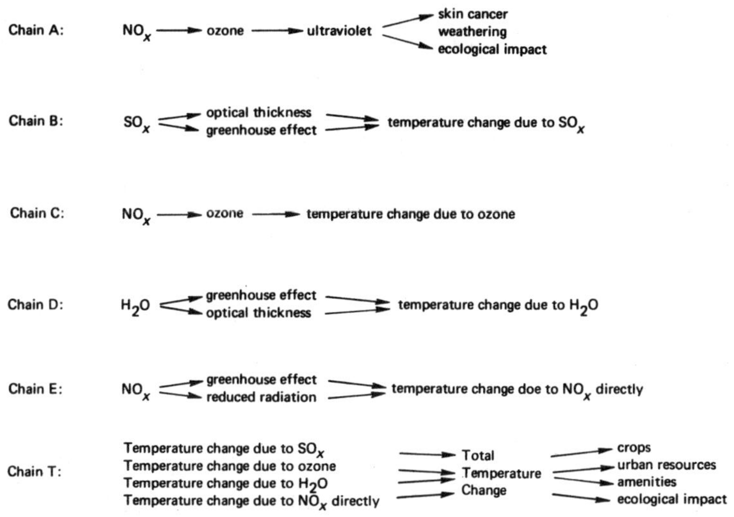

FIGURE K.1 Chains of consequences considered.

emissions in the various altitude bands are considered separately, since the residence time of NO_x (the length of time an average NO_x molecule remains in the stratosphere with an opportunity to react with ozone) is greater the higher the altitude.

The remainder of Chain A is displayed in Figure K.3. A reduction in ozone results in some degree of increased ultraviolet-B radiation. In

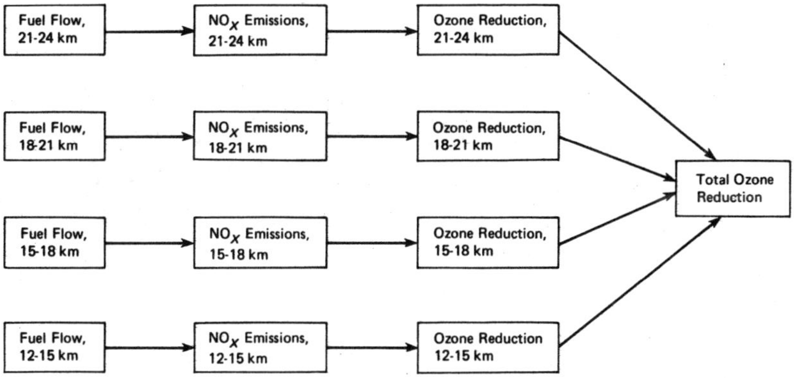

FIGURE K.2 Chain A (Part 1): $NO_x \rightarrow O_3$ reduction $\rightarrow$ uv increase; Chain C (Part 1): $NO_x \rightarrow O_3$ reduction $\rightarrow$ climate.

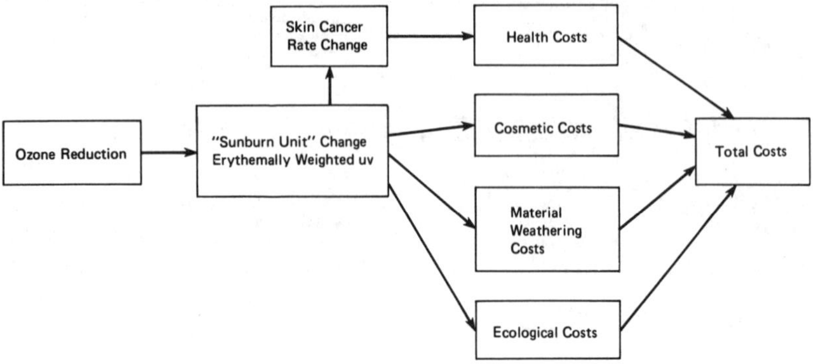

FIGURE K.3 Chain A (Part 2): $NO_x \rightarrow O_3$ reduction $\rightarrow$ uv increase.

Figure K.3, this change is labeled "sunburn unit" change, by which is meant that the wavelengths of uv radiation are weighted according to an index describing their contribution to sunburn. This increase in weighted uv radiation, in turn, is believed to result in some increase in the rate of skin cancer. In addition, the uv increase may result directly in effects on skin aging, weathering of materials, and possible ecological changes. The present analysis is concerned only, however, with the uncertainties of the impact on skin cancer.

Chain B stems from the emission of oxides of sulfur. As shown in Figure K.4, the emission of sulfur oxides at various altitudes results in the formation of aerosols of sulfuric acid, depending on the residence times at each altitude and the rate of conversion from SO_2 to H_2SO_4

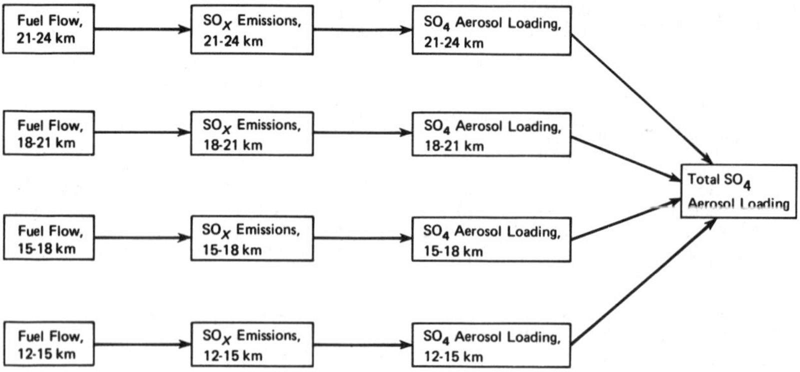

FIGURE K.4 Chain B (Part 1): $SO_x \rightarrow$ aerosol $\rightarrow$ climate.

FIGURE K.5 Chain B: $SO_x \rightarrow$ aerosol $\rightarrow$ climate.

aerosol. The next step, perhaps the most uncertain in the whole picture, is shown in Figure K.5. The effect of these aerosols on surface temperature is generally believed to be in the downward direction, because of the increased optical thickness of the stratosphere and a decrease in visible radiation. But the magnitude of this effect is in question, and, moreover, there is a possibility that the greenhouse effect of the aerosols (allowing less radiation to escape from the atmosphere) may have a *warming* result that may negate or outweigh the *cooling* effect.

Chain C, like Chain A, begins with the effect of NO_x on ozone, as shown in Figure K.2. But the next step is the little-understood effect of ozone reduction on temperature displayed in Figure K.6. Climatologists disagree as to the direction of this effect, although most are confident that ozone reduction results in warming, thus offsetting in part the likely cooling effects of SO_x aerosols.

Chain D begins with emission of water vapor in the stratosphere at various altitudes. As shown in Figure K.7, this results in an increased loading of water vapor in the stratosphere, as determined by residence times. The effect of this water vapor on temperature (Figure K.8) is again uncertain, with both the greenhouse effect (warming) and radiation decrease effect (cooling) possible.

Chain E begins with emission of NO_x in the stratosphere at various altitudes. As shown in Figure K.9, this results in an increased concentration of NO_x in the stratosphere. This, in turn, affects temperature in an uncertain way, with either cooling or warming possible, depending on whether the greenhouse effect outweighs the direct cooling effect.

The temperature changes from different causes have to be added, a process represented in Figure K.1 by Chain T.

The final link in Chain T is displayed in Figure K.10. The effects of temperature change are seen in effects on crops (positive or negative), including especially wheat, corn, cotton, and rice; effects on urban resources, including energy consumption, insulation of materials, snow removal, health, and aesthetics; and effects on natural resources and ecological balance. For purposes of the present consequence analysis,

FIGURE K.6 Chain C (Part 2): $NO_x \rightarrow O_3$ reduction $\rightarrow$ climate.

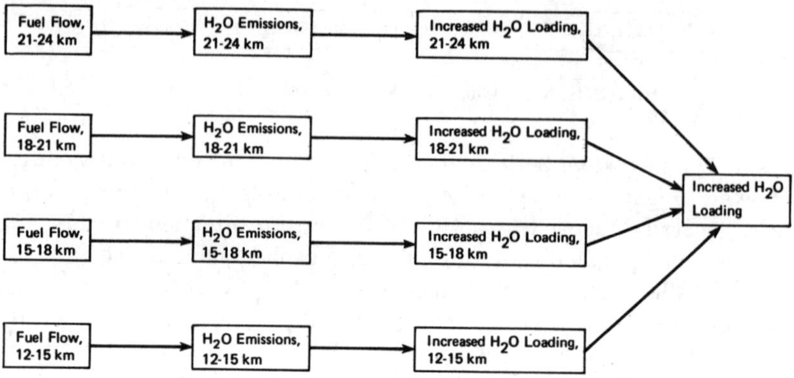

FIGURE K.7 Chain D (Part 1): $H_2O \rightarrow$ climate.

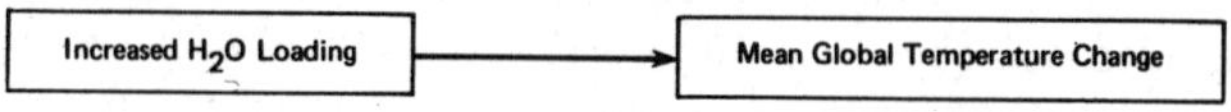

FIGURE K.8 Chain D (Part 2): $H_2O \rightarrow$ climate.

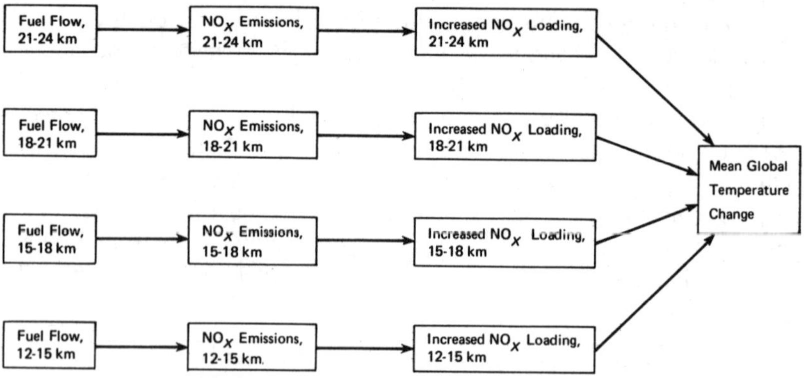

FIGURE K.9 Chain E: $NO_x \rightarrow$ climate.

however, this last link is omitted, total temperature change being considered the final outcome to be assessed.

It should be noted that effects on precipitation, independent of temperature, have not been considered. A more sophisticated analysis would treat precipitation separately.

It should also be noted that the chemical effect of H_2O in reducing NO_x (and thereby offsetting the effects of Chains A and C) is ignored because this reduction in NO_x is very small compared with total NO_x.

III. Kinds of Uncertainty

Before getting involved in the details of the uncertainty analysis, it is useful to distinguish several kinds of uncertainty that should not be confused. Each uncertainty has a special significance in suggesting appropriate policy responses.

The first, and most basic, type of uncertainty is our present *uncertainty about the magnitudes* of the climatic impacts to be expected. This includes both uncertainty due to true randomness in the universe and uncertainty due to our own lack of knowledge about climatic phenomena. We are not, however, in a state of total ignorance. We do have information from theories of climate and simulation models based on those theories. And we can observe in history the climatic impacts

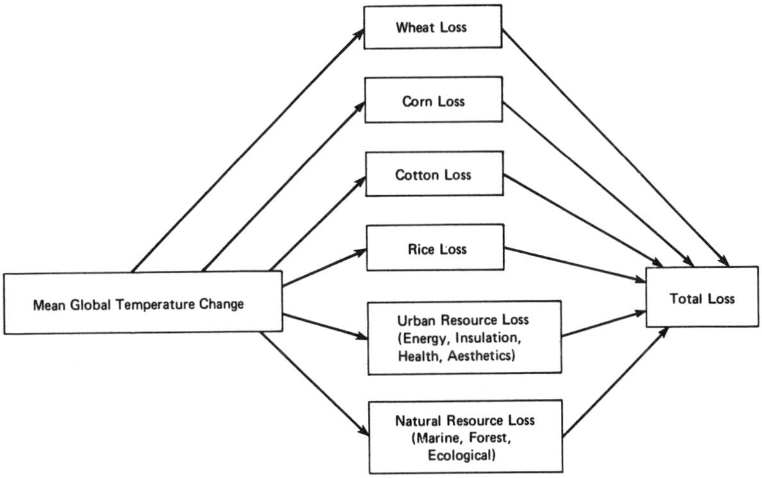

FIGURE K.10 Chain T: Climate → economic loss.

that resulted from volcanoes and other sources of large quantities of stratospheric contamination.

Another quite different kind of uncertainty that one must bear in mind involves the *diurnal variation* in weather. Even if we knew the overall climatic effects, there would still be variation around these climatic means. The importance of such diurnal variation can be seen by considering the impact of a decline in mean temperature. If there is not only a drop in the mean temperature but also in the whole distribution of day-to-day temperatures around the mean, then the impact on crops may be far more severe than the drop in mean temperature would lead one to believe. This would be the case, for example, if the probability of at least one frost during a critical month increased from 5 to 25 percent, while the mean temperature changed by only 0.1 °C.

A third kind of uncertainty, one that is really quite different from the other two, is *measurement error*. This is not really uncertainty in the same sense as the other two, since it is the *actual* impact, and not the *observed* impact, that is important. An unobservable change in stratospheric ozone *may* result in more than a few cases of skin cancer. Lack of observability does not imply unimportance.

IV. Simplifying Assumptions

In order to perform a meaningful and relatively simple quantitative analysis, three basic simplifying assumptions were made.

First, the entire consequence analysis is based on a *steady-state* level of emissions and effects. Thus, for example, the changes in stratospheric ozone and associated cases of skin cancer might not occur for several decades if emissions were to commence today. Furthermore, the analysis assumes a constant annual level of emissions up to that steady-state equilibrium.

Second, the consequence analysis deals only with *global means* and *annual means*.* This has the general effect of understating impacts because it is typically the extremes, and not the means, that have the greatest stakes attached to them. It is the marginal wheat fields in Russia and Canada that may experience the greatest losses from small temperature changes. If crop losses occur in drought-stricken West Africa, the effects would be far more severe than if they occur in nations without a severe food shortage. In short, the global distribution of climatic changes may be an important determinant of total loss. Similarly, the global distribution of ozone may have implications for the

* In the ozone reduction case, the mean is taken over the northern hemisphere.

rate of skin cancer, considering that different races have different degrees of resistance to ultraviolet radiation. If the changes were concentrated in areas where mostly dark-skinned people live, then the increase in skin-cancer incidence would be less than if the changes were concentrated in areas where mostly light-skinned people live.

Finally, the probability analysis assumes probabilistic independence among the transfers along the chains. Thus, for example, the probability distribution for temperature change per unit increase in SO_x aerosol loadings is assessed independently from residence times for SO_x (and, therefore, independently from the actual level of aerosol loading). The only exception to this assumption is that the probability distribution for fractional uv increase per fractional ozone reduction is allowed to depend on the fractional ozone reduction itself and, therefore, on the probability distributions leading to the determination of the ozone reduction.

V. Quantitative Relationships Employed in the Analysis

The following are the parameterized quantitative relationships corresponding to Chains A, B, C, D, and E. (Refer to Figures K.2 through K.9 as applicable.)

Chain A: $NO_x \rightarrow O_3$ Reduction $\rightarrow$ uv Increase $\rightarrow$ Skin Cancer

1. Fuel Flow $\rightarrow O_3$ Reduction

   Specify fuel flow (g/yr) in each altitude band:

$$FF_{1215} = \text{fuel flow in 12–15 km},$$
$$FF_{1518} = \text{fuel flow in 15–18 km},$$
$$FF_{1821} = \text{fuel flow in 18–21 km},$$
$$FF_{2124} = \text{fuel flow in 21–24 km}.$$

   Specify emission index (g/kg) for NO_x emissions:

$$EINO_x = \text{emission index for } NO_x.$$

Both of the above are input as flight/emission scenario (see below), depending on demand for flights as well as on the regulatory options chosen to limit flights in various altitude bands or to restrict emissions.

Specify probability distributions for the unknown parameters $\tilde{B}_{1215}$, $\tilde{B}_{1518}$, $\tilde{B}_{1821}$, and $\tilde{B}_{2124}$ in the following equation relating NO_x emissions (g/yr) at each altitude to northern hemisphere fractional ozone reduction ($\Delta O_3/O_3$):

$$(\Delta O_3/O_3)_i = \tilde{B}_i \cdot (FF)_i \cdot EINO_x \qquad (i = \text{altitude band}),$$

where the left-hand side is the relative reduction of ozone at all levels due to injection of a unit NO_x emission at level *i*. A linear relationship was used here, rather than a log-linear relationship, because the log-linear fit obtained from data by Chang yielded an exponent *A* of very close to unity in the equation

$$\log(\Delta O_3/O_3) = \log B_i + A_i \log(NO_x)_i.$$

Thus, the equation for fractional ozone change is

$$\Delta O_3/O_3 = (FF_{1215}\cdot\tilde{B}_{1215} + FF_{1518}\cdot\tilde{B}_{1518} + FF_{1821}\cdot\tilde{B}_{1821} + FF_{2124}\cdot\tilde{B}_{2124})EINO_x. \qquad (K.A1)$$

X 2. ~~Increased SO_x Loading → Increased Aerosol Loading~~

Specify a probability distribution for the unknown parameter $\tilde{K}$ in the following equation relating fractional increase in weighted uv radiation to fractional ozone change:

$$\Delta uv/uv = \tilde{K}\cdot\Delta O_3/O_3. \qquad (K.A2)$$

Allow for different probability distributions for *K* depending on whether $\Delta O_3/O_3 \gtrless 0.2$, an arbitrary cutoff point.

3. Weighted uv Increase → Skin Cancer Rate Increase

Specify a probability distribution for the unknown parameter $\tilde{a}$ in the following equation relating fractional increase in weighted uv radiation to the fractional increase in the incidence of skin cancer (including the fatal variety, melanoma):

$$\Delta CA/CA = \tilde{a}\cdot\Delta uv/uv. \qquad (K.A3)$$

Chain B: SO_x → Aerosol → Climate

1. Fuel Flow → Increased Aerosol Loading

Specify fuel flow by altitude band as in Chain A.
Specify emission index (g/kg) for SO_x emissions:

$$EISO_x = \text{emission index for } SO_x.$$

Both of the above are input as a flight/emission scenario (see discussion under Chain A).

Specify probability distributions for the residence times (years) $\tilde{RTS}_{1215}$, $\tilde{RTS}_{1518}$, $\tilde{RTS}_{1821}$, and $\tilde{RTS}_{2124}$ in each altitude band. Then SO_x loadings ΔSX(g) at each altitude are determined as

$$(\Delta SX)_i = (FF)_i\cdot EISO_x\cdot\tilde{RTS}_i \qquad (i = \text{altitude band}).$$

Then, total SO_x loadings are given by

$$\Delta SX = (FF_{1215}\cdot \widetilde{RTS}_{1215} + FF_{1518}\cdot \widetilde{RTS}_{1518} + FF_{1821}\cdot \widetilde{RTS}_{1821} + FF_{2124}\cdot \widetilde{RTS}_{2124})\cdot EISO_x. \qquad (K.B1)$$

2. Increased SO_x Loading → Increased Aerosol Loading

Specify a probability distribution for $\widetilde{KS}$, the fraction of SO_x that is converted to aerosol. Then, total aerosol loading is given by

$$\Delta S = 1.5\cdot \Delta SX \cdot \widetilde{KS}, \qquad (K.B2)$$

where the factor 1.5 reflects the molecular transformation from SO_2 to H_2SO_4.

3. Increased Aerosol Loading → Mean Global Surface Temperature Change

Convert aerosol loading ΔS from g to $\mu g/m^3$ by multiplying by 4×10^{-13}.

Specify a probability distribution for the unknown parameter $\tilde{C}$ in the following equation relating aerosol loading change (in $\mu g/m^3$) to mean global surface temperature change ΔT_{SB}(°C):

$$\Delta T_{SB} = \tilde{C}\cdot \Delta S. \qquad (K.B3)$$

Chain C: NO_x → O_3 Reduction → Climate

Begin with Equation (K.A1) from Chain A.

Specify a probability distribution for the unknown parameter $\tilde{Z}$ in the following equation relating fractional ozone change to mean global surface temperature change:

$$\Delta T_{SC} = \tilde{Z}\cdot \Delta O_3/O_3. \qquad (K.C1)$$

Chain D: H_2O → Climate

1. Fuel Flow → Increased Water Vapor Loading

Specify fuel flow by altitude band as in Chains A and B.

Specify emission index (g/kg) for H_2O emissions:

$$EIH_2O = \text{emission index for } H_2O.$$

Specify probability distributions for residence times $\widetilde{RTH}_{1215}$, $\widetilde{RTH}_{1518}$, $\widetilde{RTH}_{1821}$, and $\widetilde{RTH}_{2124}$. Then water vapor loadings ΔHLOAD (g) are given by

$$\Delta HLOAD = (FF_{1215}\cdot \widetilde{RTH}_{1215} + FF_{1518}\cdot \widetilde{RTH}_{1518} + FF_{1821}\cdot \widetilde{RTH}_{1821} + FF_{2124}\cdot \widetilde{RTH}_{2124})EIH_2O. \qquad (K.D1)$$

TABLE K.2 Aircraft Characteristics

Aircraft	Altitude (km)	Fuel Burnup (kg/h)	Flight Time (h/day)	Fuel Flow (kg/yr)	NO_x (g/kg)	SO_x (g/kg)	H_2O (g/kg)
Present wide-body subsonic jet (1974)	9–12 12–15	10,200	7 [a]	2.61×10^7	15 (3, 0.9, 0.18)	1 (0.1, 0.05, 0.01)	1250
Projected wide-body subsonic jet (1990–2005)	9–12 12–15	9,500	7 [a]	2.43×10^7	7 (3, 0.9, 0.18)	1 (0.1, 0.05, 0.01)	1250
Present SST	15–18	19,100	5 [b]	4.88×10^7	18 (3, 0.9, 0.18)	1 (0.1, 0.05, 0.01)	1250
Formerly projected U.S. SST	15–18 18–21	50,000	5 [b]	1.28×10^8	18 [c] (3, 0.9, 0.18)	1 (0.1, 0.05, 0.01)	1250

[a] CIAP uses 5.6 hours per day based on an analysis of CAB data that were prepared for a different purpose and that are not ideal for the study. For international operations, a CAB October 1974 report on the wide-body jet shows that, in the 12 months ending June 30, 1974, U.S. operated B-747's had 9.45 revenue hours per aircraft per day with an average stage length of 2293 miles.

[b] CIAP uses 4.5 hours per day, not significantly different.

[c] No engine was ever completed for the U.S. formerly projected SST; some tests on a related engine designed for the RB-70 suggested that the NO_x emission index would be closer to 11 during the time of flight at cruising altitude. However, this design used afterburning; without it, a value of 22 would not be implausible. A correction can easily be made in the following results; but, as will be seen, much more drastic reductions are needed to get to satisfactory levels.

2. Increased Water Vapor Loading → Mean Global Surface Temperature Change

Convert water vapor loading ΔHLOAD from g to fractional change $\Delta H_2O/H_2O$ by dividing by the baseline loading, 2×10^{15} g.

Specify a probability distribution for the unknown parameter $\tilde{H}$ in the following equation relating fractional change in water vapor loading to mean global surface temperature change:

$$\Delta T_{SD} = \tilde{H} \cdot \Delta H_2O/H_2O. \qquad \text{(K.D2)}$$

Chain E: NO_x → Climate

1. Fuel Flow → Increased NO_x Loading

Specify fuel flow by altitude band as in Chain A.

Specify emission index (g/kg) for NO_x emissions as in Chain A.

Specify probability distributions for the parameters $\widetilde{NTX}_{1215}$, $\widetilde{NTX}_{1518}$, $\widetilde{NTX}_{1821}$, and $\widetilde{NTX}_{2124}$ in the following equation relating NO_x emissions (g/yr) at each altitude to fractional change in NO_x concentration:

$$\Delta NO_x/NO_x = (\widetilde{NTX}_{1215} \cdot FF_{1215} + \widetilde{NTX}_{1518} \cdot FF_{1518} + \widetilde{NTX}_{1821} \cdot FF_{1821} + \widetilde{NTX}_{2124} \cdot FF_{2124}) \cdot EINO_x. \qquad \text{(K.E1)}$$

2. Increased NO_x Loading → Mean Global Surface Temperature Change

Specify a probability distribution for the parameter $\tilde{Y}$ relating fractional change in NO_x concentration to mean global surface temperature change in the equation

$$\Delta T_{SE} = \tilde{Y} \cdot \Delta NO_x/NO_x. \qquad \text{(K.E2)}$$

Chain T: Temperature Changes from Chains B, C, D, and E → Total Temperature Change

Compute total mean global surface temperature change ΔT_S from Equations (K.B3), (K.C1), (K.D2), and (K.E2):

$$\Delta T_S = \Delta T_{SB} + \Delta T_{SC} + \Delta T_{SD} + \Delta T_{SE}. \qquad \text{(K.T1)}$$

This is the sum of four random variables, which will be regarded as independent.

VI. Flight and Emission Models Considered

Based on the assumptions about the fuel flow per year discussed in Table K.2, the fuel flow per year is calculated as the product of the

TABLE K.3 Scenarios 1 and 1R

Altitude (km)	Planes	kg/yr/Plane	Fuel Flow (kg/yr)	$EINO_x$	$EISO_x$	EIH_2O
Scenario 1: 1990 Upperbound CIAP Fleet						
12–15	1050 Subsonic	2.43×10^7	2.55×10^{10}	7 (3, 0.9, 0.18)	1 (0.1, 0.05, 0.01)	1250
15–18	17 U.S. SST	0.91×10^8	1.54×10^9	18 (3, 0.9, 0.18)	1 (0.1, 0.05, 0.01)	1250
18–21	323 U.S. SST	0.91×10^8	2.96×10^{10}	18 (3, 0.9, 0.18)	1 (0.1, 0.05, 0.01)	1250
Scenario 1R: 1990 Upperbound CIAP Fleet Size Restricted to Subsonic						
12–15	1293 Subsonic	2.43×10^7	3.14×10^{10}	7 (3, 0.9, 0.18)	1 (0.1, 0.05, 0.01)	1250

TABLE K.4 Scenarios 2 and 2R

Altitude (km)	Planes	kg/yr/Plane	Fuel Flow (kg/yr)	$EINO_x$	$EISO_x$	EIH_2O
Scenario 2: 2005 Upperbound CIAP Fleet						
12–15	4550 Subsonic	2.43×10^7	1.10×10^{11}	7 (3, 0.9, 0.18)	1 (0.1, 0.05, 0.01)	1250
15–18	85 U.S. SST	0.91×10^8	7.81×10^9	18 (3, 0.9, 0.18)	1 (0.1, 0.05, 0.01)	1250
18–21	1635 U.S. SST	0.91×10^8	1.50×10^{11}	18 (3, 0.9, 0.18)	1 (0.1, 0.05, 0.01)	1250
Scenario 2R: 2005 Upperbound CIAP Fleet Size Restricted to Subsonic						
12–15	5779 Subsonic	2.43×10^7	1.40×10^{11}	7 (3, 0.9, 0.18)	1 (0.1, 0.05, 0.01)	1250

fuel flow per hour while at cruising altitude, the number of hours of flight per day (assumed to be 7 or 5 depending on whether it is subsonic or supersonic), and a number of days per year (365).

The first NO_x emission index in each row is essentially the known current value (but see footnote [c] to Table K.2). The others are alternative possible values whose effects are to be studied.

The number of hours per day, unlike the other figures, is an estimate. It may be argued that an airplane could have higher utilization although there are scheduling difficulties, but on the other hand there will be many shorter flights on which the ratio of cruising time in the stratosphere to total time may be small or zero. Actually, for a given usage in passenger miles, changes in this figure are presumably offset by the number of airplanes projected; this can be a major factor in the economics of the planes but not in total emissions.

VII. Flight and Emission Scenarios Considered

Seven basic flight scenarios are evaluated. These are made up of various combinations of the aircraft whose characteristics are assumed to be as shown in Table K.2. The prototype aircraft are (1) present wide-bodied subsonic jet, (2) subsonic wide-bodied jet in 1990–2005, (3) Concorde/Tupolev-144 type SST, referred to as "present SST," (4) an SST with many of the characteristics of that being developed by the United States until 1971, referred to as "U.S. SST." The base emission rates for NO_x, SO_x, and H_2O are shown in the exhibit, along with possible restricted emission rates given in parentheses.

Scenario 1, described in Table K.3, is the "upperbound" 1990 CIAP fleet. It consists of 1050 subsonics at 12–15 km,* 17 U.S. SST's † at 15–18 km, and 323 U.S. SST's at 18–21 km. Scenario 1R, described in the same table, reallocates all supersonic flights to subsonics at 12–15 km.

Scenario 2, described in Table K.4, is the "upperbound" 2005 CIAP fleet. It consists of 4550 subsonics at 12–15 km, 85 U.S. SST's at 15–18 km, and 85 U.S. SST's at 18–21 km. Scenario 2R is analogous

* The number of subsonics at 9–12 km is omitted because the climatic effects are negligible at that altitude.

† CIAP actually hypothesizes a smaller plane (256 passengers rather than 298) operating at 4.5 h per day rather than 5. The numbers of large advanced supersonic transports postulated in CIAP were converted to equivalent numbers of U.S. SST's by multiplying by the factor (256/298)·(4.5/5).

TABLE K.5 Scenarios 3, 4, and 5

Altitude (km)	Planes	kg/yr/Plane	Fuel Flow (kg/yr)	$EINO_x$	$EISO_x$	EIH_2O
Scenario 3: 100 Concorde/Tupolev						
15–18	100 Present SST	3.63×10^7	3.63×10^9	18	1	1250
Scenario 4: 100 U.S. SST						
18–21	100 U.S. SST	0.91×10^8	0.91×10^{10}	18	1	1250
Scenario 5: 1000 Subsonics (1974)						
12–15	1000 Subsonic (1974)	2.61×10^7	2.61×10^{10}	15	1	1250

to scenario 1R in that all supersonic flights are reallocated to subsonics at 12–15 km.

Scenario 3 consists of 100 present SST's at 15–18 km, Scenario 4 consists of 100 U.S. SST's at 18–21 km, and as a baseline Scenario 5 consists of 1000 present-day, wide-bodied subsonics at 12–15 km. These scenarios are described in Table K.5.

For each of Scenarios 1, 1R, 2, and 2R, the implications of emissions reduction are considered by evaluating the consequences for emission rates of 18, 3, 0.9, and 0.18 g/kg for NO_x (7, 3, 0.9, and 0.18 for subsonics), and 1, 0.1, 0.05, and 0.01 g/kg for SO_x. The reductions for SO_x would be achieved by sulfur removal from fuel, while the reductions for NO_x would require substantial engine modification. Thus four subscenarios, corresponding to the four levels of emission control (for NO_x and SO_x together) are evaluated under Scenarios 1, 1R, 2, and 2R.

VIII. Probability Estimation

The uncertain quantities to be assessed are listed in Table K.6. Probability distributions for each of these were obtained by assessment and pooling of expert opinion.

Assessment was conducted by the fractile method. Experts were asked to respond to a questionnaire (Addendum A), in which approximations to their probability distributions for the uncertain quantities identified above were elicited. For each quantity, the 0.01, 0.25, 0.50, 0.75, and 0.99 fractiles were assessed. An explanation of the fractile method is given in the preface to the questionnaire (Addendum A).

TABLE K.6 Uncertain Quantities to Be Assessed

B_{1215} =	Fractional Ozone Change per annual NO_x emission (g) at 12–15 km
B_{1518} =	Fractional Ozone Change per annual NO_x emission (g) at 15–18 km
B_{1821} =	Fractional Ozone Change per annual NO_x emission (g) at 18–21 km
K_1 =	Fractional uv Change per fractional ozone change ($-\Delta O_3/O_3 < 0.2$)
K_2 =	Fractional uv Change per fractional ozone change ($-\Delta O_3/O_3 > 0.2$)
A =	Fractional change in incidence of skin cancer per fractional uv change
RTS_{1215} =	Residence time (yr) for SO_x at 12–15 km
RTS_{1518} =	Residence time (yr) for SO_x at 15–18 km
RTS_{1821} =	Residence time (yr) for SO_x at 18–21 km
KS =	Fraction of SO_x converted to SO_4 aerosol
C =	Change in mean surface temperature (°C) per increased aerosol loading ($\mu g/m^3$)
Z =	Change in mean surface temperature (°C) per fractional ozone change
RTH_{1215} =	Residence time (yr) for H_2O at 12–15 km
RTH_{1518} =	Residence time (yr) for H_2O at 15–18 km
RTH_{1821} =	Residence time (yr) for H_2O at 18–21 km
H =	Change in mean surface temperature (°C) per fractional change in H_2O
NTX_{1215} =	Fractional change in NO_x loading per annual NO_x emission (g) at 12–15 km
NTX_{1518} =	Fractional change in NO_x loading per annual NO_x emission (g) at 15–18 km
NTX_{1821} =	Fractional change in NO_x loading per annual NO_x emission (g) at 18–21 km
Y =	Change in mean surface temperature (°C) per fractional change in NO_x

Put simply, the 0.50 fractile is the subjective *median* of the distribution; it is chosen so that the true value is believed to be equally likely to lie above or below the 0.50 fractile. The 0.25 and 0.75 fractiles divide the intervals below and above the median into equally likely subintervals. The 0.01 and 0.99 fractiles are "surprise" levels such that the true value is given just a 1 percent chance of lying below the 0.01 level or above the 0.99 level.

In addition, each expert was asked to rate his relative level of expertise on a 0–10 scale for each question. This rating was used in the pooling procedure described below.

In all, eight questionnaires were returned by members of the Climatic Impact Committee. Each respondent answered only questions in those fields in which he had expertise, so that for each question there were from two to five responses.

The responses are tabulated in Addendum B, with the respondents identified by letters.

IX. Pooling of Distributions

The 0.01, 0.25, 0.50, 0.75, and 0.99 fractiles of each individual assessor were converted into a cumulative distribution as follows. The cumulative distribution function was approximated by a piecewise linear interpolation and extrapolation from the given fractiles. The internal segments (from 0.01 to 0.25, 0.25 to 0.50, 0.50 to 0.75, 0.75 to 0.99) were interpolated linearly. The external segments (from 0 to 0.01, 0.99 to 1.0) were derived either by (a) linear extrapolation at a slope of one half that of the neighboring internal segment (e.g., the 0.75–0.99 slope extrapolated from 0.99 to 1.0) or (b) linear extrapolation to a natural upper or lower bound on the uncertain quantity (e.g., zero is a lower bound for any quantity that must be nonnegative), if it exists. A sample interpolation/extrapolation is illustrated in Figure K.11. The uncertain quantity is assumed to have a natural lower bound of zero.

Pooling of the derived cumulative distribution assessments for each parameter was obtained by weighting the distributions probabilistically according to the self-rating index of the assessor. If the (linearized) cumulative distribution for assessor j is $F_j(q)$, and the self-rating for

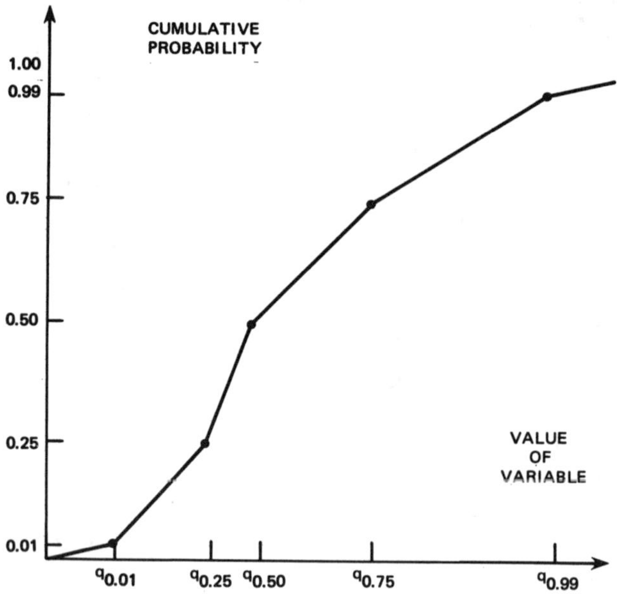

FIGURE K.11 Sample linearization of a cumulative distribution function from fractile estimates.

TABLE K.7 Pooled Fractile Data Used in the Analysis [a]

	Fractile				
	0.01	0.25	0.50	0.75	0.99
B_{1215}	−7.5 E–13	−2 E–13	−2.5 E–14	−3 E–15	−8 E–16
B_{1518}	−1 E–12	−3.3 E–13	−1.1 E–13	−3.7 E–14	−1 E–14
B_{1821}	−1 E–12	−4 E–13	−2 E–13	−1 E–13	−4 E–14
K_1	−3.2	−2.6	−2.2	−1.8	−1.4
K_2	−5.5	−4.5	−3.5	−2.5	−2.0
A	0.028	0.8	1.4	2.8	36.9
RTS_{1215}	0.053	0.31	0.6	1.1	2.75
RTS_{1518}	0.11	0.79	1.35	2.3	4.6
RTS_{1821}	0.25	1.37	2.07	3.0	5.0
KS	0.06	0.59	0.93	0.99	1.0
C	−19.2	−4.05	−1.29	−0.33	+0.25
Z	−5	−3.2	−1.4	−0.29	+4.6
RTH_{1215}	0.5	1	2	3	5
RTH_{1518}	1	1.5	3	5	8
RTH_{1821}	2	3	5	7	10
H	0	0	0.11	0.35	0.95
$\mathrm{NT}X_{1215}$	2 E–14	1 E–13	2 E–13	8 E–13	2 E–12
$\mathrm{NT}X_{1518}$	3 E–13	1.5 E–12	3 E–12	6 E–12	2 E–11
$\mathrm{NT}X_{1821}$	5 E–13	2.5 E–12	5 E–12	1 E–11	3 E–11
Y	−0.02	−0.002	0	+0.013	+0.025

[a] Definitions of variables are given in Table K.6.

that assessor is r_j, then the aggregate cumulative distribution function for the quantity $\tilde{Q}$ is derived as

$$\text{Prob}\ \{\tilde{Q}<q\}=F(q)=\sum_j(r_j/\Sigma r_j)F_j(q).$$

The 0.01, 0.25, 0.50, 0.75, and 0.99 fractiles of the pooled distribution can then be ascertained. The data derived and pooled from the question-naire sample of CIC experts, and used in the analysis, are shown in Table K.7.

X. Method of Uncertainty Analysis and Computation

The resulting distributions for each uncertain parameter were discretized as follows. The 0.50 fractile was taken as a representative value of the region from the 0.33 fractile (not assessed) to the 0.67 fractile (not assessed) and assigned mass probability 0.34. The 0.25 fractile was taken as representative of the region from 0.03 to 0.33 and assigned mass probability 0.30. Similarly, the 0.75 fractile was taken as representative of the region from 0.67 to 0.97 and assigned mass probability 0.30. The 0.01 and 0.99 fractiles were taken as representative of the tails of the distribution outside the 0.03 and 0.97 fractiles and were each assigned mass probability 0.03.

Within each chain (A, B, C, D, and E) all probability distributions were assumed to be independent, with the following exception. The altitude-dependent parameter groups (B_{1215}, B_{1518}, B_{1821}), (RTS_{1215}, RTS_{1518}, RTS_{1821}), (RTH_{1215}, RTH_{1518}, RTH_{1821}), and (NTX_{1215}, NTX_{1518}, NTX_{1821}) were assumed to be dependent in the sense that the various fractile values of each member of the triplet were assumed to occur together.

A Fortran computer program * was used to combine the discretized probability distributions for Chains A, B, C, D, and E by a simple convolution procedure. Consider the calculation for Chain C as an example. Five represented combinations of values for B_{1215}, B_{1518}, and B_{1821} are specified, corresponding to the 0.01, 0.25, 0.50, 0.75, and 0.99 fractiles, respectively, for the ratio of fractional ozone reduction to NO_x emissions in each altitude band. Since these possible sets of values are assigned mass probabilities of 0.03, 0.30, 0.34, 0.30, and 0.03, respectively, by applying Eq. (K.A1), the five resulting values of $\Delta O_3/O_3$ are each assigned probabilities 0.03, 0.30, 0.34, 0.30, and 0.03, respectively. This is the first link of the chain. Then, each of the five values for the constant Z, the ratio of temperature change to ozone reduction, is combined in Eq. (K.C1) with the five values of $\Delta O_3/O_3$ derived from the first link to yield 25 possible values for ΔT_C. The product of the lowest $\Delta O_3/O_3$ and the lowest C has probability (0.03) (0.03) = 0.0009; the product of the median $\Delta O_3/O_3$ and the median C has probability (0.34)(0.34) = 0.1156; and so forth for the other 23 combinations. This gives a final probability distribution for temperature change due to ozone. Note that in Chains A and B this process of "convoluting" distributions is repeated three times, while in Chains C, D, and E it is repeated twice.†

The computer program is given in Addendum C. The outputs of the program, for any fuel flow and emissions scenario, are cumulative probability distributions for fractional change in skin-cancer incidence

* The computer programming for this analysis was performed by Thomas E. Shemo, Jr.

† The distribution of total temperature change is the convolution of the distributions of the temperature changes in Chains B, C, D, and E.

(as well as for uv increase and ozone reduction); cumulative distributions for the temperature change associated with SO_x, ozone, H_2O, and NO_x; and for the total or any subtotal. Results for the scenarios defined for analysis are given below.

The program was verified by checking it against sample computations done by hand.

XI. Results of Consequence Analysis

Cumulative probability distributions for fractional increase in skin cancer and mean global temperature change are displayed for each fleet and emissions scenario. Since these are the results for which the whole analysis is presented, they are presented in Tables K.8 and K.9 in abbreviated form. In these tables, we give the median (the value for which

TABLE K.8 Percentage Change in Skin-Cancer Incidence for Alternative Scenarios and NO_x Emission Indices

Scenario	NO_x Emission Index (g/kg)	Median	Distinctly Possible [a]	Conceivable [b]
100 Present SST	18	2	12	82
100 U.S. SST	18	11	36	340
CIAP upperbound 1990	18 for SST 7 for subsonic	40	145	1300
	3	7	32	240
	0.9	2	10	70
	0.18	0.4	2	14
CIAP upperbound 2005	18 for SST 7 for subsonic	200	730	6600
	3	35	200	1200
	0.9	10	50	360
	0.18	2	10	70
Extrapolated current	15	4	40	260
CIAP upperbound 1990	7	2	24	150
All subsonic	3	1	10	60
	0.9	0.2	3	20
CIAP upperbound 2005	7	10	110	660
All subsonic	3	4	50	280
	0.9	1	14	85

[a] Probability of being exceeded = 0.10.
[b] Probability of being exceeded = 0.01.

TABLE K.9 Changes in Mean Surface Temperature for Alternative Scenarios, NO_x Emission Indices, and SO_x Emission Indices

Scenario	NO_x Emission Index (g/kg)	SO_x Emission Index (g/kg)	Decrease in Mean Surface Temperature (°C): Median	Distinctly Possible [a]	Conceivable [b]
100 Present SST	18	1	0.03	0.05	0.2
100 U.S. SST	18	1	0.1	0.18	0.6
CIAP upperbound 1990	18 for SST 7 for subsonic	1	0	0.10	0.95
	3	0.1	0	0.01	0.1
	0.9	0.05	0	0.01	0.05
CIAP upperbound 2005	18 for SST 7 for subsonic	1	0.30	1.50	5.0
	3	0.1	0.02	0.20	0.6
	0.9	0.05	0.01	0.08	0.3
Extrapolated current	5	1	0.01	0.02	0.2
CIAP upperbound 1990 All subsonic	7	1	0.01	0.13	0.2
CIAP upperbound 2005 All subsonic	7	1	0.06	0.15	1.0

[a] Probability of being exceeded = 0.10.
[b] Probability of being exceeded = 0.01.

there is a 50 percent chance of being exceeded, and therefore of which there is a 50 percent chance of falling short), the uppermost decile (the point that would be exceeded with 10 percent probability), and the uppermost percentile (the point exceeded with 1 percent probability). For easy reference, we refer to the uppermost decile as the "distinctly possible" value and the uppermost percentile as the "conceivable" value. By "exceed," we mean "greater in the direction usually thought of as bad": in the case of skin cancer, an increase of incidence; in the case of mean global temperature, a fall. To illustrate, the distinctly possible value for skin-cancer incidence is the number such that the probability of that value or higher for skin-cancer incidence is 10 percent; the conceivable value for temperature change is the number such that the probability of that fall or greater is 1 percent.

We also display graphically entire cumulative distributions for fractional increase in skin cancer and mean global temperature change for each fleet and emissions scenario.

The following is a complete index of the results presented in Figures K.12–K.32.

Scenario 1: Upperbound CIAP 1990

- NO_x Emissions = 18 g/kg, SO_x Emissions = 1.0 g/kg
  - Fractional increase in skin cancer / Mean global temperature change } Figure K.12
  - Fractional increase in uv / Fractional ozone reduction } Figure K.13
  - Temperature change (sulfur only) / Temperature change (ozone only) } Figure K.14
  - Temperature change (water only) / Temperature change (nitrogen only) } Figure K.15
- NO_x Emissions = 3 g/kg, SO_x Emissions = 0.1 g/kg
  - Fractional increase in skin cancer / Mean global temperature change } Figure K.16
  - Temperature change (sulfur only) / Temperature change (ozone only) } Figure K.17
- { NO_x Emissions = 0.9 g/kg, SO_x Emissions = 0.05 g/kg / NO_x Emissions = 0.18 g/kg, SO_x Emissions = 0.01 g/kg }
  - Fractional increase in skin cancer / Mean global temperature change } Figure K.18

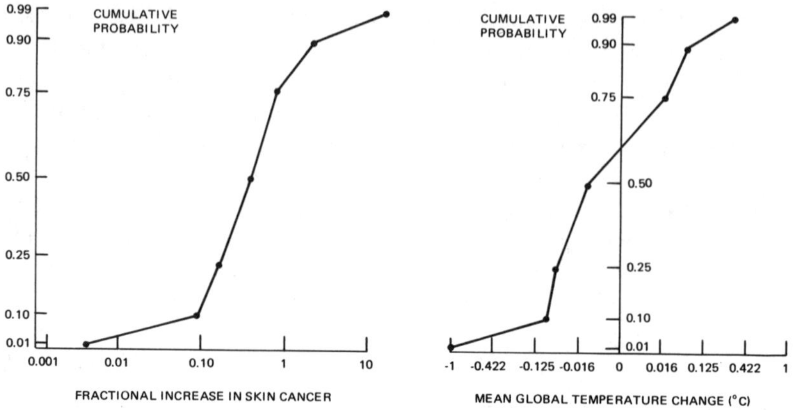

FIGURE K.12 CIAP upperbound 1990, no regulation.

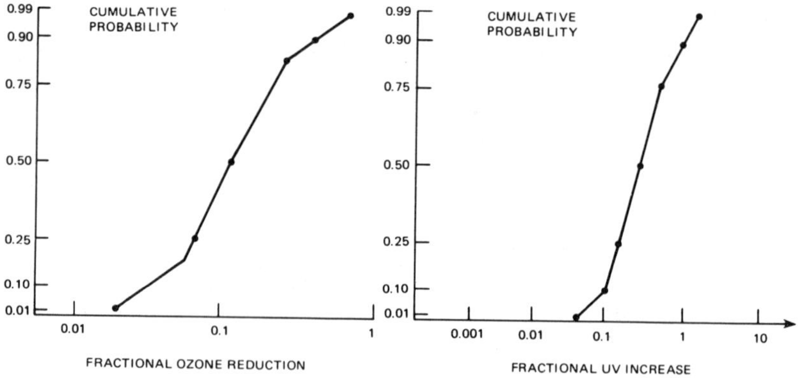

FIGURE K.13 CIAP upperbound 1990, no regulation.

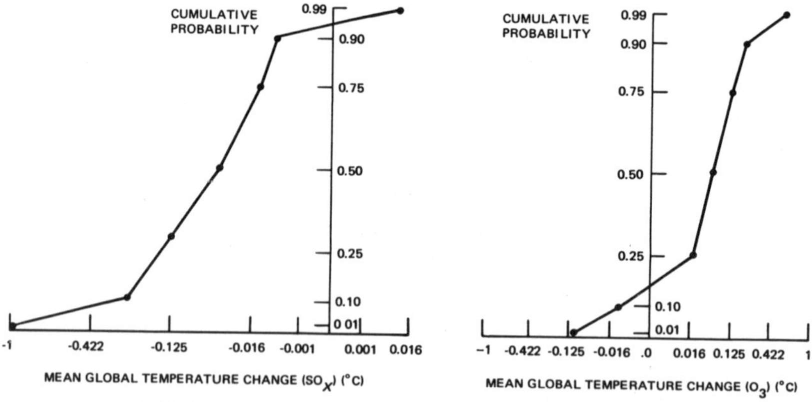

FIGURE K.14 CIAP upperbound 1990, no regulation.

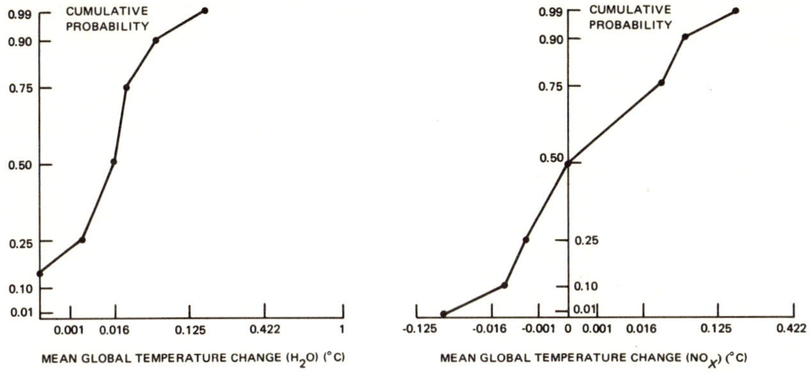

FIGURE K.15 CIAP upperbound 1990, no regulation.

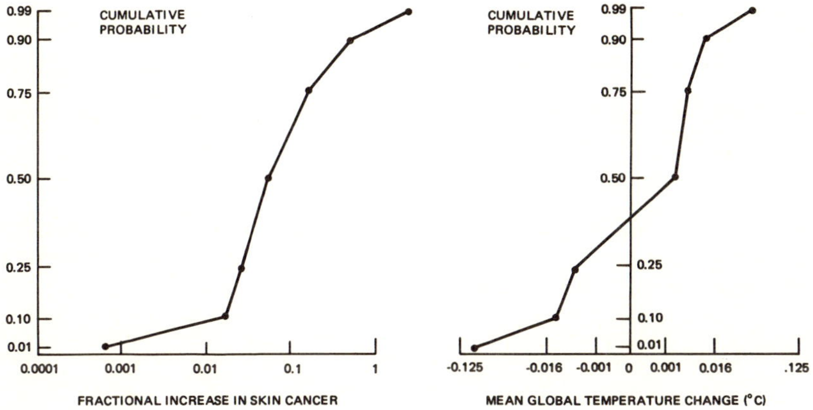

FIGURE K.16 CIAP upperbound 1990: NO_x emissions = 3 g/kg, SO_x emissions = 0.1 g/kg.

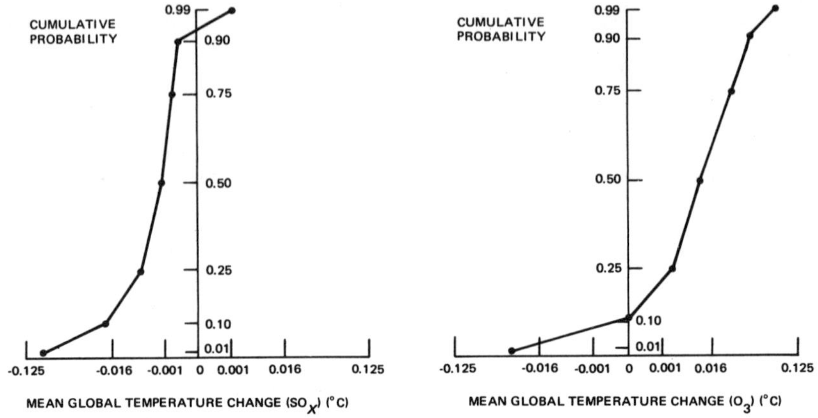

FIGURE K.17 CIAP upperbound 1990: NO_x emissions = 3 g/kg, SO_x emissions = 0.1 g/kg.

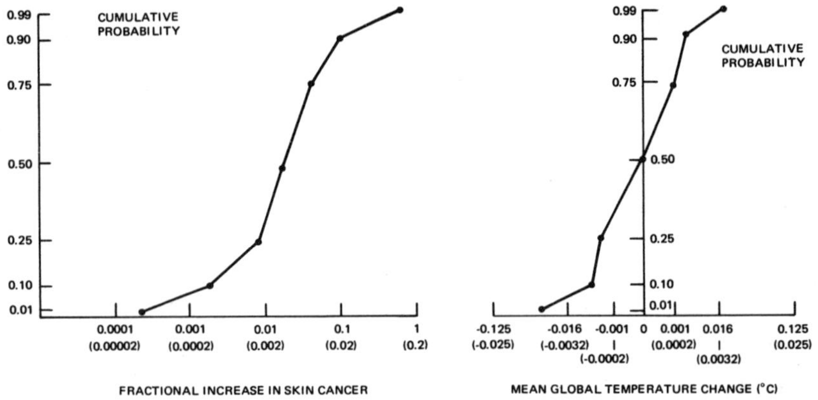

FIGURE K.18 CIAP upperbound 1990: NO_x emissions = 0.9 g/kg, SO_x emissions = 0.05 g/kg; NO_x emissions = 0.18 g/kg, SO_x emissions = 0.01 g/kg. Numbers in parentheses refer to the lower emission levels.

Scenario 1R: Upperbound CIAP 1990, All Subsonic

NO_x Emissions = 18 g/kg, SO_x Emissions = 1.0 g/kg

Fractional increase in skin cancer } Figure K.19
Mean global temperature change }

{ NO_x Emissions = 3 g/kg
{ NO_x Emissions = 0.9 g/kg

Fractional increase in skin cancer   Figure K.20

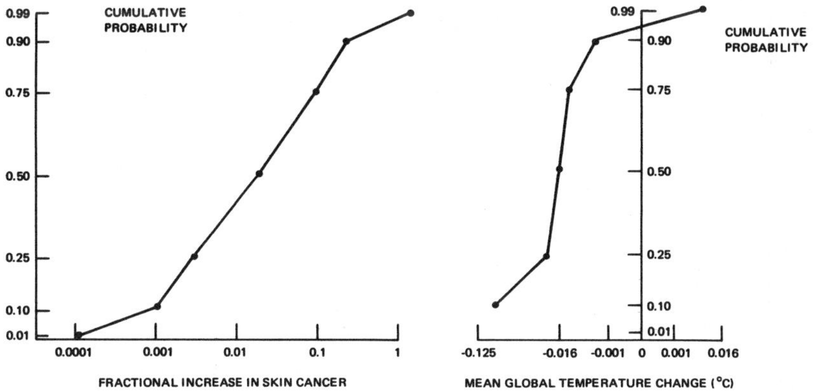

FIGURE K.19 CIAP upperbound 1990, all flights subsonic, no regulation.

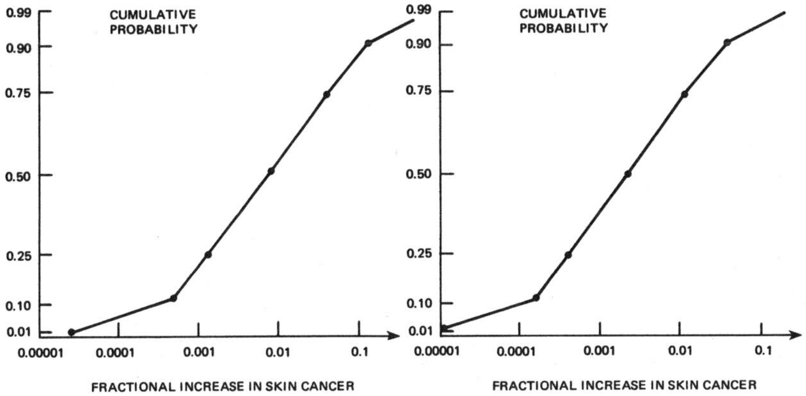

FIGURE K.20 CIAP upperbound 1990, all flights subsonic. Left: NO_x emissions $=3$ g/kg; right: NO_x emissions$=0.9$ g/km.

Scenario 2: Upperbound CIAP 2005

NO_x Emissions = 18 g/kg, SO_x Emissions = 1.0 g/kg

Fractional increase in skin cancer } Figure K.21
Mean global temperature change }

Fractional ozone reduction } Figure K.22
Fractional increase in uv }

Temperature change (sulfur only) } Figure K.23
Temperature change (ozone only) }

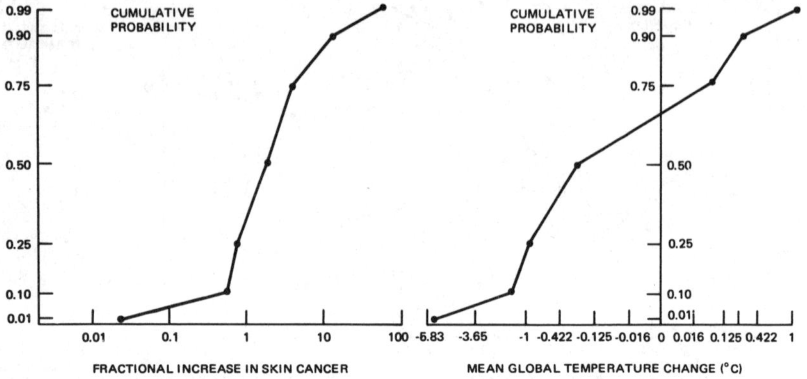

FIGURE K.21 CIAP upperbound 2005, no regulation.

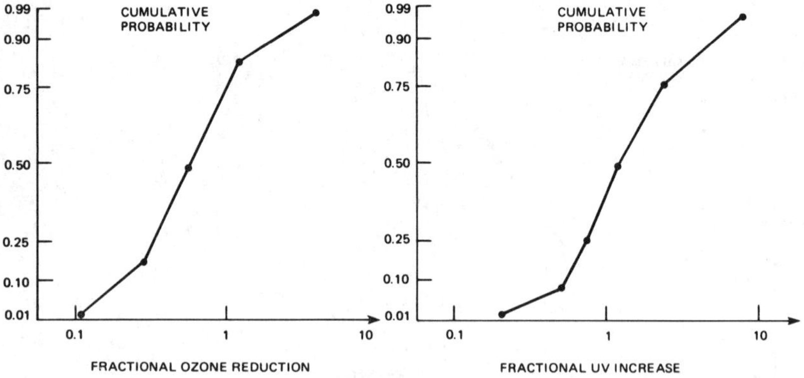

FIGURE K.22 CIAP upperbound 2005, no regulation.

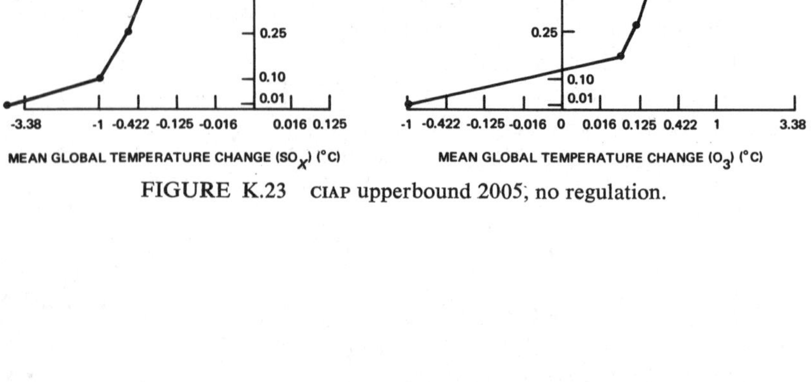

FIGURE K.23 CIAP upperbound 2005; no regulation.

Temperature change (water only)
Temperature change (nitrogen only) } Figure K.24

NO_x Emissions = 3 g/kg, SO_x Emissions = 0.1 g/kg

Fractional increase in skin cancer
Mean global temperature change } Figure K.25

Temperature change (sulfur only)
Temperature change (ozone only) } Figure K.26

{ NO_x Emissions = 0.9 g/kg, SO_x Emissions = 0.05 g/kg
{ NO_x Emissions = 0.18 g/kg, SO_x Emissions = 0.01 g/kg

Fractional increase in skin cancer
Mean global temperature change } Figure K.27

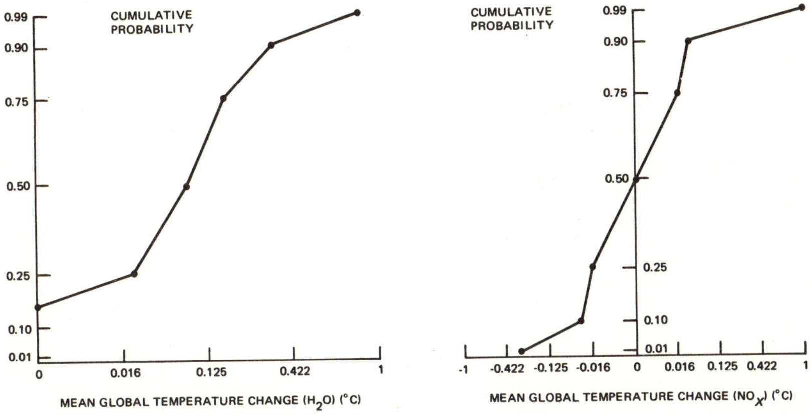

FIGURE K.24 CIAP upperbound 2005, no regulation.

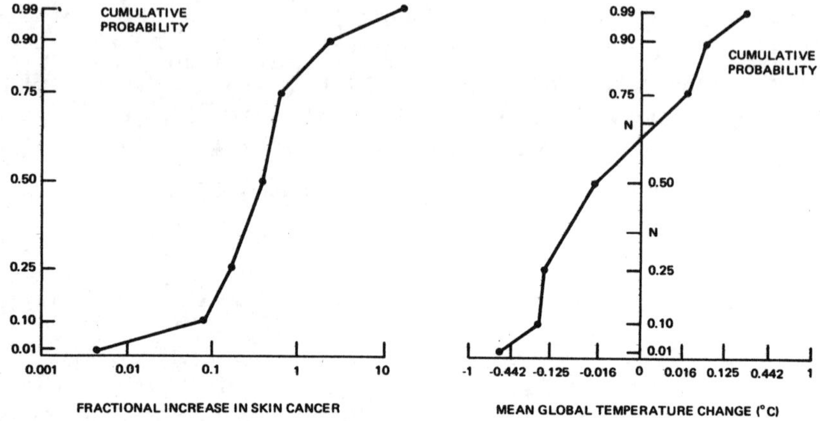

FIGURE K.25 CIAP upperbound 2005: NO_x emissions=3 g/kg, SO_x emissions =0.1 g/kg.

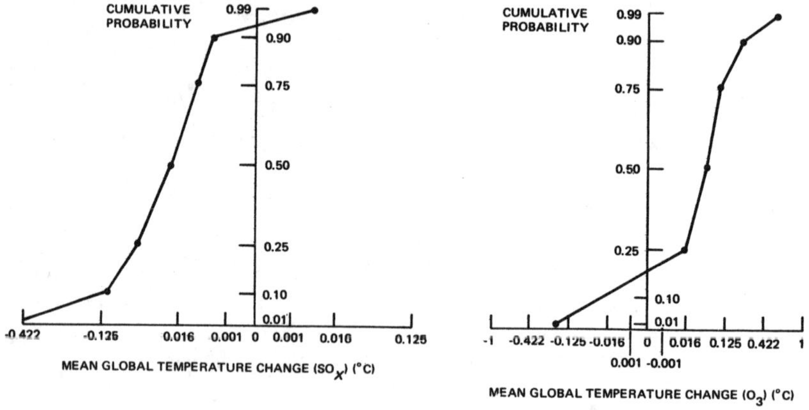

FIGURE K.26 CIAP upperbound 2005: NO_x emissions=3 g/kg, SO_x emissions =0.1 g/kg.

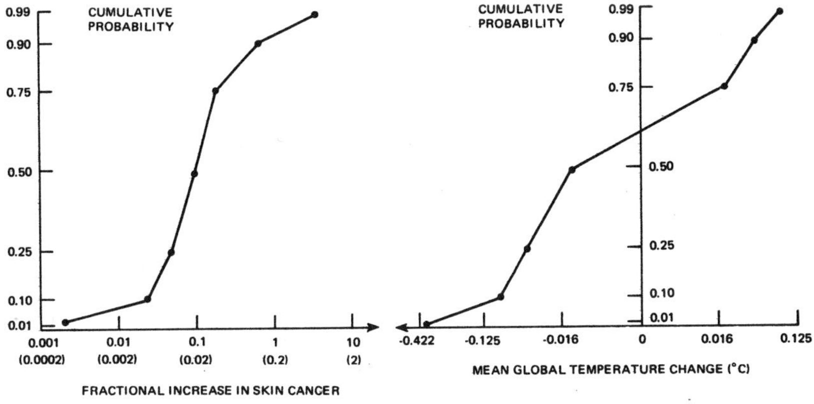

FIGURE K.27 CIAP upperbound 2005: NO_x emissions = 0.9 g/kg, SO_x emissions = 0.05 g/kg; NO_x emissions = 0.18 g/kg, SO_x emissions = 0.01 g/kg. Numbers in parentheses refer to the lower emission levels.

Scenario 2R: Upperbound CIAP 2005, All Subsonic

NO_x Emissions = 18 g/kg, SO_x Emissions = 1.0 g/kg

Fractional increase in skin cancer } Figure K.28
Mean global temperature change }

{ NO_x Emissions = 3 g/kg
{ NO_x Emissions = 0.9 g/kg

Fractional increase in skin cancer   Figure K.29

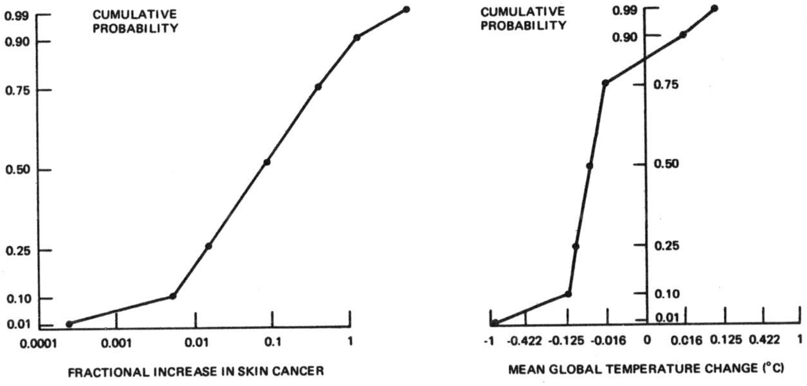

FIGURE K.28 CIAP upperbound 2005, all flights subsonic, no regulation.

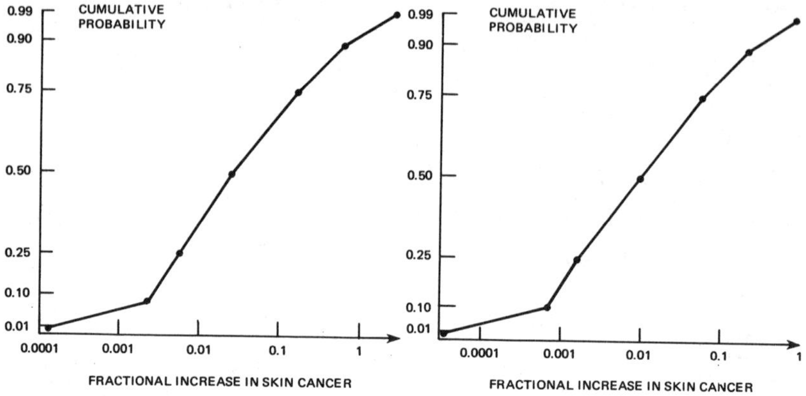

FIGURE K.29 CIAP upperbound 2005, all flights subsonic. Left: NO_x emissions $=3$ g/kg; right: NO_x emissions$=0.9$ g/kg.

Scenario 3: 100 Concorde-Tupolev

Fractional increase in skin cancer } Figure K.30
Mean global temperature change }

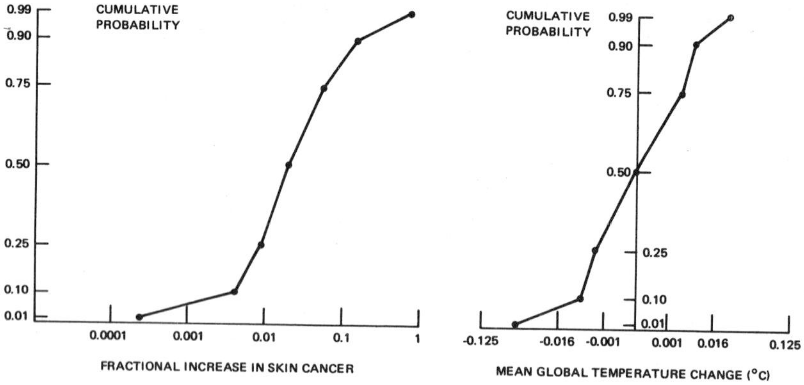

FIGURE K.30 100 Concorde/Tupolev-144, no regulation.

Scenario 4: 100 U.S. SST's

Fractional increase in skin cancer
Mean global temperature change } Figure K.31

Scenario 5: 1000 Subsonic (1974)

Fractional increase in skin cancer
Mean global temperature change } Figure K.32

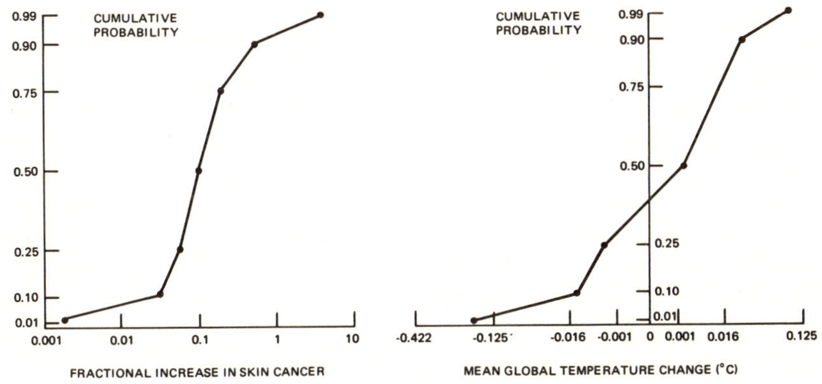

FIGURE K.31 100 U.S. SST's, no regulation.

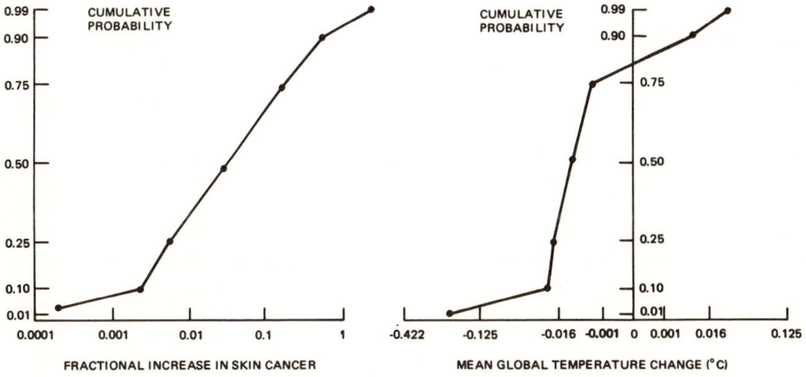

FIGURE K.32 1000 subsonic (1974), no regulation.

Addendum A: Questionnaire for Probability Assessment

PREFACE TO QUESTIONNAIRE

There is general agreement among the participants in CIAP and the CIC of the National Research Council that considerable uncertainty surrounds any point estimates of the links in the chains relating stratospheric emissions to changes in radiation and climate. This uncertainty reflects more than uncertainties in model parameters; it reflects basic gaps in our understanding of the cause–effect relationships. In many instances, not only the magnitudes, but indeed the *signs* of various partial effects are subject to uncertainty.

In order to guide rationally the decisions government will have to make regarding stratospheric flight, we must do our best to assess these gross uncertainties. The decision analysis approach requires that you, as experts in your scientific fields, perform the uncomfortable task of assessing in a quantitative way the extent of the uncertainty you honestly feel, deep down inside.

The initial goal is to derive a set of judgmental probability *distributions* for the key links in the causal chains. It is *not* the purpose of this study to use these probability estimates in a full-blown decision analysis for stratospheric flight, since that would require measurement of the benefits derived from such flight and consideration of value judgments regarding the relative weighting of risks and benefits. It *is* our purpose ultimately to provide the decision-makers in government with an assessment of the probabilities of possible consequences of stratospheric flight, in the form of honest judgmental probability distributions.

This questionnaire is designed to elicit your own expert evaluation of the uncertainties about some of the basic parameters that enter at various stages into the determination of the climatic impacts of stratospheric travel. We ask that you respond to this questionnaire on the basis of all the information available to you including that conveyed in the CIAP and CIC documents. This is an "open book" examination, dated September 1974.

The parameters examined are drawn from a wide variety of subject matters—atmospheric physics, chemistry, biology, and technology. Each respondent will have his own domain of expertise. There may be some questions on which you have done some research directly. There may be others in a field in which you are a scholar but which you have not directly studied. Do not hesitate to answer; your opinion, based on general knowledge of the field, on examination of the models developed in the CIAP or CIC studies, on knowledge of other materials, and on perceptions about the scholars who have done work in the area, is invaluable.

To account for the degree of your relation to any given question, we have attached to each question a "Self-Rating Index" box. Please enter a number from 0 to 10 in increasing order of self-judged expertise; 0 signifies no knowledge beyond a layman's. You are free not to answer any question about which you feel no special competence.

The questionnaire deals with elements of five causal chains in climatic impact.

Chain A: NO_x Emissions ⟶
Ozone Reduction ⟶
uv Increase ⟶
Skin Cancer

Chain B: SO_x Emissions ⟶
Change in Aerosol Loading ⟶
Change in Mean Surface Temperature

Chain C: Ozone Reduction ⟶
Change in Mean Surface Temperature

Chain D: NO_x Emissions ⟶
Change in Mean Surface Temperature

Chain E: H_2O Emissions ⟶
Change in Mean Surface Temperature

In addition, there are questions about the possibilities of engine development to reduce NO_x emissions.

The questionnaire guides you in assessing your subjective probability distributions for the following uncertain quantities, taken as steady-state values where relevant:

1. Residence times for sulfate aerosol at various altitudes
2. Fraction of sulfur emissions that convert to aerosols
3. Incremental change in mean surface temperature (°C) per unit change in aerosol loading
4. Background concentration of NO_x (ppb) at various altitudes
5. Incremental change in NO_x concentration (ppb) per 10^9 g/yr NO_x emitted at various altitudes
6. Fractional ozone reduction per fractional change in NO_x concentration (ppb) at various altitudes
7. Fractional ozone reduction per 10^9 g/yr NO_x emitted at various altitudes
8. Incremental change in mean surface temperature (°C) per unit change in NO_x concentration (ppb) (direct effect)
9. Incremental change in mean surface temperature (°C) per fractional change in ozone

10. Residence time for H_2O vapor at various altitudes

11. Incremental change in mean surface temperature (°C) per fractional change in H_2O burden

12. Fractional change in incidence of skin cancer per fractional change in weighted uv radiation

13. Incremental cost (in 1974 dollars) of development of subsonic engine to achieve reduction of NO_x emissions to 6, 2, or 0.2 g/kg

14. Incremental cost (in 1974 dollars) of development of large advanced supersonic engine to achieve reductions of NO_x emissions to 6, 2, or 0.2 g/kg *if* subsonic engine NO_x emission has been reduced to 2 g/kg

15. Incremental cost (in 1974 dollars) of development of large advanced supersonic engine to achieve reductions of NO_x emissions to 6, 2, or 0.2 g/kg in the absence of development leading to reduced NO_x emission on subsonic engines

For each of these uncertain quantities, you will be asked to assess the following:

(a) Your median, or 50% fractile. This is the value such that you think it equally likely that the true value of the uncertain quantity lies above or below this value. To test yourself, you might play the following game. Select your median. Imagine you have the choice between staking your scientific reputation on the proposition that the true value of the uncertain quantity lies above the announced median and staking it on the proposition that the true value lies below. If you do not care which choice you have made (i.e., are willing to let your worst enemy choose for you), then you have selected your honest median. If you do not prefer this first choice, raise your median until you are indifferent.

(b) Your upper quartile, or 75% fractile. This is the value such that you think it equally likely that the true value of the uncertain quantity lies between your median and this value as that it lies above this value. In other words, the 75% estimate cuts the upper half of your distribution into two equally likely parts.

(c) Your lower quartile, or 25% fractile. This is the value that cuts the lower half of your distribution into two equally likely parts.

(d) Your upper "surprise" level, or 99% fractile. This is hard to assess, but it should represent the value of the uncertain quantity such that you think the chance that the true value lies above this level is exactly 1/100.

(e) Your lower "surprise" level, or 1% fractile. This is analogous to the upper "surprise" estimate at the low end of your distribution.

In assessing the 50%, 75%, 25%, 99%, and 1% values in the questionnaire, please heed the following:

1. The effects considered are long-run *steady-state* effects (in questions 1–12).
2. It is assumed implicitly that relationships are *linear,* at least in the range of interest. If you believe linearity to be an unusually bad assumption in any relationship, please so indicate in the space provided and also indicate the magnitude and direction of the nonlinearity (i.e., how does the proportionality constant vary with the level of the independent variable?).
3. The effects considered are *incremental* effects, superimposed upon any background changes that may occur. Furthermore, the several effects are assumed to be *noninteractive* for purposes of this analysis.
4. Unless otherwise stated, all relationships refer to average effects in the northern hemisphere. Please feel free to make explicit any assumptions you need to make in order to think clearly about some parameter.
5. CAVEAT! In making your assessments consider all the variability you would consider if you actually had to make a judgment. Therefore, e.g., treat seriously the *possibility* that SO_x aerosols cause net warming via greenhouse effects, or that O_3 reduction cools rather than warms.

We may not have asked uncertainties about all the parameters we should have. At the end of the questionnaire, please insert any further similar questions, and answer them if you wish.

QUESTIONNAIRE

Q1a. Assess the residence time for SO_x emitted at 0–3 km above the tropopause.

______	______	______	______	______
1%	25%	50%	75%	99%

Self-Rating Index (0–10)__________

Q1b. Assess the residence time for SO_x emitted at 3–6 km above the tropopause.

______	______	______	______	______
1%	25%	50%	75%	99%

Self-Rating Index (0–10)__________

Q1c. Assess the residence time for SO_x emitted at 6–9 km above the tropopause.

______	______	______	______	______
1%	25%	50%	75%	99%

Self-Rating Index (0–10)__________

Q1d. Assess the residence time for SO_x emitted at 9–12 km above the tropopause.

1%	25%	50%	75%	99%

Self-Rating Index (0–10)_______________

Q2. Assess the percentage of SO_x emissions that eventually convert into aerosols.

1%	25%	50%	75%	99%

Do you think this percentage is altitude-dependent (please specify)?________

Self-Rating Index (0–10)_______________

Q3. Assess the constant C in the following equation relating increased aerosol loading ΔS ($\mu g/m^3$) to change in mean global surface temperature ΔT_S (°C):

$$\Delta T_S(°\mathrm{C}) = C \cdot \Delta S$$

1%	25%	50%	75%	99%

Linearity?_______________

Self-Rating Index (0–10)_______________

Q4a. Assess the concentration of NO_x (in ppb) in the absence of emissions by aircraft at 0–3 km above the tropopause.

1%	25%	50%	75%	99%

Self-Rating Index (0–10)_______________

Q4b. Assess the concentration of NO_x (in ppb) in the absence of emissions by aircraft at 3–6 km above the tropopause.

1%	25%	50%	75%	99%

Self-Rating Index (0–10)_______________

Q4c. Assess the concentration of NO_x (in ppb) in the absence of emissions by aircraft at 6–9 km above the tropopause.

1%	25%	50%	75%	99%

Self-Rating Index (0–10)_______________

Q4d. Assess the concentration of NO_x (in ppb) in the absence of emissions by aircraft at 9–12 km above the tropopause.

______	______	______	______	______
1%	25%	50%	75%	99%

Self-Rating Index (0–10)______________

Q5a. Assess the constant α in the following equation relating NO_x emissions E at 0–3 km above the tropopause (in 10^9 g/yr) to change in steady-state NO_x concentration (ppb):

$$\Delta NO_x = \alpha \cdot E$$

______	______	______	______	______
1%	25%	50%	75%	99%

Linearity?______________

Self-Rating Index (0–10)______________

Q5b. Assess the constant α in the following equation relating NO_x emissions E at 3–6 km above the tropopause (in 10^9 g/yr) to change in steady-state NO_x concentration (ppb):

$$\Delta NO_x = \alpha \cdot E$$

______	______	______	______	______
1%	25%	50%	75%	99%

Linearity?______________

Self-Rating Index (0–10)______________

Q5c. Assess the constant α in the following equation relating NO_x emissions E at 6–9 km above the tropopause (in 10^9 g/yr) to change in steady-state NO_x concentration (ppb):

$$\Delta NO_x = \alpha \cdot E$$

______	______	______	______	______
1%	25%	50%	75%	99%

Linearity?______________

Self-Rating Index (0–10)______________

Q5d. Assess the constant α in the following equation relating NO_x emissions E at 9–12 km above the tropopause (in 10^9 g/yr) to change in steady-state NO_x concentration (ppb):

$$\Delta NO_x = \alpha \cdot E$$

______	______	______	______	______
1%	25%	50%	75%	99%

Linearity?______________

Self-Rating Index (0–10)______________

Q6a. Assess the constant β in the following equation relating fractional change in NO_x concentration (ppb) at 0–3 km above the tropopause to fractional ozone reduction:

$$\Delta O_3/O_3 = -\beta \cdot \Delta NO_x/NO_x$$

______	______	______	______	______
1%	25%	50%	75%	99%

Linearity?______________

Self-Rating Index (0–10)______________

Q6b. Assess the constant β in the following equation relating fractional change in NO_x concentration (ppb) at 3–6 km above the tropopause to fractional ozone reduction:

$$\Delta O_3/O_3 = -\beta \cdot \Delta NO_x/NO_x$$

______	______	______	______	______
1%	25%	50%	75%	99%

Linearity?______________

Self-Rating Index (0–10)______________

Q6c. Assess the constant β in the following equation relating fractional change in NO_x concentration (ppb) at 6–9 km above the tropopause to fractional ozone reduction:

$$\Delta O_3/O_3 = -\beta \cdot \Delta NO_x/NO_x$$

______	______	______	______	______
1%	25%	50%	75%	99%

Linearity?______________

Self-Rating Index (0–10)______________

Q6d. Assess the constant β in the following equation relating fractional change in NO_x concentration (ppb) at 9–12 km above the tropopause to fractional ozone reduction:

$$\Delta O_3/O_3 = -\beta \cdot \Delta NO_x/NO_x$$

______	______	______	______	______
1%	25%	50%	75%	99%

Linearity?______________

Self-Rating Index (0–10)______________

Q7a. Assess the constant B in the following equation relating NO_x emissions E at 0–3 km (in 10^9 g/yr, NO_2) to fractional O_3 reduction:

$$-\Delta O_3/O_3 = B \cdot E$$

________	________	________	________	________
1%	25%	50%	75%	99%

Linearity?____________

Self-Rating Index (0–10)____________

Q7b. Assess the constant B in the following equation relating NO_x emissions E at 3–6 km (in 10^9 g/yr, NO_2) to fractional O_3 reduction:

$$-\Delta O_3/O_3 = B \cdot E$$

________	________	________	________	________
1%	25%	50%	75%	99%

Linearity?____________

Self-Rating Index (0–10)____________

Q7c. Assess the constant B in the following equation relating NO_x emissions E at 6–9 km (in 10^9 g/yr, NO_2) to fractional O_3 reduction:

$$-\Delta O_3/O_3 = B \cdot E$$

________	________	________	________	________
1%	25%	50%	75%	99%

Linearity?____________

Self-Rating Index (0–10)____________

Q7d. Assess the constant B in the following equation relating NO_x emissions E at 9–12 km (in 10^9 g/yr, NO_2) to fractional O_3 reduction:

$$-\Delta O_3/O_3 = B \cdot E$$

________	________	________	________	________
1%	25%	50%	75%	99%

Linearity?____________

Self-Rating Index (0–10)____________

Q8. Assess the constant δ in the following equation relating change in the steady-state level of NO_x concentration (ppb) in the atmosphere to change in mean surface temperature, T_S (°C):

$$\Delta T_S = \delta \cdot \Delta NO_x$$

(This is the direct effect due to absorption of sunlight by NO_x, apart from any effect through ozone reduction.)

________	________	________	________	________
1%	25%	50%	75%	99%

Linearity?____________

Self-Rating Index (0–10)____________

Q9. Assess the constant β in the following equation relating fractional ozone reduction to change in mean surface temperature:

$$\Delta T_S = \beta \cdot \Delta O_3 / O_3$$

1%	25%	50%	75%	99%

Linearity?________________

Self-Rating Index (0–10)________________

Q10a. Assess the residence time for H_2O emitted at 0–3 km above the tropopause.

1%	25%	50%	75%	99%

Self-Rating Index (0–10)________________

Q10b. Assess the residence time for H_2O emitted at 3–6 km above the tropopause.

1%	25%	50%	75%	99%

Self-Rating Index (0–10)________________

Q10c. Assess the residence time for H_2O emitted at 6–9 km above the tropopause.

1%	25%	50%	75%	99%

Self-Rating Index (0–10)________________

Q10d. Assess the residence time for H_2O emitted at 9–12 km above the tropopause.

1%	25%	50%	75%	99%

Self-Rating Index (0–10)________________

Q11. Assess the constant β in the following equation relating fractional change in stratospheric H_2O burden $\Delta H/H$ to change in mean surface temperature ΔT_S:

$$\Delta T_S = \beta \cdot \Delta H / H$$

1%	25%	50%	75%	99%

Linearity?________________

Self-Rating Index (0–10)

Q12. Assess the constant α in the following equation relating fractional change in weighted uv radiation to fractional change in skin cancer incidence. (The weighting of the uv radiation is a fifty–fifty mixture of that appropriate for sunburn and that appropriate for DNA damage.)

$$\Delta Ca/Ca = \alpha \cdot \Delta uv/uv$$

_____	_____	_____	_____	_____
1%	25%	50%	75%	99%

Linearity?__________

Self-Rating Index (0–10)__________

Q13a. Given that a new engine will be developed over a normal development cycle for subsonic aircraft, what is the incremental cost (in 1974 dollars) to reduce the NO_x emission index from 18 down to 6 g/kg?

_____	_____	_____	_____	_____
1%	25%	50%	75%	99%

Self-Rating Index (0–10)__________

Q13b. Given that a new engine will be developed over a normal development cycle for subsonic aircraft, what is the incremental cost (in 1974 dollars) to reduce the NO_x emission index from 18 down to 2.0 g/kg?

_____	_____	_____	_____	_____
1%	25%	50%	75%	99%

Self-Rating Index (0–10)__________

Q13c. Given that a new engine will be developed over a normal development cycle for subsonic aircraft, what is the incremental cost (in 1974 dollars) to reduce the NO_x emission index from 18 down to 0.2 g/kg?

_____	_____	_____	_____	_____
1%	25%	50%	75%	99%

Self-Rating Index (0–10)__________

Q14a. Given that the subsonic engine above has been developed (with an Emission Index of 2 g/kg) and that a new large advanced supersonic engine will be developed over a normal development cycle, what is the incremental cost (in 1974 dollars) to reduce the NO_x emissions from 18 down to 6 g/kg?

_____	_____	_____	_____	_____
1%	25%	50%	75%	99%

Self-Rating Index (0–10)__________

Q14b. Given that the subsonic engine above has been developed (with an Emission Index of 2 g/kg) and that a new large advanced supersonic engine will be developed over a normal development cycle, what is the incremental cost (in 1974 dollars) to reduce the NO_x emissions from 18 down to 2.0 g/kg?

______	______	______	______	______
1%	25%	50%	75%	99%

Self-Rating Index (0–10)______________

Q14c. Given that the subsonic engine above has been developed (with an Emission Index of 2 g/kg) and that a new large advanced supersonic engine will be developed over a normal development cycle, what is the incremental cost (in 1974 dollars) to reduce the NO_x emissions from 18 down to 0.2 g/kg?

______	______	______	______	______
1%	25%	50%	75%	99%

Self-Rating Index (0–10)______________

Q15a. In absence of the above new subsonic engine development, if a new large advanced supersonic engine is to be developed over a normal development cycle, what is the incremental cost (in 1974 dollars) to reduce the NO_x emissions from 18 down to 6 g/kg?

______	______	______	______	______
1%	25%	50%	75%	99%

Self-Rating Index (0–10)______________

K15b. In absence of the above new subsonic engine development, if a new large advanced supersonic engine is to be developed over a normal development cycle, what is the incremental cost (in 1974 dollars) to reduce the NO_x emissions from 18 down to 2.0 g/kg?

______	______	______	______	______
1%	25%	50%	75%	99%

Self-Rating Index (0–10)______________

Q15c. In absence of the above new subsonic engine development, if a new large advanced supersonic engine is to be developed over a normal development cycle, what is the incremental cost (in 1974 dollars) to reduce the NO_x emissions from 18 down to 0.2 g/kg?

______	______	______	______	______
1%	25%	50%	75%	99%

Self-Rating Index (0–10)______________

Please fill in below any questions (and the corresponding answers) like those above which you feel should be added to this questionnaire.

QI . . .

______	______	______	______	______
1%	25%	50%	75%	99%

Linearity?______________

Self-Rating Index (0–10)______________

QII . . .

______	______	______	______	______
1%	25%	50%	75%	99%

Linearity?______________

Self-Rating Index (0–10)______________

QIII . . .

______	______	______	______	______
1%	25%	50%	75%	99%

Linearity?______________

Self-Rating Index (0–10)______________

QIV . . .

______	______	______	______	______
1%	25%	50%	75%	99%

Linearity?______________

Self-Rating Index (0–10)______________

Addendum B: Summary of Responses to Questionnaire

The questions are given the same numbers as in the questionnaires. For each question, the responses of each respondent to that question are given, together with his self-rating index. The respondents are identified by letter, from A to H.

Question	Respondent	Probability Levels					Self-Rating Index
		1%	25%	50%	75%	99%	
1a (yr)	A	0.17	0.5	0.67	1.3	2	10
	B	0.04	0.25	0.5	1.0	2.5	5
	C	0.1	0.4	0.7	1.5	0.3	5
	D	0.05	0.1	0.15	0.3	0.8	4
	E	0.2	0.4	0.7	1.0	1.5	5
1b (yr)	A	0.5	0.8	1.5	2.5	3	10
	B	0.08	0.75	1	1.5	4	5
	C	0.5	1	2	3	5	5
	D	0.1	0.2	0.5	1.0	2	4
	E	0.7	1.2	1.7	2.4	3	6
1c (yr)	A	1	1.5	2	3	5	10
	B	0.25	1.2	2	3	5	5
	C	0.6	1.2	2.5	3.5	5	5
	D	0.2	0.5	1	1.5	3	4
	E	1.5	2.0	2.5	3	4	6
1d (yr)	A	1.5	2	2.5	4	6	10
	B	0.5	2	3	4	6	5
	C	0.6	1.2	2.5	3.5	5	5
	D	0.3	0.7	1.5	2.5	5	4
	E	1.8	2.2	2.8	3.5	4	6

2	A	0.1	0.5	0.9	0.95	1	5
	B	0.05	0.2	0.5	0.9	0.995	4
	E	0.5	0.99	1	1	1	8
3	A	−20	−6	−3	−0.3	+0.3	3
	B	− 0.2	−0.15	−0.1	−0.05	−0.01	4
	D	−10	−2	−1	−0.5	0	4
4a	A	0.5	2	3	5	10	8
	B	0.05	1	3	5	12	3
	C	0.1	1	3	10	20	2
	E	0.2	0.35	0.5	0.75	1.5	8
4b	A	3	6	8	10	20	8
	B	1	2	6	9	12	3
	C	0.1	1	3	10	20	2
	E	0.5	1	1.5	2	5	8
4c	A	5	7	10	15	20	8
	B	1	2	4	10	12	3
	C	0.1	1	3	10	20	2
	E	1	2	3	4	10	8
4d	A	7	10	13	17	20	8
	B	0.05	2	5	8	12	3
	C	0.1	1	3	10	20	2
	E	2	4	6	8	12	8
5a	A	2×10^{-5}	1×10^{-4}	2×10^{-4}	8×10^{-4}	2×10^{-3}	10
	C	5×10^{-4}	2×10^{-3}	3×10^{-3}	6×10^{-3}	1×10^{-2}	5
5b	A	3×10^{-4}	1.5×10^{-3}	3×10^{-3}	6×10^{-3}	1×10^{-2}	10
	C	2×10^{-3}	4×10^{-3}	1×10^{-2}	2×10^{-2}	5×10^{-2}	5

Addendum B Continued

Question	Respon-dent	Probability Levels 1%	25%	50%	75%	99%	Self-Rating Index
5c	A	5×10^{-4}	2.5×10^{-3}	5×10^{-3}	1×10^{-2}	3×10^{-2}	10
	C	2×10^{-3}	6×10^{-3}	1×10^{-2}	2×10^{-2}	5×10^{-2}	5
5d	A	1×10^{-3}	5×10^{-3}	9×10^{-3}	2×10^{-2}	5×10^{-2}	10
	C	2×10^{-3}	6×10^{-3}	1×10^{-2}	2×10^{-2}	5×10^{-2}	5
6a	B	0	1	2	3	4	4
	E	0.1	0.15	0.2	0.3	0.4	8
6b	B	1	2	3	5	7	4
	E	0.1	0.15	0.2	0.3	0.4	8
6c	B	1	3	5	8	10	4
	E	0.1	0.15	0.2	0.3	0.4	8
6d	B	2	5	7	9	12	4
	E	0.1	0.15	0.2	0.3	0.4	8
7a	A	-2.8×10^{-16}	-1.4×10^{-15}	-2.8×10^{-15}	-1.1×10^{-14}	-2.8×10^{-14}	9
	D	-5×10^{-16}	-1.5×10^{-15}	-3×10^{-15}	-6×10^{-15}	-2.0×10^{-14}	4
	E	-1×10^{-15}	-3×10^{-15}	-4×10^{-15}	-7×10^{-15}	-1.2×10^{-14}	8
7b	A	-4×10^{-15}	-2×10^{-14}	-4×10^{-14}	-8×10^{-14}	-3×10^{-13}	9
	D	-2×10^{-16}	-2×10^{-15}	-5×10^{-15}	-1×10^{-14}	-3×10^{-14}	4
	E	-3×10^{-15}	-6×10^{-15}	-8×10^{-15}	-1.2×10^{-14}	-2×10^{-14}	8
7c	A	-7×10^{-15}	-3.5×10^{-14}	-7×10^{-14}	-1.4×10^{-13}	-4×10^{-13}	9
	D	-5×10^{-15}	-1.5×10^{-14}	-3×10^{-14}	-6×10^{-14}	-2×10^{-13}	4
	E	-8×10^{-15}	-1×10^{-14}	-1.2×10^{-14}	-1.6×10^{-14}	-2.4×10^{-14}	8

7d	A	-1.4×10^{-14}	-7×10^{-14}	-1.3×10^{-13}	-3×10^{-13}	-7×10^{-13}	9
	D	-2×10^{-15}	-2×10^{-14}	-5×10^{-14}	-1×10^{-13}	-3×10^{-13}	4
	E	-1×10^{-14}	-1.2×10^{-14}	-1.6×10^{-14}	-2×10^{-14}	-3×10^{-14}	8
8	A	−0.02	−0.006	−0.003	−0.001	+0.001	6
	B	0	0	+0.01	+0.02	+0.025	8
9	A	−5	−2	−1	−0.5	+0.5	4
	B	−5	−4	−2	−1	0	8
	H	−5	−2	0	+0.2	+5	4
10a (yr)	A	0.5	1	2	3	5	10
	D	0.05	0.1	0.15	0.3	0.8	4
	E	0.3	0.5	1	1.5	2	5
10b (yr)	A	1	1.5	3	5	8	10
	D	0.1	0.2	0.5	1	2	4
	E	1	1.5	2	3	4	8
10c (yr)	A	2	3	5	7	10	10
	D	0.2	0.5	1	1.5	3	4
	E	2	2.5	3	3.5	5	8
10d (yr)	A	3	5	8	11	15	10
	D	0.3	0.7	1.5	2.5	5	4
	E	2.5	3	3.5	4	5	7
11	B	0	0	0	0.3	0.5	9
	D	0.02	0.1	0.2	0.5	1	4

Addendum B Continued

Question	Respon-dent	Probability Levels					Self-Rating Index
		1%	25%	50%	75%	99%	
12	A	0.5	0.8	1	2	3	5
	C	0	0.4	1	2	10	8
	D	0	1	2	4	10	4
	E	0.5	0.8	1.0	1.2	2	2
	F	0.6	1.7	2.2	6	60	9

NOTE: There were no responses to questions 13, 14, or 15. The answers to questions 4 and 6 were not used in subsequent analysis.

Addendum C: Computer Program for Uncertainty Analysis

```
      REAL HOOHAH(5)
      REAL NTX12(5),NTX15(5),NTX18(5),NTX21(5),Y(5),DNOX(5),DE(5,5)
      DATA IY/'YES'/
      REAL TP(2401),T(2401)
      REAL RTS12(5),RTS15(5),RTS18(5),RTS21(5),C(5),DS(5),DTB(5,5,5),
     &PROB2(5,5),CBS(9),Z(5),TC(5,5),CCS(9),EIH2O,RTH12(5),RTH15(5),
     &RTH18(5),RTH21(5),H(5),HLOAD(5),TD(5,5),CDS(9),DT(7,7,7,7),PROB3
     &(7,7,7,7),P1(7),P2(7),P3(7),RW,RK,RC,RR,RU,RN,DS1(5,5),KS(5)
      INTEGER PRINT1(6),PRINTT(4),NX(4),PR(6)
      LOGICAL PRI(6)
      INTEGER PR83(4)
      DATA PR83/'B','C','D','E'/
      DATA PR/'A','B','C','D','T','E'/
      INTEGER IV(2401)
      REAL B12(5),B15(5),B18(5),B21(5),K1(5),K2(5),A(5),OZRED(5),
     &UVINC(5,5),CA(5,5,5),PROB(5,5,5),PC(5),CES(9),P4(7)
      NAMELIST/INPUT/B12,B15,B18,B21,K1,K2,A,FF12,FF15,
     &FF18,FF21,ENOX12,ENOX15,ENOX18,ENOX21
     &,RTS12,RTS15,RTS18,RTS21,C,EISOX,Z,EIH2O,RTH12,RTH15,RTH18,
     &RTH21,H,KS,NTX12,NTX15,NTX18,NTX21,Y
      DATA PC/.03,.3,.34,.3,.03/
      DO 41 I=1,5
      DO 41 J=1,5
41    PROB2(I,J)=PC(I)*PC(J)
      DO 40 I=1,5
      DO 40 J=1,5
      DO 40 K=1,5
40    PROB(I,J,K)=PC(I)*PC(J)*PC(K)
100   WRITE(5,91)
91    FORMAT(//' INPUT FROM FILE? ',/)
      READ(5,92) IANS
92    FORMAT(A3)
      IF(IANS.NE.IY) GOTO 93
      CALL IFILE(20,'MILT')
      READ(20,INPUT)
      CALL RELEAS(20)
93    WRITE(5,97)
97    FORMAT(//' LIST DATA? ',/)
      READ(5,92) IANS
      IF(IANS.NE.IY) GOTO 94
      WRITE(5,INPUT)
94    WRITE(5,3)
3     FORMAT(//' READY FOR INPUT'//)
      READ(5,INPUT)
      WRITE(5,95)
95    FORMAT(//' STORE DATA ON FILE? ',/)
      READ(5,92) IANS
      IF(IANS.NE.IY) GOTO 96
      CALL OFILE(20,'MILT')
      WRITE(20,INPUT)
      CALL RELEAS(20)
96    CONTINUE
      DO 11 I=1,6
11    PRI(I)=.FALSE.
      WRITE(5,12)
12    FORMAT(//' ENTER PRINT CODES (A,B,C,D,E,T)'/)
      READ(5,13) PRINT1
13    FORMAT(6A1)
      DO 14 I=1,6
      DO 14 J=1,6
14    IF(PRINT1(I).EQ.PR(J)) PRI(J)=.TRUE.
      DO 10 I=1,5
      OZRED(I)=FF12*ENOX12*B12(I)+FF15*ENOX15*B15(I)+FF18*ENOX18*B18(I
     &)+FF21*ENOX21*B21(I)
      IF(OZRED(I).LT.-1.0)OZRED(I)=-1.0
```

```
10      HOOHAH(I)=OZRED(I)
        DO 20 I=1,5
        DO 20 J=1,5
        IF(OZRED(I).GT.-0.2) GOTO 21
        UVINC(I,J)=OZRED(I)*K2(J)
        GOTO 20
21      UVINC(I,J)=OZRED(I)*K1(J)
20      CONTINUE
        DO 30 I=1,5
        DO 30 J=1,5
        DO 30 K=1,5
30      CA(I,J,K)=UVINC(I,J)*A(K)
        IF(.NOT.PRI(1))GOTO 50
        CALL SORT(HOOHAH,5,IV)
        CALL COMP(HOOHAH,PC,5,N2,TP,IV,T)
        WRITE(5,2100)
2100    FORMAT(///' FRACTIONAL OZONE CHANGE  PROBABILITY     CUMULA',
     &'TIVE PROBABILITY'//)
        CALL OUT (HOOHAH,N2,IV,TP)
        CALL SORT(UVINC,25,IV)
        CALL COMP(UVINC,PROB2,25,N2,TP,IV,T)
        WRITE(5,2002)
2002    FORMAT(///' FRACTIONAL UV CHANGE         PROBABILITY',
     &'      CUMULATIVE PROBABILITY'//)
        CALL OUT2(UVINC,N2,IV,TP)
        WRITE(5,1)
1       FORMAT(///' CANCER RATE INCREASE      PROBABILITY      ',
     &'CUMULATIVE PROBABILITY'//)
        CALL SORT(CA,125,IV)
        CALL COMP(CA,PROB,125,N2,TP,IV,T)
        CALL OUT2(CA,N2,IV,TP)
50      IF((.NOT.PRI(2)).AND.(.NOT.PRI(5)))GOTO 60
        DO 51 I=1,5
51      DS(I)=(FF12*EISOX*RTS12(I)+FF15*EISOX*RTS15(I)
     &+FF18*EISOX*RTS18(I)+FF21*EISOX*RTS21(I))
        DO 52 I=1,5
        DO 52 J=1,5
52      DS1(I,J)=1.5*KS(I)*DS(J)
        DO 2001 I=1,5
        DO 2001 J=1,5
        DO 2001 K=1,5
2001    DTB(I,J,K)=C(J)*DS1(I,K)*4.0E-13
        CALL SORT(DTB,125,IV)
        IF(PRI(2))WRITE(5,53)
53      FORMAT(///' SOX TEMPERATURE CHANGE   PROBABILITY      ',
     &'CUMULATIVE PROBABILITY'//)
        CALL COMP(DTB,PROB,125,N2,TP,IV,T)
        IF(PRI(2)) CALL OUT2(DTB,N2,IV,TP)
        IF(PRI(5))CALL TRANS(DTB,TP,N2,CBS,IV)
60      IF((.NOT.PRI(3)).AND.(.NOT.PRI(5)))GOTO 70
        DO 61 I=1,5
        DO 61 J=1,5
61      TC(I,J)=Z(J)*OZRED(I)
        CALL SORT(TC,25,IV)
        IF(PRI(3))WRITE(5,62)
62      FORMAT(///' OZONE TEMP CHANGE          PROBABILITY      ',
     &'CUMULATIVE PROBABILITY'//)
        CALL COMP(TC,PROB2,25,N2,TP,IV,T)
        IF(PRI(3)) CALL OUT2(TC,N2,IV,TP)
        IF(PRI(5))CALL TRANS(TC,TP,N2,CCS,IV)
70      IF((.NOT.PRI(4)).AND.(.NOT.PRI(5))) GOTO 2003
        DO 71 I=1,5
71      HLOAD(I)=FF12*EIH2O*RTH12(I)+FF15*EIH2O*RTH15(I)+FF18*EIH2O*
     &RTH18(I)+FF21*EIH2O*RTH21(I)
        DO 72 I=1,5
        DO 72 J=1,5
72      TD(I,J)=H(J)*HLOAD(I)/2.0E15
        CALL SORT(TD,25,IV)
        IF(PRI(4))WRITE(5,73)
73      FORMAT(///' H2O TEMPERATURE CHANGE   PROBABILITY      ',
```

```
      &'CUMULATIVE PROBABILITY'//)
       CALL COMP(TD,PROB2,25,N2,TP,IV,T)
       IF(PRI(4)) CALL OUT2(TD,N2,IV,TP)
       IF(PRI(5)) CALL TRANS(TD,TP,N2,CDS,IV)
2003   IF((.NOT.PRI(6)).AND.(.NOT.PRI(5)))GOTO 80
       DO 2004 I=1,5
2004   DNOX(I)=FF12*ENOX12*NTX12(I)+FF15*ENOX15*NTX15(I)+
      &FF18*ENOX18*NTX18(I)+FF21*ENOX21*NTX21(I)
       DO 2005I=1,5
       DO 2005 J=1,5
2005   DE(I,J)=Y(J)*DNOX(I)
       CALL SORT(DE,25,IV)
       IF(PRI(6)) WRITE(5,2006)
2006   FORMAT(//' NOX TEMPERATURE CHANGE   PROBABILITY
      &'CUMULATIVE PROBABILITY'//)
       CALL COMP (DE,PROB2,25,N2,TP,IV,T)
       IF(PRI(6))CALL OUT2(DE,N2,IV,TP)
       IF(PRI(5))CALL TRANS(DE,TP,N2,CES,IV)
80     IF(.NOT.PRI(5)) GOTO 200
       DO 81 I=2,6
       P1(I)=PC(I-1)
       P2(I)=PC(I-1)
       P4(I)=PC(I-1)
81     P3(I)=PC(I-1)
       P1(1)=CBS(8)
       P2(1)=CCS(8)
       P4(1)=CES(8)
       P3(1)=CDS(8)
       P1(7)=CBS(9)
       P2(7)=CCS(9)
       P4(7)=CES(9)
       P3(7)=CDS(9)
       P1(2)=P1(2)-P1(1)
       P2(2)=P2(2)-P2(1)
       P4(2)=P4(2)-P4(1)
       P3(2)=P3(2)-P3(1)
       P1(6)=P1(6)-P1(7)
       P2(6)=P2(6)-P2(7)
       P4(6)=P4(6)-P4(7)
       P3(6)=P3(6)-P3(7)
       WRITE(5,3001)
3001   FORMAT(//' ENTER CODES FOR TOTAL (BCDE)'/)
       READ(5,3002) PRINTT
3002   FORMAT(4A1)
       DO 3003 I=1,4


       PRI(I)=.FALSE.
       DO 3003 J=1,4
3003   IF(PRINTT(J).EQ.PR83(I)) PRI(I)=.TRUE.
       NIN=1
       DO 3004 I=1,4
       NX(I)=1
       IF(.NOT.PRI(I)) GOTO 3004
       NX(I)=7
       NIN=NIN*7
3004   CONTINUE
       IF(PRI(1))GOTO 3005
       P1(1)=1.0
       CBS(1)=0.0
3005   IF(PRI(2))GOTO 3006
       P2(1)=1.0
       CCS(1)=0.0
3006   IF(PRI(3)) GOTO 3007
       P3(1)=1.0
       CDS(1)=0.0
3007   IF(PRI(4))GOTO 3008
       P4(1)=1.0
       CES(1)=0.0
3008   CONTINUE
```

```
      CALL ZAP(PROB3,2401)
      DO 82 I=1,NX(1)
      DO 82 J=1,NX(2)
      DO 82 K=1,NX(3)
      DO 82 L=1,NX(4)
      PROB3(I,J,K,L)=P1(I)*P2(J)*P3(K)*P4(L)
      DT(I,J,K,L)=CBS(I)+CCS(J)+CDS(K)+CES(L)
82    CONTINUE
      CALL ZC(DT,PROB3,2401,NIN,T,TP)
      CALL SORT(DT,NIN,IV)
      WRITE(5,83)
83    FORMAT(///' TOTAL TEMP CHANGE          PROBABILITY
     &'CUMULATIVE PROBABILITY'//)
      CALL COMP(DT,PROB3,NIN,N2,TP,IV,T)
      CALL OUT2(DT,N2,IV,TP)
200   PAUSE 'REENTER PARAMETERS?'
      GOTO 100
      END
C
C
C
C
C
      SUBROUTINE SORT(A,N,I)
      REAL A(N)
      INTEGER I(N),Z
      LOGICAL A1
      DO 10 K=1,N
10    I(K)=K
      Z=2
32    A1=.FALSE.
      J=N+1
31    J=J-1
      IF(A(J).GE.A(J-1))GOTO 30
      HOLD=A(J)
      A(J)=A(J-1)
      A(J-1)=HOLD
      IHOLD=I(J)
      I(J)=I(J-1)
      I(J-1)=IHOLD
      A1=.TRUE.
30    IF(J.NE.Z) GOTO 31
      Z=Z+1
      IF(Z.EQ.(N+1))RETURN
      IF(A1)GOTO 32
      RETURN
      END
C
C
C
      SUBROUTINE OUT(A,N,I,P)
      DIMENSION A(N),I(N),P(N)
      CUM=0.0
      DO 10 K=1,N
      CUM=CUM+P(I(K))*0.5
      WRITE(5,1) A(K),P(I(K)),CUM
10    CUM=CUM+P(I(K))*0.5
1     FORMAT(9X,G11.5,5X,G11.5,16X,G11.5)
      RETURN
      END
C
C
C
C
C
      SUBROUTINE TRANS(A,P,N,O,IV)
      DIMENSION A(N),P(N),O(9),C(5),IV(N)
      DATA C/.01,.25,.5,.75,.99/
      K=1
      O(1)=A(1)
```

```
      O(7)=A(N)
100   PC=0.0
      DO 10 I=1,N
      PC=PC+P(IV(I))
10    IF(C(K).LE.PC) GOTO 20
      I=N
20    IF(I.EQ.1) GOTO 30
      PL=PC-P(IV(I-1))
      D=C(K)-PL
      COF=D/(PC-PL)
      O(K+1)=(A(I)-A(I-1))*COF+A(I-1)
      GOTO 35
30    O(K+1)=A(I)
35    IF(K.EQ.5) GOTO 200
      K=K+1
      GOTO 100
200   O(8)=P(IV(1))
      O(9)=P(IV(N))
      RETURN
      END
C
C
C
C
C
      SUBROUTINE OUT2(A,N,I,P)
      DIMENSION A(N),I(N),P(N),Z(9)
      DATA Z/.001,.01,.10,.25,.50,.75,.90,.99,.999/
      CUM=P(I(1))*0.5
      J=1
      WRITE(5,1) A(1),P(I(1)),CUM
1     FORMAT(9X,G11.5,5X,G11.5,16X,G11.5)
      CUM=CUM*2.0
      DO 10 K=2,N-1
      CUM=CUM+P(I(K))*0.5
      IF(J.GE.10) GOTO 10
      IF(CUM.LT.Z(J)) GOTO 10
      J=J+1
      WRITE(5,1) A(K),P(I(K)),CUM
10    CUM=CUM+P(I(K))*0.5
      CUM=CUM+P(I(N))*0.5
20    WRITE(5,1)A(N),P(I(N)),CUM
      RETURN
      END
C
C
C
C
      SUBROUTINE COMP (A,P,N1,N2,P2,I,A2)
      DIMENSION A(N1),A2(N1),P(N1),P2(N1),I(N1)
      N2=0
      K=1
      KK=2
40    T=P(I(K))
50    IF(A(K).NE.A(KK))GOTO 10
      T=T+P(I(KK))
      KK=KK+1
      IF(KK.GT.N1) GOTO 30
      GOTO 50
10    N2=N2+1
      A2(N2)=A(K)
      P2(N2)=T
      K=KK
      IF(KK.EQ.N1) T=P(I(KK))
      IF(KK.EQ.N1) GOTO 30
      KK=K+1
      GOTO 40
30    N2=N2+1
      A2(N2)=A(K)
      P2(N2)=T
```

```
      DO 60 L=1,N2
      I(L)=L
60    A(L)=A2(L)
      RETURN
      END
C
C
C
C
C
      SUBROUTINE ZAP(A,N)
      DIMENSION A(N)
      DO 10 I=1,N
10    A(I)=0.0
      RETURN
      END
C
C
C
C
      SUBROUTINE ZC(A,B,N,N2,T,TP)
      REAL A(N),B(N),T(N),TP(N)
      J=1
      DO 10 I=1,N
      IF(B(I).EQ.0.0) GOTO 10
      T(J)=B(I)
      TP(J)=A(I)
      J=J+1
10    CONTINUE
      IF(J.NE.N2+1) PAUSE 'VERY BAD NEWS IN ZC'
      DO 20 I=1,N2
      A(I)=TP(I)
20    B(I)=T(I)
      RETURN
      END
```